Lecture Notes in Control and Information Sciences

Edited by M. Thoma and A. Wyner

For information about Vols. 1–61 please contact your bookseller or Springer-Verlag.

Lecture Notes in Control and Information Sciences

Edited by M. Thoma and A. Wyner

126

N. Christopeit, K. Helmes
M. Kohlmann (Editors)

Stochastic Differential Systems

Proceedings of the 4th Bad Honnef Conference,
June, 20–24, 1988

Springer-Verlag Berlin Heidelberg GmbH

Editors
Norbert Christopeit
Institut für Ökonometrie
und Operations Research
der Universität Bonn
Ökonometrische Abteilung
Adenauerallee 24-42
D-5300 Bonn

Michael Kohlmann
Fakultät für Wirtschafts-
wissenschaften und Statistik
Universität Konstanz
Postfach 5560
D-7750 Konstanz

Kurt Helmes
Department of Mathematics
University of Kentucky
Lexington, KY 40506
USA

ISBN 978-3-540-51299-8 ISBN 978-3-540-46188-3 (eBook)
DOI 10.1007/978-3-540-46188-3

Library of Congress Cataloging in Publication Data

Stochastic differential systems : proceedings of the 4th Bad Honnef conference,
June 20–24, 1988
/ N. Christopeit, K. Helmes, M. Kohlmann, eds.
(Lecture notes in control and information sciences ; 126)
Papers of the 4th Bad Honnef Conference on Stochastic Differential Systems.

1. Stochastic systems – Congresses. 2. Differentiable dynamical systems – Congresses.
I. Christopeit, N. II. Helmes, K. (Kurt) III. Kohlmann, M. (Michael)
IV. Bad Honnef Conference on Stochastic Differential Systems (4th : 1988) V. Series.
QA402.S846 1989
003–dc20 89-11454

Originally published by Springer-Verlag Berlin Heidelberg New York in 1989.

Offsetprinting: Mercedes-Druck, Berlin

2161/3020-543210

PREFACE

This volume contains the major part of contributions to the 4th Bad Honnef Conference on Stochastic Differential Systems held at Bad Honnef, West Germany, June 20–24, 1988.Following the tradition of the preceding Bad Honnef Conferences, the meeting was intended to highlight recent advances in the areas of stochastic control and filter theory as well as stochastic analysis.

As sort of thematic "domains of attraction", special emphasis has been given to two rather active fields of current research: the use of adaptive methods in stochastic systems analysis and the theory of random fields. In view of the overwhelming flood of information accumulated in these two areas in recent years, the most that could be hoped for at this conference was to offer a glimpse of the status of research and to inspire interest and discussion in these fields. Several survey lectures were intended to provide some introduction to the more mature parts of the theory for those less acquainted with the subject, complemented by contributions that should give some taste of the diversifying issues of current research both in theory and practice. We leave it to the reader to judge how well this goal has been achieved.

It is a privilege of the organizing committee to express its gratitude to the Deutsche Forschungsgemeinschaft, whose generous support made this conference possible. We are also indebted to the members of the International Advisory Committee; assisting us with many valuable suggestions concerning program and speakers they have a substantial share in the success of the conference. Last, but not least, our thanks go to the staff of the Elly–Hölterhoff–Stift for their kind hospitality as well as to G. Nöldeke and A. Schütt for their skilfull job in data processing for the conference and in preparing this volume.

Bonn, February 1989

N. Christopeit
K. Helmes
M. Kohlmann

LIST OF PARTICIPANTS

S. Albeverio
Fakultät und Institut für Mathematik
Ruhr-Universität Bochum
Postfach 10 21 48
4630 Bochum 1
F.R.G.

A.N. Al-Hussaini
Dept. of Statistics
University of Alberta
Edmonton T6G 2G1
Canada

S.A. Anulova
Institute of Control Sciences
Profsoyusnaya 65
117 806Moscow
U.S.S.R.

A. Bensoussan
I.N.R.I.A.
Domaine de Voluceau-Rocquencourt
B.P. 105
78135 Le Chesnay Cedex
France

T. Bielecki
Institute of Mathematics
Polish Academy of Sciences
Sniadeckich 8
00-950 Warszawa
Poland

K.J. Bierth
Institut für Angewandte Mathematik
Universität Bonn
Wegelerstr. 6
5300 Bonn 1
F.R.G.

P.E. Caines
Dept. of Electrical Engeneering
McGill University
3480 University
Montreal, PQ H3A 2A7
Canada

R.J. Chitashvili
Dept. of Probability Theory and Math. Statistics
Institute of Mathematics
Georgian Academy of Sciences
150 a Plekhanov Ave.
380012 Tbilisi
U.S.S.R.

N. Christopeit
Institut für Ökonometrie und Operations Research
Universität Bonn
Adenauerallee 24-42
5300 Bonn 1
F.R.G.

M.H.A. Davis
Department of Electrical Engeneering
Imperial College of Science and Technology
Exhibition Road
London SW7 2BT
Great Britain

T. E. Duncan
Dept. of Mathematics
University of Kansas
Lawrence, Kansas 66045
U.S.A.

R.J. Elliott
Dept. of Statistics and Probability
University of Alberta
Edmonton, T6G2G1
Canada

H.J. Engelbert
Sektion Mathematik
Friedrich-Schiller-Universität
6900 Jena
G.D.R.

L. Gerencser
Department of Electrical Engeneering
Mc Gill University
Montreal, Quebec, H3A 2A7
Canada

J. Glas
Hochgrafenstraße 65
7750 Konstanz
F.R.G.

B. Goldys
Institute of Mathematics
Polish Academy of Sciences
00-950 Warszawa
Sniadeckich 8, P.O.Box 137
Poland

G.L. Gomez
Abt. für Mathematik
Universität Erlangen-Nürnberg
8920 Erlangen
F.R.G.

B. Grigelionis
Academy of Sciences of the Lithuanian SSR
Institute of Mathematics and Cybernetics
232600, Vilnius, 54
U.S.S.R.

A. Heinricher
Dept. of Mathematics
University of Kentucky
Lexington, KY 40506
U.S.A.

K. Helmes
Dept. of Mathematics
University of Kentucky
Lexington, KY 40506
U.S.A.

A. Hilbert
Fakultät und Institut für Mathematik
Ruhr-Universität Bochum
Postfach 10 21 48
4630 Bochum 1
F.R.G.

H. Hui
Institut für Angewandte Mathematik
Universität Bonn
Wegelerstr. 6
5300 Bonn 1
F.R.G.

K. Iwata
Fakultät und Institut für Mathematik
Ruhr-Universität Bochum
Postfach 10 21 48
4630 Bochum 1
F.R.G.

A. Jakubowski
Institut für Mathematische Stochastik
Universität Göttingen
Lotzestr. 13
3400 Göttingen
F.R.G.

M. Jerschow
FB 6 Mathematik
Gesamthochschule Essen
Universitätsstr. 2
4300 Essen
F.R.G.

W. Kirsch
Fakultät und Institut für Mathematik
Ruhr-Universität Bochum
Postfach 10 21 48
4630 Bochum 1
F.R.G.

D. Köhnlein
Institut für Angewandte Mathematik
Universität Bonn
Wegelerstr. 6
5300 Bonn 1
F.R.G.

M. Kohlmann
Fakultät für Wirtschaftswissenschaften und Statistik
Postfach 5560
7750 Konstanz 1
F.R.G.

T. Kottmann
Institut für Ökonometrie und Operations Research
Universität Bonn
Adenauerallee 24-42
5300 Bonn 1
F.R.G.

P. Kröger
Mathematisches Institut
Universität Erlangen-Nürnberg
8521 Erlangen
F.R.G.

P.R. Kumar
Decision and Control Laboratory
University of Illinois at Urbana-Champaign
1101 West Springfield Avenue
Urbana, IL 61801
U.S.A.

H.J. Kushner
Dept. of Applied Mathematics
Lefschetz Center for Dynamical Systems
Brown University
Providence, R.I. 02912
U.S.A.

S. Kusuoka
Research Institute for Mathematical Science
Kyoto University
606 Kyoto
Japan

T.L. Lai
Department of Statistics
Sequoia Hall
Stanford University
Stanford, CA 94305-4065
U.S.A.

P. Mandl
Dept. of Probability and Mathematical Statistics
Charles University
Sokolovska 83
186 Prag 8
Czechoslovakia

S.P. Meyn
Department of Systems Engineering
Australian National University
Canberra, ACT
Australia

M. Musiela
School of Mathematics
University of New South Wales
P.O. Box 1
Kensington, New South Wales
Australia

J. Parisi
Lehrstuhl für Experimentalphysik II
Universität Tübingen
7400 Tübingen 1
F.R.G.

B. Pasik-Duncan
Dept. of Mathematics
University of Kansas
Lawrence, Kansas 66045
U.S.A.

N.I. Portenko
Institute of Mathematics
Ukrainian Academy of Sciences
Repin Str. 3
Kiev
U.S.S.R.

F. Russo
BIBOS
Universität Bielefeld
Postfach 8640
4800 Bielefeld 1
F.R.G.

M. Schäl
Institut für Angewandte Mathematik
Universität Bonn
Wegelerstr. 6
5300 Bonn 1
F.R.G.

K.-U. Schaumlöffel
Universität Bremen
Institut für Dynamische Systeme
Bibliothekstraße 1
Postfach 330 440
2800 Bremen 33
F.R.G.

K. Schilling
Hannoversche Str. 68
3400 Göttingen
F.R.G.

S.E. Shreve
Dept. of Mathematics
Carnegie-Mellon University
Pittsburgh, Penn. 15213
U.S.A.

P. Spreij
Dept. of Economics
Free University P.O. Box 7161
1007 MC Amsterdam
Netherlands

L. Stettner
Institute of Mathematics
Polish Academy of Sciences
Sniadeckich 8
00-950 Warszawa
Poland

D. Surgailis
Institute of Mathematics and Cybernetics
Academy of Sciences of the Lithuanian SSR
Akademijos 4
232600, Vilnius
U.S.S.R.

T.A. Toronjadze
Dept. of Probability Theory and Mathematical Statistics
Institute of Mathematics
Georgian Academy of Sciences
150 a Plekhanov Ave.
Tbilisi 380012
U.S.S.R.

N. van Thu
Goetheinstitut
7800 Freiburg i.Br.
F.R.G.

E. Wong
Department of Electrical Engineering
and Computer Sciences
University of California at Berkeley
Berkeley, CA 94720
U.S.A.

K. Yamada
Institute of Information Sciences and
Electronics
University of Tsukuba
Tsukuba-shi, Ibaraki 305
Japan

M. Zhiming
Fakultät und Institut für Mathematik
Ruhr-Universität Bochum
Postfach 10 21 48
4630 Bochum 1
F.R.G.

CONTENTS

Some Results on Newton Equation with an Additional Stochastic Force

S. Albeverio#
A. Hilbert

Fakultät für Mathematik, Ruhr-Universität-Bochum (FRG)
BiBoS–Research Centre, SFB 237 Bochum–Essen–Düsseldorf, CERFIM Locarno (CH)

Abstract

In this report, based on joint work with E. Zehnder, we discuss stochastic perturbations of classical Hamiltonian systems by a white noise force. We give existence and uniqueness results for the solutions of the equation of motion, allowing for forces growing stronger than linearly at infinity. We prove that Lebesgue measure in phase space is a σ-finite invariant measure. Moreover we give a Girsanov formula relating the solutions for a nonlinear force to those for a linear force.

1. Introduction

The study of stochastic perturbations of classical dynamical systems has been developed in recent years, however, the case of stochastic perturbations of classical Hamiltonian systems has been much less investigated on a mathematical basis, despite its great interest in applications (celestial mechanics, vibrations in mechanical systems, wave propagation in solid state physics ...). One reason for this is the fact that the structure of the classical flows themselves is much more complicated. Orbits of very different long time behaviour, in general cannot be separated in finite time intervals, stable and unstable behaviour being mixed, see e.g. [Ar], [Mo], and [Mo-Ze]. The nature of the orbits depends on the dimension of the system. For degrees of freedom exceeding three the behaviour can vary between periodic motion, and Arnold diffusion. Often the difficult mathematically rigorous investigation of the longtime behaviour, has been replaced by heuristic numerical approaches, see e.g. [Li-Lie].
In case the perturbation is stochastic, hence typically non smooth, and under the restriction of one degree of freedom, Potter [Po] (see also McKean [McK]) analysed nonlinear oscillators perturbed by a white noise force, described by the equations

$$\begin{aligned} \dot{x} &= v \\ \dot{v} &= K(x) + \dot{w} \end{aligned} \tag{1.1}$$

where K is the deterministic force, $\cdot$ means time derivative, $\dot{w}$ is white noise (the derivative of Brownian motion $w(t)$). Under assumptions on the force $K(x) = -V'(x)$, $V \in C^1(\mathbb{R})$ being attracting towards the origin, i.e. $x \cdot K(x) \leq 0$, Potter proved the existence of global solutions and results about recurrence and the invariance of Lebesgue measure $dxdv$ under the flow given in (1.1). Some of these results have been recently extended by Markus and Weerasinghe [MaW] who also studied winding numbers associated with the solution process (x, v) around the origin, see also [AGQ].

Existence and uniqueness for solutions of higher (but finite) dimensional second order Ito equations, as the systems of the type given in (1.1) have been called by Borchers [Bo], have furthermore been deduced by Goldstein [Go] for systems with globally Lipschitz continuous force K, and Narita [Na] in case there exists a function, decreasing along the paths analogously to a Ljapunov function in the deterministic theory.

The present paper is based on joint work with E. Zehnder [AHZ], to which we refer for more details and further discussions. We shall study equations of the form (1.1) in the case where x, and v run in $\mathbb{R}^d$. In Section 2 we establish existence and uniqueness results for strong solutions of the equations, under assumptions on K which are of the type $K(x) = -\nabla V(x)$ for some $V \in C^1(\mathbb{R}^d)$,with either a condition of the form $V(x)$ quadratic or such that $x \cdot K(x) \leq 0$ for $|x|$ sufficiently large. Then the solution process possesses the Markov property and continuous sample paths, furthermore it depends continuously on the initial conditions.

In Section 3 we compare the solutions of the nonlinear system (1.1) with the ones of a corresponding linear system given by

$$\begin{aligned} dx &= vdt \\ dv &= -\gamma x dt + dw \end{aligned} \tag{1.2}$$

(w as above, and γ a constant $d \times d$-matrix with positive eigenvalues). This is done by establishing a Cameron - Martin - Maruyama - Girsanov type formula for the Radon Nikodym derivative of the probability measures. This result can be applied to prove some properties which hold with probability one for the nonlinear system by exploiting their validity for the associated linear system (see [AHZ],[H]).

In Section 4 we exhibit some features of the behaviour of the solution process of the nonlinear system for large times. In particular, we give estimates for the energy functional for the process. We introduce the generator of the diffusion, solving (1.1), and ensure its hypoellipticity (in the sense that the occurring vector fields span the tangent space to phase space). By a Hörmander's type theorem we demonstrate absolute continuity of the transition probability w.r. to Lebesgue measure without further restrictions but continuity of the coefficient functions. Moreover we show that Lebesgue measure is a σ-finite invariant measure.

2. Existence and Uniqueness of the Solution

We consider a Hamiltonian System with corresponding Newton equation

$$\frac{d}{dt}x(t) = v(t), \quad \frac{d}{dt}v(t) = K(x) \tag{2.1}$$

where $t \in \mathbb{R}_+$ is time, $x(t)$ is position in $\mathbb{R}^d$ at time t, $v(t)$ is velocity at time t, and $K(x)$ is the deterministic force. The initial conditions $x(0) = x_0, \quad v(0) = v_0, \quad (x_0, v_0) \in \mathbb{R}^{2d}$ are given. Adding a white noise force $\dot{w}$, we arrive at a system of stochastic differential equations in the phase space random variable $Y \equiv (y(t) \in \mathbb{R}^{2d} | y(t) = (x(t), v(t)), \ t \geq 0)$ of the form

$$dy(t) = \beta(y(t))dt + \sigma d\tilde{w}_t, \qquad y(0) = y_0 = \begin{pmatrix} x_0 \\ v_0 \end{pmatrix} \tag{2.2}$$

with

$$\beta(y(t)) \equiv \begin{pmatrix} v(t) \\ K(x(t)) \end{pmatrix}, \quad \sigma = \begin{pmatrix} 0 & 0 \\ 0 & 1 \end{pmatrix}, \quad \tilde{w} = \begin{pmatrix} b_t \\ w_t \end{pmatrix},$$

where $((b_t, w_t), \mathcal{A}_t \otimes \mathcal{F}_t, \ t \geq 0)$ is a Brownian motion in $\mathbb{R}^{2d}$ issued from 0 at time 0, with independent families of σ-algebras $(\mathcal{A}_t)$, and $(\mathcal{F}_t)$.

<u>Theorem 2.1</u>

Each of the following conditions is sufficient for the existence of pathwise unique solutions of (2.2) for all $t \in \mathbb{R}_+$:
Let $\alpha, \beta \in \mathbb{R}^d$, and $R \geq 0$ then there exist constants $C_1, C_2 \geq 0$, where C_1 depends on R, such that

a) 1) $|K(\alpha) - K(\beta)| \leq C_1 |\alpha - \beta| \qquad \forall |\alpha|, |\beta| < R$
   2) $|K(\alpha)| \leq C_2(1 + |\alpha|) \qquad \forall \alpha \in \mathbb{R}^d$.

b) 1) $\alpha \mapsto K(\alpha)$ is a locally Lipschitz continuous function. Moreover,
   2) For $d \geq 1$: $K(\alpha) = -\nabla V(\alpha)$ for some $V \in C^1(\mathbb{R}^d)$
   3) $(\alpha - x_0)K(\alpha) \leq 0 \qquad \forall \alpha \in \mathbb{R}^d$, for some $x_0 \in \mathbb{R}^d$.

<u>Proof:</u>

a) Statement a) is proven by a stochastic version of the Picard-Lindelöf method of iteration, see e.g. [Mc K] (Cor 6.3.4).

b) For $d = 1$ the statement (b) is a special case of a result of Potter[Po] see e.g. [Mc K], [Na 1].

We give a proof valid for $d \geq 1$ which uses a process which adopts the part played by the Ljapunov function in the deterministic theory. Let us introduce the energy functional

$$W(y) = \frac{1}{2} |v|^2 + V(x) - V(x_0). \tag{2.3}$$

From condition b)2), i.e. $(\alpha - x_0) \cdot K(\alpha) = -|\alpha - x_0| \frac{\partial V(\alpha)}{\partial |\alpha|}$, where $\frac{\partial V}{\partial |\alpha|}$ is the derivative of V along the direction $(\alpha - x_0)$, we conclude

$$V(\alpha) = V(x_0) - \int_0^{|\alpha - x_0|} \frac{1}{|\beta|} (\beta \cdot K(\beta + x_0))\, d|\beta| \geq V(x_0) \tag{2.4}$$

where we also used assumption b)3).
For the energy functional in (2.3) this implies

$$W(y) \geq \frac{1}{2} |v|^2 , \tag{2.5}$$

and

$$|\sigma \cdot \nabla_y W(y)|^2 = |v|^2 \leq 2W(y).$$

Since β is locally Lipschitz continuous, according to [Na 2], and [Ik-Wa], there exist local solutions Y.

Let us introduce stopping times

$$\sigma_n = \inf\{\, t \geq 0 |\, y(t) \geq n\} \wedge n$$

of the process $Y \equiv (y(t), t \geq 0)$, $t < \sigma_n$, and define the <u>explosion time</u> $e(y_0)$ <u>of Y</u> for given initial condition $y(0) = y_0$, by

$$e(y_0) = \sup_{n \in \mathbb{N}} \sigma_n \qquad \mathbb{P} \text{ a.e.} \tag{2.6}$$

For $n \to \infty$ the local solutions converge a.e. to the solution of (2.2), cf [Na 3]. The existence of a global solution Y with initial condition $y(t_0) = y_0$ is equivalent to an infinite explosion time $e(y_0)$, i.e.

$$\mathbb{P}(N_{y_0}) = \mathbb{P}(\{e(y_0) < \infty\}) = 1 \tag{2.7}$$

with

$$N_{y_0} = \{e(y_0) < \infty, \text{ and } \lim_{t \nearrow e(y_0)} |y(t)| = +\infty\}). \tag{2.8}$$

At times before any explosion can possibly occur we can reexpress the energy functional W (2.3), applying Ito's formula to its differential, by

$$W(y(t)) = W(y_0) + \int_0^t v(s) \cdot dw(s) + d\frac{t}{2}\,. \tag{2.9}$$

Let us set

$$\tau(t) \equiv \int_0^t |v(s)|^2 \, ds \tag{2.10}$$

and

$$a(t) \equiv \int_0^{\tau^{-1}} v(s) \cdot dw(s). \tag{2.11}$$

Then $a(t)$, with filtration $\mathcal{F}_{\tau(t)}$ and with a clock running according to the time τ is a new Brownian motion. Under the assumption of Theorem 2.1b) a global solution of (2.2) is established due to (2.7) by the following

<u>Lemma 2.2</u>

Under the hypothesis of Theorem 2.1b) we have

$$\mathbb{P}(e(y_0) = +\infty) = 1.$$

<u>Proof:</u>

The proof of the higher dimensional statement can be reduced to the one for the one dimensional case in [Po] with y being replaced by $|y|$. The proof is by construction, distinguishing the cases $\tau(e(y_0)) < \infty$ and $\tau(e(y_0)) = \infty$, and using the sample path properties of the Brownian Motion $a(\cdot)$. Thus the a.s. finiteness of $|y(t)|$, where $t_0 \leq t \leq e(y_0)$, in deduced, which yields the contradiction. ■

We are left with the proof of uniqueness

a) This case is covered e.g. by [Fr].

b) This case follows from [Ik-Wa], (Theorem 3.1), since the coefficients of the equation (2.2) are in particular locally Lipschitz continuous. This yields uniqueness for $t \in [0, e(y_0)(w)]$. Since $e(y_0)(w) = \infty$ a.s. , by the first part of this theorem, pathwise uniqueness holds for all $t \geq 0$. ■

<u>Remark 2.3</u>

The statement of the theorem holds for $t \geq t_0$ with initial condition $y(t_0)$ given. Furthermore, the condition $b3$) may be generalized to

b3′) There exists a constant $r > 0$ s.t. $(\alpha - x_0) \cdot K(\alpha) \leq 0$ for $|\alpha - x_0| \geq r$

In fact the result on pathwise uniqueness is left untouched, in case we are able to establish the existence of a global solution. Proceeding as in the proof by contradiction of Lemma 2.2, assume that with probability different from zero $e(y_0) < \infty$. Then for w s.t. $e(y_0)(w) < \infty$, and $\tau(e(y_0))(\omega) \leq \infty$, the part showing boundedness of the configuration variable x(t) for $t \in [0, e(y_0)]$ does not involve the force K, and therefore remains unchanged. Splitting the integral in the expression for $V(x(t))$ in (2.4), and inserting the solution of (2.2) we receive a nonnegative contribution,

$$\int_r^{|x(t)-x_0|} |\beta|^{-1} (\beta \cdot K(\beta + x_0)) d|\beta| \tag{2.12}$$

plus a term

$$C_K \equiv \int_0^{\tilde{\gamma}} |\beta|^{-1} (\beta \cdot K(\beta + x_0)) d|\beta| \tag{2.13}$$

with $\tilde{\gamma} \equiv |x(t) - x_0| \wedge r$ which is bounded for $t \in [0, e(y_0)]$, since K is a continuous function. We can estimate the norm of the momentum v, using the energy process, by

$$|v(t)|^2 \leq 2(W(y_0) + d\frac{t}{2} + C_K + a(\tau(t))) \tag{2.14}$$

This yields the boundedness of the process Y in case the random time τ stays finite. In case $\tau(e(o, y_0)) = \infty$ the force is not involved in proving boundedness of the phase space process Y. So the same argument as in the proof of Lemma 2.2 applies. ■

Remark 2.4

Let us look at some analytical properties of the solution of (2.2). We observe that by setting $z \equiv v(t) - w(t)$ we achieve a C^1 - (in time) solution (x, z) of the system of stochastic differential equations

$$\begin{aligned} \dot{x} &= z(t) + w(t) \\ \dot{z} &= K(x(t)) \end{aligned} \tag{2.15}$$

which is equivalent to the system given in (2.2). Since the Brownian motion $(w(t))$ possesses Hölder continuous paths of index $< \frac{1}{2}$ we have local in time solutions to which the theory of systems of ordinary differential equations can be applied. Finally we easily find continuous dependence of the initial data for the solution of (2.2).

Theorem 2.5

Let $K(\alpha, \mu)$ be a continuous function of α, μ satisfying a local Lipschitz condition in α, uniformly in μ, for μ a parameter in some open connected domain in $\mathbb{R}^k$ of the form $\{\mu \in \mathbb{R}^k | \; |\mu - \mu_0| < c\}$ for some $c > 0$, $\mu_0 \in \mathbb{R}^k$. Consider the unique solution (2.2) for $t \geq t_0$ in a bounded domain D of $\mathbb{R}^{2d}$ with initial condition $y_0 = y(t_0)$, and coefficient $K(\alpha, \mu)$. Let us call $y(t, \mu, t_0, y_0)$ the solution, then the mapping $\mu \longmapsto y(t, \mu, t_0, y_0)$ is a.s. continuous. We also have

$$\lim_{\mu \to \mu_0} y(t, \mu, t_0, y_0) = y(t, \mu_0, t_0, y_0)$$

uniformly over every compact (t, y)-set.
The mappings $y_0 \mapsto y(t, \mu, t_0, y_0)$, and $t_0 \mapsto y(t, \mu, t_0, y_0)$ are locally Lipschitz continuous.

Proof:

Since there exists a pathwise (unique) solution of (2.2) this follows from results on ordinary differential equations, see e.g. [Am] (Theorem II.8).

3. A Girsanov Formula

In this section we shall answer the question whether almost sure properties about the solution to non linear stochastic differential equations can be reduced to ones of some Gaussian process. More precisely, we shall investigate whether the probability measure associated with the solution $Y = (y(t) = (x(t), v(t)),\ t \geq 0)$, $y(t_0) = y_0$ of the stochastic differential equation (2.2), is absolutely continuous with respect to the probability measure associated with the process given by the stochastic differential equation

$$d\eta(t) = a(\eta)\, dt + \sigma d\tilde{w}_t \tag{3.0}$$

with $\eta = \binom{z}{u}$, $\quad a(\eta) = \binom{u}{-\gamma z}$, where γ is a positive constant $2d \times 2d$ matrix, and vice versa. This amounts to deriving a (Cameron-Martin-Maruyama-) Girsanov formula relating the probability measures.

Lemma 3.1

Let K, V be as in Theorem 2.1 and let Y be the corresponding solution of (2.2). Then for W defined as in (2.3) we have

$$\text{i) } \mathbb{E}(W(y(t))) = \mathbb{E}(W(y_0)) + d\frac{t}{2}$$

$$\text{ii) } \mathbb{E}\left(W(y(t))^2\right) = \mathbb{E}\left(W(y_0)^2\right) + \frac{d}{2}\int_0^t \mathbb{E}(W(y(s)))\,ds + \frac{1}{2}\int_0^t \mathbb{E}\left(|v(s)|^2\right)ds$$

Proof:

Statement $i)$ follows from 2.9 by taking expectation and using that $\int_0^t v(s)\cdot dw(s)$ is a martingale with expectation zero.

ii) For any $F \in C^2(\mathbb{R})$, we have, using Ito formula successively, for all $t \geq 0$

$$F(W(y(t))) = F(W(y_0)) + \int_0^t F'(W(y(s)))\,v(s)\cdot dw(s) + \frac{1}{2}\int_0^t \left(dF'(W(y(s))) + |v(s)|^2 F''(W(y(s)))\right) ds$$

Inserting $F(\lambda) = \lambda^2, \in \mathbb{R}$, we get the equation $ii)$ of the lemma.

■

Lemma 3.2

Let K be as in Theorem 2.1 and let $x(t)$ be the space component of the solution $y(t) = (x(t), v(t))$ of (2.2). In the case of hypothesis 2.1b assume furthermore that for all $\alpha \in \mathbb{R}^n$ and some constant $C > 0$:

$$|K(\alpha)|^2 \leq C\left(|V(\alpha) - V(x_0)|^2 + 1\right).$$

Then the stochastic integral

$$\int_0^t K(x(s))\cdot dw(s)$$

defines a square integrable martingale w.r. to the filtration $(\mathcal{G}_t)_{t\geq 0} \equiv (\mathcal{A}\otimes\mathcal{F}_t)_{t\geq 0}$, which has zero expectation.

Proof:

On Wiener space Ω over $\mathbb{R}^{2d}$ we are given the stochastic integral

$$\int_0^t \beta(y(s))\cdot\sigma d\tilde{w}(s) = \int_0^t K(x(s))\cdot dw(s). \tag{3.1}$$

a) For the assumption of Theorem 2.1a), i.e. for a linear growth condition, the statement of the Lemma follows by Gronwall's theorem, see [Am] (see also e.g. [Ik-Wa],[Lip-S],[McK]).

b) Now we turn to the case of assumption 2.1b in Theorem 2.1. Using $V(x(t)) - V(x_0) \geq 0$, and the definition of the energy function W (2.3), we get according to Lemma 3.1

$$\mathbb{E}\left(|V(x(t)) - V(x_0)|^2\right) \leq \mathbb{E}\left(\frac{|(v_0)|^2}{2}\right) + \frac{d+1}{2}\int_0^t \mathbb{E}(W(y(s)))\,ds \tag{3.3}$$

From the growth condition on K and (3.3) we deduce

$$\mathbb{E}\left(|K(x(t))|^2\right) \leq C\left[C_1 + \frac{d+1}{2}\int_0^t \mathbb{E}\left(W(y(s))\right) ds\right] \tag{3.4}$$

with $C_1 \equiv (\mathbb{E}\left(\frac{|v_0|^2}{2}\right)$. Inserting in (3.4) the expectation $\mathbb{E}(W(y(s)))$ expressed in Lemma 3.1 we find

$$\mathbb{E}\left(\int_0^t |K(x(s))|^2 ds\right) \leq p_1 t + p_2 t^2 + p_3 t^3 \equiv P_3(t) \tag{3.5}$$

where

$$p_1 = C\mathbb{E}(W(y_0)), \quad p_2 = C\frac{d+1}{4}\mathbb{E}(W(y_0)), \quad p_3 = C\frac{d(d+1)}{24} .$$

As in part *i*) the statement of the lemma follows since by (3.1) combined with the estimate in (3.5), $\int_0^t K(x(s)) \cdot dw_s$ represents the stochastic integral of a nonanticipative function (in $L^2(\Omega, \mathbb{P})$),see e.g [Ik-Wa] (p. 48) or [Lip] (p. 97). In particular we have a martingale, with zero expectation (as seen from its form).

∎

In order to prove the existence of a density of the probability on path space for the nonlinear solution w.r.t. the one corresponding to the linear system, we need the following special case of [Do].

Lemma 3.3

Under the assumptions of Lemma 3.2 the following estimate holds:

$$\mathbb{E}\left(\int_0^t |\gamma x(t) + K(x(t))|^2\right) dt < \infty,$$

where γ is a $d \times d$ matrix.

Proof:

a) We treat separately the cases a), b) corresponding to the assumptions 2.1a, 2.1b in Theorem 2.1. By Schwarz's inequality and the growth assumption on K

$$|\gamma x(t) + K(x(t))|^2 \leq 2\left(2C_2^2 + (2C_2^2 + C_\gamma)|x(t)|^2\right) \tag{3.7}$$

with C_γ the norm of γ.

Replacing $x(t) = x_0 + \int_0^t v(s)ds$, it follows again by Schwarz's inequality

$$|x(t)|^2 \leq 2\left(|x_o|^2 + t\int_0^t |v(s)|^2 ds\right). \tag{3.8}$$

Using $v(s) = v_0 + \int_0^s K(x(u))\,du + \int_0^s dw_u$ and the independence of v_0 of the rest, we get applying Ito's formula, taking into account a2) and the fact that the martingale in Lemma 3.2 has zero expectation:

$$\mathbb{E}\left(|v(s)|^2\right) \leq \mathbb{E}\left(|\int_0^s K(x(u))\,du|^2\right) + s^2 + \mathbb{E}\left(|v_0|^2\right). \tag{3.9}$$

Combining (3.8) with (3.9) and applying Schwarz's inequality we arrive at the following estimate for the r.h.s. of (3.7)

$$\begin{aligned}\mathbb{E}\left(|x(t)|^2\right) &\leq 2\left(\mathbb{E}\left(|x_0|^2\right) + 2\mathbb{E}\left(|v_0|^2\right)t^2 + C_2^2 t^3\,\mathbb{E}\left(\int_0^t 2(1+|x(u)|^2)du\right) + \frac{1}{3}t^4\right)\\ &\leq C_1(t) + C_2(t)\mathbb{E}\left(\int_0^t |x(u)|^2 du\right)\end{aligned}$$

with $C_1(t) \equiv 2\mathbb{E}\left(|x_0|^2\right) + 2\mathbb{E}\left(|v_0|^2\right) t^2 + \frac{1}{3}t^4 + 2C_2^2 t^4$, and $C_2(t) \equiv 2C_2^2 t$
Using Gronwall-Lemma (see e.g. [Am]) in the variable $\mathbb{E}(|x(t)|^2)$ we arrive at

$$\mathbb{E}\left(|x(t)|^2\right) \leq C_1(t) + C_3(t) \tag{3.10}$$

with $C_3(t)$ a smooth function of t, bounded for $t \geq 0$.
From (3.9) and (3.10) we deduce the estimate required in statement a) of the Lemma, namely for $0 \leq t < \infty$:

$$\mathbb{E}\left(\int_0^t |\gamma x(s) + K(x(s))|^2 ds\right) \leq \tilde{C}(t) < \infty \tag{3.11}$$

with $\tilde{C}(t) \equiv 4C_2^2 t + 2(2C_2^2 + C_\gamma)\int_0^t [C_1(s) + C_3(s)]\, ds$.

b) Let K fulfill the conditions of Theorem 2.1b). Analogously to the first step in (3.7), we split the squared norm by Schwarz's inequality. Integrating, and taking expectation in a second and third step yields

$$\mathbb{E}\left(\int_0^t |\gamma x(s) + K(x(s))|^2 ds\right) \leq 2\int_0^t \left(\mathbb{E}\left(|\gamma x(s)|^2 + \mathbb{E}\left(|K(x(s))|^2\right)\right)\right) ds. \tag{3.12}$$

We have to estimate the linear and nonlinear term separately. Since the explosion time is a.s. infinite, combining (2.8) and (3.8) we get

$$|\mathbb{E}\left(x_0 \int_0^t v(s)ds\right)| \leq \int_0^t \left(\mathbb{E}|x_0 v(s)|\right) ds \leq \left(\mathbb{E}|x_0|^2\right)^{\frac{1}{2}} \int_0^t \left(\mathbb{E}\left(W(s)\right)\right)^{\frac{1}{2}} ds = P(t) \tag{3.13}$$

where $P(t)$ is a polynomial of degree 3 in $t^{\frac{1}{2}}$.
Then for all t

$$\mathbb{E}\int_0^t |x(s)|^2 ds \leq C_\gamma t\mathbb{E}\left(|x_0|^2\right) + 2\int_0^t \left[s\int_0^s \left(\mathbb{E}\left(W\left(y_0\right)\right) + \frac{d}{2}u\right) du + 2P(s)\right] ds,$$

which is a polynomial $P_4(t)$ of degree 4 in t. Collecting the estimates given in (3.4) and (3.13) we find

$$\mathbb{E}\left(\int_0^t |\gamma x(s) + K(x(s))|^2 ds\right) \leq 2\int_0^t \left(P_4(s) + P_3(s)\right) ds < \infty. \tag{3.14}$$

This finishes the proof.

■

Remark 3.4

It looks tempting to try to use the estimate in Lemma 3.3, and the assumptions on K to check Novikov's condition $\mathbb{E}(exp(\frac{1}{2}\int_0^t |\gamma x(s) + K(x(s))|^2)) < \infty$, sufficient for Girsanov's theorem below, to hold. In reality, this is not possible in our situation. However, it is possible to check that other types of sufficient (and necessary) conditions for Girsanov's theorem [Lip-S] hold under the conditions of Theorem 2.1, see [AHZ],[H] for a detailed proof.

Theorem 3.5

Let Y with initial data y_0 be the global solution of the nonlinear stochastic differential equation (2.2) with K satisfying the assumptions of the existence and uniqueness Theorem 2.1, and the growth condition of Lemma 3.2.

Let $\eta = (\eta(t) = (z(t), u(t)), t \geq 0)$ be the solution of the linear stochastic differential equation (3.0). Then the process η is equivalent to Y, in the sense that the probability measures μ_η and μ_y constructed on path space Ω are equivalent. The density of μ_y w.r.t. μ_η is given by:

$$\frac{d\mu_y}{d\mu_\eta}(\eta) = \exp\left(\int_0^T (K(z(t)) + \gamma z(t)) \cdot dw(t) - \frac{1}{2}\int_0^T |K(z(t)) + \gamma z(t)|^2 dt\right).$$

Proof:

For arbitrary fixed $t \in [0,T]$, and $a(y)$, $\beta(y)$, σ as in Lemma 3.2 the system of algebraic equations in x over $\mathbb{R}^{2d}$ admits a measurable solution $\bar{\alpha}(x)$

$$\sigma\bar{\alpha}(y) = \beta(y) - a(y) \tag{3.15}$$

Denote by $\bar{\alpha}_1$, $\bar{\alpha}_2$ the first resp. last d components of $\bar{\alpha}(x)$. $\bar{\alpha}_1$ is left undefined by (3.15), for convenience we choose $\bar{\alpha}_1(y) = 0$. $\bar{\alpha}_2$ is determined by (3.15) as

$$\bar{\alpha}_2(y) \equiv \alpha(x) \equiv K(x) + \gamma x \quad . \tag{3.16}$$

The above Lemma 3.3 yields in particular that almost surely

$$\int_0^T |\gamma x(s) + K(x(s))|^2 ds < \infty \qquad .$$

This implies (extending the result [Lip-S](Theorem 7.19) for the case d=1) that μ_y is absolutely continuous with respect to μ_η, and the Radon-Nikodym derivative

$$\begin{aligned}\frac{d\mu_\eta}{d\mu_y}(y) &= \exp\left(-\int_0^T \alpha(x(t)) \cdot dw(t) - \frac{1}{2}\int_0^T |\alpha(x(t))|^2 dt\right) \\ &\equiv \zeta(-\alpha)_T\end{aligned} \tag{3.17}$$

is a martingale with expectation one. This implies, as is well known, that for

$$\hat{w}_t = w_t + \int_0^t \alpha(x(s)) \cdot dw(s) \tag{3.18}$$

$((b_t, \hat{w}_t), \mathcal{G}_t)$, $0 \leq t < T$, constitutes an $2n$-dimensional Brownian motion $\check{w}$ on the probability space $(\Omega, \mathcal{G}_t, \check{\mathbb{P}})$, with $\check{\mathbb{P}}$ the measure on $(\Omega, \mathcal{G})$ defined by

$$\check{\mathbb{P}} \equiv \zeta(-\alpha)_T \, d\mathbb{P}. \tag{3.19}$$

On the other hand the sufficient condition for the linear process in (3.0)

$$\int_0^t |\gamma z(s) + K(z(s))|^2 \, ds < \infty$$

holds almost surely, using the Lipschitz condition on K, and the estimate $\mathbb{E}[\exp(\lambda \max_{0\leq s\leq t} |w(s)|)] \leq 2^d \exp(d\lambda^2 t)$, see e.g.[Y](Theorem 13.13),[AHZ]. Then using the d-dimensional version of the result by [Lip-S](Theorem 7.19), one has that μ_η is absolutely continuous with respct to μ_y. Combining the above results, induces equivalence of μ_y, and μ_η, and the theorem is proven. ■

Remarks

a) In case the condition b3 of Theorem 2.1 is valid outside a ball of radius r only, we introduce the stopping times $\tau_r \equiv \inf\{t \geq 0 | \, |x(t) - x_0| \geq r\}$. Then we may estimate

i) $\int_0^{\tau_r} |\beta(\psi(s))|^2 ds$, and

ii) $\int_{\tau_r}^t |\beta(\psi(s))|^2 ds$

separately, by the Lipschitz condition for i), and by the condition on K given in Lemma 3.2 for ii).

b) Theorem 3.5 has been proven before for $d = 1$ under a sublinearity assumption on K in [Ma-W]. Under the assumptions of Theorem 3.5 properties of the linear process, which are valid almost surely, transfer to nonlinear diffusions Y, as given in Theorem 2.1. For example, provided $y(0) \geq 0$ a.s., there holds

$$|y(t)|^2 \geq 0 \qquad a.s. \quad .$$

This is demonstrated by adopting a result from [Mark]. For other applications of this result see [Ma-W] and [AHZ].

4. Some Additional Remarks

We can use the energy function defined in (2.3) to obtain estimates on the behaviour of the solution process Y of (2.2) with initial condition y_0 for $t \to \infty$, compare [Po], [McK]. In fact we have

Theorem 4.1

Under the assumption of Theorem 3.5 we have for the energy functional W

a) $W(y(t)) - d\frac{t}{2}$ is a martingale

b) $[W(y_0) + d\frac{t}{2}]^n \leq E([W(y(t))]^n) \leq d^n \sum_{k=0}^n \prod_{l=1}^k (2l-1)\,[W(y_0)]^{n-k} \left(d\frac{t}{2}\right)^k$ (with $\prod_{l=1}^k (2l-1) = 0$ for $k = 0$).

c) for t such that $0 \leq \lambda t d \leq 1$ there holds

$$e^{\frac{1}{2}\lambda t d} \leq E\left[e^{\lambda(W(y(t)) - W(y_0))}\right] \leq (1 - \lambda t d^2)^{-\frac{1}{2}} \exp \frac{\lambda t d^2}{(1 - \lambda t d^2)^{\frac{1}{2}}}$$

Proof:

a) Y solves the diffusion equation (2.2) so that it is $(\mathcal{A}_t \otimes \mathcal{F}_t)_{t \geq 0}$ adapted and from (2.9) we deduce that $W(y(t)) - d\frac{t}{2}$ is a martingale.

b) Applying Ito's formula to $(W(y(t)))^n$, repeatedly, and using that the martingale $\int_0^t v(s)(W(y(s)))^{n-1} \cdot dw_s$ has expectation zero, we find taking expectation

$$\begin{aligned} &E(W(y(t))^n) - E(W(y_0)^n) = \\ &= \int_0^t \left(d\frac{n}{2} E\left(W(y(s))^{n-1}\right) + \frac{n(n-1)}{2} E\left(|vW(y(s))^{n-2}|\right)\right) ds\,, \quad n \geq 2. \end{aligned} \tag{4.1}$$

From (4.1) and

$$0 \leq E\left(|v(s)|^2 W(y(s))^{n-2}\right) \leq 2E\left(W(y(s))^{n-1}\right)$$

we get the right inequality of b) by induction from

$$E(W(y(t))^n) - E(W(0)^n) \leq n(n - 1 + \frac{d}{2}) \int_0^t E\left(W(y(s))^{n-1}\right) ds. \tag{4.2}$$

One can easily see by induction that

$$\mathbb{E}\,(W(y(t))^n) \le \mathbb{E}\left(\left(\sqrt{t}z + \sqrt{W(y_0)}\right)^{2n}\right), \tag{4.3}$$

where the expectation on the r.h.s. is with respect to standard Gaussian measure with mean 0 and covariance $\frac{d}{2}$,z being the corresponding Gaussian variable. Introducing this into the following inequality, deduced from (2.9) and (4.1),

$$d\frac{n}{2}\int_0^t \mathbb{E}\left(W(y(s))^{n-1}\right)ds \le \mathbb{E}\,(W(y(t))^n) - \mathbb{E}\,(W(y_0)^n) \tag{4.4}$$

gives the l.h.s. of the inequalities in b) and therefore completes the proof of statement a).

c) Let $k > 0$. Using the inequality of the l.h.s. in b) we achieve the inequality on the left in c) since

$$e^{\lambda\left(d\frac{t}{2}+W(y_0)\right)} \le \mathbb{E}\left(e^{\lambda W(y(t))}\right).$$

Using (4.3) to estimate $\mathbb{E}(W(y)^k)$, and monotone convergence one gets

$$\mathbb{E}\left(e^{\lambda W(y(t))}\right) \le \mathbb{E}\left(\exp\left(\lambda\left(\sqrt{t}z + W(y_0)\right)^2 - z^2\right)\right)$$

An explicit computation of the r.h.s. yields the remaining bound.

■

<u>Remark:</u>

The statistical quantities expectation and variance of the process $y(t)$ are given by

$$\mathbb{E}(y(t)) = \left(v_0 + \int_0^t \mathbb{E}(K(x(s)))ds, x_0 + v_0 t + \int_0^t\int_0^s \mathbb{E}(K(x(u)))du ds\right),$$

$$\mathrm{Var}(y(t)) = \left(\mathrm{Var}\left(\int_0^t K(x(s))ds\right) + td, \mathrm{Var}\left(\int_0^t K(x(s))(t-s)ds\right) + d\frac{1}{3}t^3\right).$$

The process $Y(t)$ of Theorem 2.1 is a Markov diffusion process, since it solves the stochastic equation (2.2). The Markov kernel $P(t,a,db)$, $a,b \in \mathbb{R}^{2d}$, defined by the transition probability is then well defined. Since K is continuous by our assumptions, $P(t,a,db)$ defines a (Feller) Markov semigroup on $C_b(\mathbb{R}^{2d})$. Let L be its infinitesimal generator. Using Ito's formula see e.g. [Fr,Si] we get

$$(Lf)(a) = (\Delta_v + K(x)\nabla_v + v\nabla_x)\,f(a) \tag{4.5}$$

with $a \equiv (x,v)$. Following [Po] one can show that $P(t,a,db)$ is absolutely continuous w.r. to Lebesgue-measure db for fixed t,a. This is seen by looking at the transition probability kernels $P_n(t,a,db)$ for the approximation of (2.2) obtained by replacing $K(x)$ by $K_n(x) = K(x) \wedge K(\frac{n}{|x|}x)$. By known results on the fundamental solution of degenerate parabolic equations with globally Lipshitz coefficients we have that $P^n(t,a,db) = p^n(t,a,b)db$, with $p^n(t,a,\cdot) \in L^1_{\mathrm{loc}}(db)$. A dominated convergence argument shows that $P^n(t,a,A) \to P(t,a,A)$ and from $P^n(t,a,A) = 0$ for $|A| = 0$ it follows that $P(t,a,A) = 0$, hence the absolute continuity of $P(t,a,db)$.

Let us regard $P(t,a,db)$ as defining a Markov semigroup in the space $\mathcal{M}$ of signed measures with finite total variation, for $\mu \in \mathcal{M}$ we define

$$T_t\mu(\cdot) = \int_{\mathbb{R}^{2d}} P(t,a,\cdot)\mu(da). \tag{4.6}$$

We call μ an invariant measure for the Markov semigroup T_t, or an invariant measure for the process Y if

$$T_t\mu(A) = \mu(A) \tag{4.7}$$

for any Borel subset A of $\mathbb{R}^{2d}$ and all $t \geq 0$. We shall see that if μ is invariant then it has a density w.r. to Lebesgue measure. In fact from (4.6) by Fubini's Theorem

$$\mu(A) = \int_{\mathbb{R}^{2d}} P(t,a,A)\mu(da) = \int_A \left(\int_{\mathbb{R}^{2d}} p(t,a,y)\mu(da) \right) dy \tag{4.8}$$

where we used the absolute continuity of $P(t,a,\cdot)$ w.r. to Lebesgue measure. From (4.8) absolute continuity of the invariant measure follows.

Remark 4.2

Let $\tilde{L}$ be the infinitesimal generator of the semigroup T_t, then any invariant measure μ of the process Y of Theorem 2.1 satisfies

$$\tilde{L}\mu = 0 \, , \tag{4.9}$$

and conversely, as easily seen.

Lemma 4.3

The Lebesgue measure is an invariant measure for Y.

Proof:

We have

$$\int_{\mathbb{R}^{2d}} f(b)\tilde{L}\mu(db) = \int_{\mathbb{R}^{2d}} Lf(b)\mu(db), \qquad (\forall f \in C_o^2(\mathbb{R}^{2d}))$$

This implies using the special form (4.5) of L:

$$\tilde{L} = \Delta_v - K(x)\nabla_v - v\nabla_x \tag{4.13}$$

on measures of the form $\mu(db) = \rho(b)db$, $\rho \in C^2(\mathbb{R}^{2d})$. In particular we get

$$\tilde{L}\lambda = 0 \, . \tag{4.14}$$

■

Remark 4.4

One verifies that if K is C^∞, L is hypoelliptic in the sense that it has the form

$$L = X_1^2 + X_0 \tag{4.15}$$

with $X_1 = \nabla_v$, $X_0 = (K(x)_1 \frac{\partial}{\partial v_1} + \frac{\partial}{\partial x_1}, \ldots, K(x)_d \frac{\partial}{\partial v_d} + \frac{\partial}{\partial x_d})$, and $[X_1, X_0] = \nabla_x$ so that $\{[X_1,[X_1,X_0]]\}$ span the smooth vector fields over $\mathbb{R}^{2d}$.

Acknowledgements

This is a report based on common work with Professor Edy Zehnder, whom we would like to thank most warmly for the joy of collaboration and stimulating discussions. We are also grateful to Professors Eric Carlen, Antonella Calzolari, Gianfausto Dell'Antonio, Detlev Dürr, David Elworthy, Shigeo Kusuoka, Wilfried Loges, Gianna Nappo, for very interesting discussions.

REFERENCES

[AGQ] S. Albeverio, Gong Lu, Quian Min Ping, in preparation

[AHZ] S. Albeverio, A. Hilbert, E. Zehnder, **Hamiltonian Systems with a Stochastic Force; Nonlinear versus Linear, and a Girsanov Formula**, Preprint, Bochum

[Am] H. Amann, **Gewöhnliche Differentialgleichungen**, W. de Gruyter 1983

[Ar] V.I. Arnold, **Mathematical Methods of Classical Mechanics**, Springer-Verlag 1978

[Arn1] L. Arnold, **Stochastic Differential Equations**, J. Wiley & Sons 1971

[Arn2] L. Arnold, W. Kliemann, **On Unique Ergodicity for Degenerate Diffusions**, Bremen Report **147** (1986)

[Bor] D.R. Borchers, *Second Order Stochastic Differential Equations and related Ito Processes*, Ph.D. Thesis, Carnegie Institute of Technology 1964

[Die] R. Dieckerhoff & E. Zehnder, *An A-Priori Estimate for Nonlinear Oscillatory Differential Equations*, Ann. Scuola Norm. Pisa **14**, 79-95(1987)

[Do] C. Doléans-Dade, *Quelques applications de la Formule de changement de variables pour les semimartingales*, Zeitschr. Wahrsch. verw. Geb. **16**, 181-194 (1970)

[Fr] A. Friedman, **Stochastic Differential Equations and Applications**, Vol. 1, Academic Press, New York 1975

[Go] J.A. Goldstein, *Second Order Ito Processes*, Nagoya Math. J. **36** (1969), 27-63

[H] A. Hilbert, PhD Thesis, in preparation

[Ik-Wa] N. Ikeda, S. Watanabe, **Stochastic Differential Equations and Diffusion Processes**, North-Holland, Amsterdam 1981

[Khas] R.Z. Khas'minskiĭ, **Stochastic Stability of Differential Equations**, Sijthoff, Alphen aan den Rijn (1980)

[Ku] H.J. Kushner, **Approximation and Weak Convergence Methods for Random Processes, with Applications to Stochastic Systems Theory**, MIT Press, Cambridge 1984

[Li-Lie] A.J. Lichtenberg & M.A. Lieberman, **Regular and Stochastic Motion**, Springer Verlag (Applied Mathematical Sciences 38) 1983

[Lip-S] R.S. Lipster & A.N. Shirjaev, **Statistics of Random Processes I: General Theory**, Springer Verlag 1977

[McK] H.P. McKean, **Stochastic Integrals**, Academic Press New York 1969

[Ma-W] L. Markus & A. Weerasinghe, *Stochastic Oscillators*, J. Diff. Equ. **21**, no. 2 (1988)

[Mo] J. Moser *Stable and Random Motion in Dynamical Systems, with Special Emphasis on Celestial Mechanics* in: Ann. Math. Studies **77**, Princeton U.P., Princeton N.J. 1973

[Mo-Ze] J. Moser, E. Zehnder, book in preparation

[Na1] K. Narita, *No Explosion Criteria for Stochastic Differential Equations*, J. Math. Soc. Japan **34**, 191–203, (1982)

[Na2] K. Narita, *Explosion Time of Second-Order Ito Processes*, J. Math. Anal. Appl. **104**, 418–427, no. 2 (1984)

[Na3] K. Narita, *On explosion and growth order of inhomogeneous diffusion processes*, Yokohama Math. J. **28**, 45–57 (1980)

[Po] J. Potter, *Some Statistical Properties of the Motion of a Nonlinear Oscillator Driven by White Noise*, Ph.D. Thesis, M.I.T. (1962)

[Si] B. Simon, **Functional Integration and Quantum Physics**, Academic Press, New York 1979

[Y] J.Yeh, **Stochastic Processes and the Wiener Integral**, M.Dekker, New York 1979

[Ze] E. Zehnder, *Some Perspectives in Hamiltonian Systems*, in **"Trends and Developments in the Eighties"**, Edts. S. Albeverio, Ph. Blanchard, World Scient., Singapore, 1985

On Dirichlet forms on topological vector spaces: existence and maximality

Sergio Albeverio

Institut für Mathematik
Ruhr-Universität Bochum
4630 Bochum 1
Federal Republic of Germany;
BiBoS;
SFB 237 Bochum-Essen-Düsseldorf;
CERFIM, Locarno (CH)

Michael Röckner

Department of Mathematics
University of Edinburgh
Edinburgh EH9 3JZ
Scotland

Abstract

We recall basic problems of the theory of Dirichlet forms and symmetric Markov processes with finite or infinite dimensional state space. We discuss in particular the existence problem and give some new results concerning the structure of the space of Dirichlet forms extending a given minimal one. The results apply in particular to Dirichlet forms giving Hamiltonians of quantum mechanics and quantum field theory.

Contents

1. Introduction and framework

Symmetric Markov semigroups T_t on a Hilbert space $L^2(E;\mu) \equiv L^2(\mu)$, with E some measurable space and μ some probability measure on E, are, as well known, the basic quantities needed to construct symmetric Markov processes with state space E, see e.g. [F1],[Si]. One has $T_t = e^{+tL}$, $t \geq 0$, with L self-adjoint negative in $L^2(\mu)$ and the Markov property is expressed by $0 \leq u \leq 1 \rightarrow 0 \leq T_t u \leq 1$. A closely related but often analytically more powerful tool is the one of Dirichlet forms. We shall give below the precise definition in the topological case which interests us in this paper. Let us here shortly recall that the abstract definition of a Dirichlet form $\mathcal{E}$ is as a symmetric positive closed bilinear form on $L^2(\mu)$ which has the Dirichlet contraction property $\mathcal{E}(u^\sharp, u^\sharp) \leq \mathcal{E}(u,u)$ for all u in its domain $D(\mathcal{E})$, with $u^\sharp \equiv (u \vee 0) \wedge 1$ (see [F1],[Si]). The relation between T_t and $\mathcal{E}$ is such that $D(\mathcal{E}) = D(\sqrt{-L})$ and $\mathcal{E}(u,u) = (\sqrt{-L}u, \sqrt{-L}u)$, the latter being the scalar product of $\sqrt{-L}u$ with itself in $L^2(\mu)$. Concerning the relations of these basic analytic objects and associated processes and some basic problems concerning the entire setting let us first recall briefly the situation where E is a locally compact space with countable base for the topology. In this case T_t is a diffusion semigroup iff the Dirichlet form is a local one (this is the case if E is a manifold and L on smooth functions of compact support is a second

order elliptic differential operator). Using T_t we can define a (time reversal) symmetric diffusion process with (modified) state space in E, stationary with invariant measure μ. A corresponding stochastic calculus has been developed (see e.g. [F1],[F2],[O],[A/B/R]).
A particularly interesting situation is the one where $E = \mathbb{R}^d$ and the local Dirichlet form is a closed extension of the following minimal energy form $\tilde{\mathcal{E}}$ given by μ: $\tilde{\mathcal{E}}(u,v) \equiv \frac{1}{2}\int \nabla u \cdot \nabla v \, d\mu$, $u, v \in C_0^\infty(\mathbb{R}^d)$, with $\nabla u \equiv \left(\frac{\partial}{\partial x_1}u, \dots, \frac{\partial}{\partial x_d}u\right)$, $\nabla u \cdot \nabla v \equiv \sum_{k=1}^d \frac{\partial}{\partial x_k}u \frac{\partial}{\partial x_k}v$.
Three basic mathematical problems are encountered in such a study, all depending on μ alone:
1) Existence (i.e. closability of $\tilde{\mathcal{E}}$; namely if $\tilde{\mathcal{E}}$ is not closable it is not possible to obtain a T_t and a fortiori a process from it)
2) Structure (of the space of all possible Dirichlet forms extending the given one $\tilde{\mathcal{E}}$)
3) Uniqueness (under which circumstances is the closure $\mathcal{E}^0$ of $\tilde{\mathcal{E}}$ the only Dirichlet form extending the minimal energy form $\tilde{\mathcal{E}}$?)
These basic questions have been discussed, with applications to nonrelativistic quantum mechanics ("ground state picture", "Dirichlet quantum mechanics") (and to generalized Schrödinger operators, diffusions with singular drifts) e.g. in [A1],[F1,2],[A/HK/S],[A/F/HK /L],[A/B/R],[A/K/Rö],[A/M],[A/Rö1-4],[W] and references therein.

In the present paper we discuss these problems in the more general case where E is a (possibly infinite dimensional hence non-locally compact) topological vector space concentrating particularly on 2).
Dirichlet forms over E infinite dimensional have arisen early in work by L. Gross and have been analyzed systematically starting with [A/HK1], in a rigged space setting, see also [A/HK2,3]. Further work in Hilbert spaces is in [Pa], in a Banach space setting in [K], for more "abstract" frameworks see [Dy],[B/H]. More recently we have developed in [A/Rö1-4] a general approach in the case where E is a Souslin locally convex Hausdorff vector space, which covers all previous cases and seems most suitable for further discussions. We recall that Souslin spaces are continuous images of complete separable metrizable (i.e. polish) spaces, and include e.g. all polish spaces (and a fortiori all separable Banach spaces) and distributional spaces like $\mathcal{D}'(\mathbb{R}^d)$, $\mathcal{S}'(\mathbb{R}^d)$ (see e.g. [Sch]).
In our setting then (E, μ) is a Souslin space and μ is a Borel measure on it. We shall here first expose in detail our framework and discuss Dirichlet forms extending a minimally defined energy form $\tilde{\mathcal{E}}$ (on an equivalence class $\widetilde{\mathcal{F}C_b^\infty}$ of smooth cylinder functions): $\tilde{\mathcal{E}}(u,v) = \sum_k \int \frac{\partial}{\partial k}u \frac{\partial}{\partial k}v \, d\mu$, $u, v \in \widetilde{\mathcal{F}C_b^\infty}$, k running in a (finite or countable) set of "μ-admissible" directions in E (see below for precise definitions).
A sufficient (and in a sense also necessary) condition for the existence problem 1) (closability of $\tilde{\mathcal{E}}$) is given (Theor. 1.2) (following [A/Rö1], and extending [A/HK1-3],[K]).
Using the closure of $\tilde{\mathcal{E}}$ a diffusion process can be constructed [A/Rö3] (extending previous results of [A/HK1-3],[K]).
We also analyse more in detail the particularly important case (realized in applications) where there is a real separable Hilbert space H densely and continuously imbedded in E and containing a dense linear subspace of μ-admissible directions in E.
The closure of $\left(\tilde{\mathcal{E}}, \widetilde{\mathcal{F}C_b^\infty}\right)$ is minimal in the class of all Dirichlet forms extending $\left(\tilde{\mathcal{E}}, \widetilde{\mathcal{F}C_b^\infty}\right)$.
We clarify partially problem 2) above in this setting by introducing another Dirichlet form $(\mathcal{E}^+, D(\mathcal{E}^+))$ extending $\left(\tilde{\mathcal{E}}, \widetilde{\mathcal{F}C_b^\infty}\right)$ and showing (Theor. 2.8 below) that $(\mathcal{E}^+, D(\mathcal{E}^+))$ is maximal in a precise sense (this extends a recent result of [A/K], see also [A/Rö4]). In order to prove that $(\mathcal{E}^+, D(\mathcal{E}^+))$ is maximal a subclass of well μ-admissible directions in E is introduced, the "well-admissible directions". This concept is an extension of the one of directions k for which μ is quasi invariant under translations with a smoothness requirement on the derivatives in the direction k, as used in [A/HK1-3],[A/Rö1-3]. It turns out (cf. Th. 2.8 below) that if E has a dense subset of μ-admissible directions then $(\mathcal{E}^+, D(\mathcal{E}^+))$ is

maximal. This extends previous results in [A/K] inasmuch as no quasi invariance or strict positivity of μ is required. This result is essentially contained in [A/K/Rö], we present here a version with more detailed proofs.
The result is formulated in Section 2. Its proof involves the establishment of a "path space representation formula" (of interest in itself and extending a result of [A/K]), which is presented in Sect. 3. Further more technical details of the proof are given in Sect. 4.

Before entering into these matters let us however continue our short overview of results connected with the topics of this paper. In [A/K] a condition of "log concavity" for μ is isolated and in this case stronger results on the problem 2) above are obtained and a further step towards a uniqueness result (problem 3) above) are made. Let us recall in this connection that in the finite dimensional case $E = \mathbb{R}^n$ a strong uniqueness result has been obtained by N. Wielens [W], see also [F2] for partial extension to manifolds and [T] for extensions to measures on abstract Wiener spaces.
Examples of Dirichlet forms satisfying all postulates discussed in this paper and the papers mentioned above are provided by quantum field theory, as first remarked in [A/HK1]. In particular the free Markov field measure μ_0 on $S'(\mathbb{R}^d)$ yields a unique Dirichlet form, associated with stochastic quantization [A/Rö2]; its restriction μ_0^0 to the σ-algebra generated by time zero fields, a probability measure on $S'(\mathbb{R}^{d-1})$, yields a unique Dirichlet form associated with the free Hamiltonian (Hamiltonian of free relativistic quantum fields) [A/HK1-4],[A/HK/R] (see also e.g. [Ko],[Rö]). In these cases all problems 1) – 3) are solved, in particular in the framework provided by the present paper certain non Gaussian probability measures on $S'(\mathbb{R}^2)$ resp. $S'(\mathbb{R})$, associated with interacting quantum fields over two dimensional space-time also provide examples of interesting Dirichlet forms for which the above problems 1) – 3) are partially solved in the sense discussed above. More precisely, the measures μ on $S'(\mathbb{R}^2)$ associated with fields over $\mathbb{R}^2$ with trigonometric, exponential or polynomial interactions provide examples for all results in the present paper, cfr. also [A/Rö1-4] and the associated infinite dimensional diffusion, constructed in [A/Rö3], is the process of stochastic quantization.
The measures μ^0 obtained by restriction of μ to the σ-algebra generated by time zero fields also provide examples for all results in the present paper, cfr. [A/Rö1-3]; their associated diffusions are associated with stationary Euclidean random fields. In the case of exponential interactions (Høegh-Krohn's model) [A/HK0] the Dirichlet form $(\mathcal{E}^+, D(\mathcal{E}^+))$ mentioned above coincides with the one associated with the Euclidean Markov field given by μ (this is based on results of [Gie],[Ze] and [A/K], see also [A/K/R],[A/Rö1-4]). Finally we remark that there is a relation between our present theory of infinite dimensional Dirichlet forms and a theory of Dirichlet forms associated with white noise functionals, see [A/H/P/S],[A/H/P/R/S].

Let us now go over to a more detailed description of the framework of this paper.
Let E be a Hausdorff locally convex topological vector space over $\mathbb{R}$ which is a Souslin space (i.e. the continuous image of a Polish space). Let $\mathcal{B}(E)$ denote the system of its Borel sets and let μ be a probability measure on $(E, \mathcal{B}(E))$. We call a pair $(\mathcal{E}, D(\mathcal{E}))$ a form on $L^2(E;\mu)$ if $D(\mathcal{E})$ is a linear subspace of $L^2(E;\mu)$ and $\mathcal{E} : D(\mathcal{E}) \times D(\mathcal{E}) \to \mathbb{R}$ is a non-negative symmetric bilinear form. Given a form $(\mathcal{E}, D(\mathcal{E}))$ on $L^2(E;\mu)$ we call E its "state space" and sometimes briefly say $(\mathcal{E}, D(\mathcal{E}))$ is a "form on E" (instead of "form on $L^2(E;\mu)$"). For $\alpha > 0$ we set $\mathcal{E}_\alpha := \mathcal{E} + \alpha < , >_{L^2(E;\mu)}$, $D(\mathcal{E}_\alpha) := D(\mathcal{E})$. $(\mathcal{E}, D(\mathcal{E}))$ is called closed if the pre-Hilbert space $(D(\mathcal{E}), \mathcal{E}_1)$ is complete and closable if it has a closed extension, i.e. there exists a closed form $(\widetilde{\mathcal{E}}, D(\widetilde{\mathcal{E}}))$ on H such that $D(\mathcal{E}) \subset D(\widetilde{\mathcal{E}})$ and $\mathcal{E} = \widetilde{\mathcal{E}}$ on $D(\mathcal{E})$. Clearly, $(\mathcal{E}, D(\mathcal{E}))$ is closable if and only if the following condition is satisfied:

$$(1.0)\quad \begin{cases} \text{If } u_n \in D(\mathcal{E}),\ n \in \mathbb{N}, \text{ such that } u_n \xrightarrow[n\to\infty]{} 0 \text{ in } H \text{ and } (u_n)_{(n\in\mathbb{N})} \text{ is} \\ \mathcal{E}\text{-Cauchy (i.e. } \mathcal{E}(u_n - u_m, u_n - u_m) \xrightarrow[n,m\to\infty]{} 0\text{), then } \lim\limits_{n\to\infty} \mathcal{E}(u_n, u_n) = 0. \end{cases}$$

Furthermore, if $(\mathcal{E},D(\mathcal{E}))$ is closable it has a smallest closed extension $(\bar{\mathcal{E}},D(\bar{\mathcal{E}}))$ called its closure (cf. [F1,§1.1]).

1.1 Definition (i) A form $(\mathcal{E}, D(\mathcal{E}))$ on $L^2(E;\mu)$ is a *Dirichlet form* if it is closed, $D(\mathcal{E})$ is dense in $L^2(E;\mu)$ and every normal contraction operates on $(\mathcal{E}, D(\mathcal{E}))$, i.e. given $T : \mathbb{R} \to \mathbb{R}$ such that $T(0) = 0$ and $|T(x) - T(y)| \leq |x - y|$ for all $x, y \in \mathbb{R}$ then for every $u \in D(\mathcal{E})$, $T \circ u \in D(\mathcal{E})$ and $\mathcal{E}(T\circ u, T\circ u) \leq \mathcal{E}(u,u)$.
(ii) The unique negative definite self adjoint operator L on $L^2(E;\mu)$ satisfying

$$D\left(\sqrt{-L}\right) = D(\mathcal{E}) \quad \text{and} \quad \mathcal{E}(u,v) = \left\langle \sqrt{-L}u, \sqrt{-L}v \right\rangle_{L^2(E;\mu)}, \quad u,v \in D(\mathcal{E}),$$

is called the *generator* of the Dirichlet form $(\mathcal{E}, D(\mathcal{E}))$.

The fact that every normal contraction operates on a Dirichlet form $(\mathcal{E}, D(\mathcal{E}))$ is equivalent to the following property of its generator L (cf. [B/H, Théorème 1.1]):

$$\left\langle Lu, (u-1)^+ \right\rangle_{L^2(E;\mu)} \leq 0 \quad \text{for each } u \in D(L). \tag{1.1}$$

Now we will recall the definition of an important class of forms called classical Dirichlet forms on E. First we need a proper domain to start with. Define the linear space

$$\mathcal{F}C_b^\infty := \{u : E \to \mathbb{R} | \text{there exist } l_1, \ldots, l_m \in E' \text{ and } f \in C_b^\infty(\mathbb{R}^m) \text{ such that } u(z) = f(l_1(z), \ldots, l_m(z)),\ z \in E\}$$

where $C_b^\infty(\mathbb{R}^m)$ is the set of all infinitely differentiable functions on $\mathbb{R}^m$ such that all partial derivatives are bounded. Let $\widetilde{\mathcal{F}C_b^\infty}$ denote the associated set of classes in $L^2(E;\mu)$. Note that if $\operatorname{supp}\mu \neq E$, two different elements in $\mathcal{F}C_b^\infty$ might belong to the same class in $\widetilde{\mathcal{F}C_b^\infty}$. Now let $k \in E \setminus \{0\}$ and define for $u \in \mathcal{F}C_b^\infty$ the following Gâteaux-type derivative (in direction k) by

$$\frac{\tilde{\partial}}{\partial k} u(z) := \left.\frac{\mathrm{d}}{\mathrm{d}s} u(z + sk)\right|_{s=0}, \quad z \in E. \tag{1.2}$$

If μ has the property

$$\frac{\tilde{\partial}}{\partial k} u = \frac{\tilde{\partial}}{\partial k} v \quad \mu-\text{a.e. if } u, v \in \mathcal{F}C_b^\infty \text{ with } u = v \quad \mu-\text{a.e.}, \tag{1.3}$$

then $\frac{\tilde{\partial}}{\partial k}$ "respects μ-classes" and therefore defines a linear operator on $L^2(E;\mu)$ with domain $\widetilde{\mathcal{F}C_b^\infty}$ which we also denote by $\frac{\tilde{\partial}}{\partial k}$. In this case we define the corresponding form by

$$\widetilde{\mathcal{E}_k}(u,v) = \int \frac{\tilde{\partial}}{\partial k} u \frac{\tilde{\partial}}{\partial k} v \, \mathrm{d}\mu, \quad u, v \in D\left(\widetilde{\mathcal{E}_k}\right) := \widetilde{\mathcal{F}C_b^\infty}. \tag{1.4}$$

Since E is Souslinean, $\widetilde{\mathcal{F}C_b^\infty}$ is dense in $L^2(E;\mu)$ (cf. [A/Rö1,3.1]). Hence $\left(\widetilde{\mathcal{E}_k}, \widetilde{\mathcal{F}C_b^\infty}\right)$ is densely defined, but it is not closed. If, however, $\left(\widetilde{\mathcal{E}_k}, \widetilde{\mathcal{F}C_b^\infty}\right)$ is closable, then it is easy to see that normal contractions operate on the closure which is therefore a Dirichlet form. The same is true for finite or countable sums of forms of type (1.4). Therefore one only has to check whether (1.4) is closable. In [A/Rö1] we proved a necessary and sufficient condition (on μ, k) for the closability of (1.4). In order to restate the corresponding theorem we need some preparations.

In the sequel we denote the Borel σ-field associated with a topological space X by $\mathcal{B}(X)$. Given two measurable spaces $(X_i, \mathcal{B}_i)$, $(i = 1, 2)$, a $\mathcal{B}_1/\mathcal{B}_2$-measurable map $T : X_1 \to X_2$ and a measure μ on $(X_1, \mathcal{B}_1)$ we denote the image measure under T on $(X_2, \mathcal{B}_2)$ by $T(\mu)$. Furthermore, for notational convenience we will denote the μ-class corresponding to a $\mathcal{B}(E)$-measurable function u also by u if no confusion is possible.

Let E' be the topological dual of E, $k \in E \setminus \{0\}$ and let $l_k \in E'$ such that $l_k(k) = 1$. Define

$$\pi_k(z) := z - l_k(z)k, \quad z \in E.$$

Let $E_k := \pi_k(E)$, then E_k as a closed subspace of E is also a Souslin space. For each $z \in E$, $z = x + sk$, where $x \in E_k$, $s \in I\!R$, are uniquely determined.
Since E, E_k are Souslinean we can disintegrate μ with respect to $\pi_k : E \to E_k$ (cf. [D/M,III70] or [R,Proposition 1]), i.e. there exists a kernel $\rho_k : E_k \times \mathcal{B}(I\!R) \to [0, 1]$ such that for all $u : E \to I\!R$, u bounded, $\mathcal{B}(E)$-measurable

$$\int_E u(z)\mu(dz) = \int_{E_k}\int_{I\!R} u(x + sk)\rho_k(x, ds)\nu_k(dx), \tag{1.5}$$

where $\nu_k := \pi_k(\mu)$ and $\rho_k(\cdot, ds)$ is ν_k-a.e. uniquely determined. It is now easy to verify that $(L^2(I\!R; \rho_k(x, ds)), <\ ,\ >_x)_{(x\in E)}$ is a measurable field of Hilbert spaces over $(E_k, \mathcal{B}(E_k), \nu_k)$ (where, of course $<\ ,\ >_x$ is the usual L^2-inner product with respect to the measure $\rho_k(x, ds)$) and that

$$L^2(E; \mu) = \int^{\oplus} L^2(I\!R; \rho_k(x, ds))\nu_k(dx) \tag{1.6}$$

(cf. [Di,Chap.II§1] and [A/Rö1, Sect. 1]). Here for $u \in L^2(E; \mu)$ the corresponding field $(u_x)_{(x\in E_k)}$ of vectors in $\int^{\oplus} L^2(I\!R; \rho_k(x, ds))\nu_k(dx)$ is given by $u_x := u(x + \cdot k)$, $x \in E_k$.

Let ds denote the Lebesgue measure on $I\!R$. Given a $\mathcal{B}(I\!R)$-measurable function $\rho : I\!R \to I\!R_+$ consider the following condition which was introduced in [Ha] (see also [Ru/Sp] and [Sp]):

$$(H) \quad \begin{cases} \rho = 0 \ \ ds\text{-a.e. on } I\!R \setminus R(\rho) \\ \qquad \text{where } R(\rho) := \left\{ t \in I\!R \,\middle|\, \int_{t-\varepsilon}^{t+\varepsilon} \rho^{-1}\, ds < +\infty \text{ for some } \varepsilon > 0 \right\}. \end{cases}$$

Note that $R(\rho)$ is an open subset of $I\!R$ and that $\rho > 0$ ds-a.e. on $R(\rho)$.
(H) is a rather weak assumption. For instance, clearly every $\mathcal{B}(I\!R)$-measurable function $\rho : I\!R \to I\!R^+$ having the property that for ds-a.e. $s \in \{\rho > 0\}$,

$$\operatorname{ess\,inf}\{\rho(s') \mid s - \varepsilon \le s' \le s + \varepsilon\} > 0$$

for some $\varepsilon > 0$, satisfies (H). In particular, (H) holds for any lower semicontinuous nonnegative function on $I\!R$. On the other hand, if $C \subset I\!R$, C closed, with empty interior and strict positive, finite Lebesgue measure, then (H) does not hold for $\rho : I\!R \to I\!R_+$ defined by

$$\rho(s) := \begin{cases} 1 & ,\text{if } s \in C, \\ 2^{-n}\dfrac{s - a_n}{(b_n - a_n)^2} & ,\text{if } s \in]a_n, b_n[, \end{cases}$$

where $a_n, b_n \in I\!R$, $n \in I\!N$, $a_n < b_n$ such that $I\!R \setminus C = \bigcup_{n=1}^{\infty}]a_n, b_n[$, $]a_n, b_n[\cap]a_m, b_m[= \emptyset$ if $n \neq m$. Note that $\rho \in L^1(I\!R; ds)$ and even $\rho > 0$ on $I\!R$, but $I\!R \setminus R(\rho) = C$.

Now we are prepared to state the main result of [A/Röl], i.e. the closability criterion mentioned above.

1.2 Theorem (i) Assume that for ν_k-a.e. $x \in E_k$, $\rho_k(x, ds) = \rho_k(x, s)\,ds$ for some $\mathcal{B}(I\!R)$-measurable function $\rho_k(x, \cdot) : I\!R \to I\!R_+$ satisfying (H). Then the form

$$(1.7)\quad \begin{cases} D(\mathcal{E}_k) := \left\{ u = (u_x)_{(x\in E_k)} \in \int^{\oplus} L^2(I\!R; \rho_k(x,s)ds)\nu_k(ds) \middle| \text{for } \nu_k-\text{a.e. } x \in E_k, \right. \\ \qquad u_x \text{ has an absolutely continuous (} ds-\text{)version } \widetilde{u_x} \text{ on } R(\rho(x,\cdot)) \\ \qquad \left. \text{and } \frac{\partial}{\partial k}u := \left(\frac{d\widetilde{u_x}}{ds}\right)_{(x\in E_k)} \in \int^{\oplus} L^2(I\!R; \rho_k(x,s)ds)\nu_k(dx) \right\}, \end{cases}$$

$$(1.8)\qquad \mathcal{E}_k(u,v) := \int \frac{\partial}{\partial k}u \frac{\partial}{\partial k}v\, d\mu, \qquad u, v \in D(\mathcal{E}_k),$$

is closed, or equivalently the operator $\frac{\partial}{\partial k}$ (defined in (1.7)) with domain $D(\mathcal{E}_k)$ is closed. Furthermore, (1.3) is satisfied and $\frac{\partial}{\partial k}$ is an extension of $\frac{\tilde{\partial}}{\partial k}$. In particular, the form $\left(\widetilde{\mathcal{E}_k}, \widetilde{\mathcal{F}C_b^\infty}\right)$ is closable.

(ii) If μ satisfies (1.3) and the form $\left(\widetilde{\mathcal{E}_k}, \widetilde{\mathcal{F}C_b^\infty}\right)$ is closable, then for ν_k-a.e. $x \in E_k$, $\rho_k(x, ds) = \rho_k(x, s)\,ds$ for some $\mathcal{B}(I\!R)$-measurable function $\rho_k(x, \cdot) : I\!R \to I\!R^+$ satisfying (H). In particular, $(\mathcal{E}_k, D(\mathcal{E}_k))$ defined by (1.7),(1.8) is a closed extension of $\left(\widetilde{\mathcal{E}_k}, \widetilde{\mathcal{F}C_b^\infty}\right)$.

Note that our assumptions in 1.2(ii) do not involve l_k (or E_k), hence 1.2 is independent of the choice of l_k (or E_k).

1.3 Definition Let $k \in E$. k is called *μ-admissible* if $k = 0$ or if for ν_k-a.e. $x \in E_k$, $\rho_k(x, ds) = \rho_k(x, s)\,ds$ for some $\mathcal{B}(I\!R)$-measurable function $\rho(x, \cdot) : I\!R \to I\!R_+$ satisfying (H) or equivalently (cf. 1.2) if (1.3) is satisfied and $\left(\widetilde{\mathcal{E}_k}, \widetilde{\mathcal{F}C_b^\infty}\right)$ is closable.

1.4 Corollary Let K_0 be a finite or countable set of admissible elements in E. Let

$$(1.9)\quad \begin{cases} D(\mathcal{E}) := \left\{ u \in \bigcap_{k\in K_0} D(\mathcal{E}_k) \middle| \sum_{k\in K_0} \mathcal{E}_k(u,u) < +\infty \right\} \\ \mathcal{E}(u,v) = \sum_{k\in K_0} \mathcal{E}_k(u,v), \quad u, v \in D(\mathcal{E}), \end{cases}$$

and let $(\widetilde{\mathcal{E}}, D(\widetilde{\mathcal{E}}))$ be defined correspondingly with $\left(\widetilde{\mathcal{E}_k}, \widetilde{\mathcal{F}C_b^\infty}\right)$ replacing $(\mathcal{E}_k, D(\mathcal{E}_k))$. Then $(\mathcal{E}, D(\mathcal{E}))$ is a closed extension of $(\widetilde{\mathcal{E}}, D(\widetilde{\mathcal{E}}))$. If, in addition

$$(1.10)\qquad \sum_{k\in K_0} |l(k)|^2 < +\infty \qquad \text{for all } l \in E'$$

then $D(\widetilde{\mathcal{E}}) = \widetilde{\mathcal{F}C_b^\infty}$, hence both $(\mathcal{E}, D(\mathcal{E}))$ and the closure of $(\widetilde{\mathcal{E}}, D(\widetilde{\mathcal{E}}))$ are densely defined. Moreover, both these closed forms are Dirichlet forms.

For the proofs of 1.2, 1.4 we refer to [A/Röl]. The Dirichlet forms of the type appearing in 1.4 are called *classical Dirichlet forms* (in accordance with the case where $E = \mathbb{R}^d$). Given an admissible k in $E\setminus\{0\}$ and $u \in D(\mathcal{E}_k)$ we have defined $\frac{\partial u}{\partial k} \in L^2(E;\mu)$. $\frac{\partial u}{\partial k}$ can be considered as a *μ-stochastic partial derivative of* u (w.r.t. k). Of course, one can also study the concept of a "total" μ-stochastic derivative in the sense of Gâteaux. To this end we need to introduce a suitable Hilbert space H that will play the role of a tangent space to E at each point (cf. [K]). ((1.10) above can be considered as a first step in this direction.) Then we are able to define the "coordinate free" classical Dirichlet forms similar to those introduced in [K] in the case where E is a separable Banach space and μ is quasi-invariant (cf. 2.7 below).

Suppose that there exists a real separable Hilbert space $(H, <\ ,\ >_H)$ densely and continuously imbedded in E. Identifying H with its dual we obtain that E' is densely imbedded in H; in this sense

$$E' \subset H \subset E, \tag{1.11}$$

and $<\ ,\ >_H$ restricted to $E' \times H$ coincides with the dualisation ${}_{E'}<\cdot,\cdot>_E$ between E' and E. Suppose furthermore, that we can find a dense linear subspace K of $(H, <\ ,\ >_H)$ consisting of ($\mu-$)admissible elements in E.

Let $(\mathcal{E}_k, D(\mathcal{E}_k))$, $k \in K$, be defined as in 1.2. Define the linear space

$$\text{(1.12)}\quad \left\{\begin{aligned} S := \Big\{u \in \bigcap_{k\in K} D(\mathcal{E}_k) \,\Big|\, & \text{ there exists a } \mathcal{B}(E)/\mathcal{B}(H)\text{-measurable function} \\ & \nabla u : E \to H \text{ such that, for each } k \in K, \quad \langle \nabla u(z), k\rangle_H = \frac{\partial u}{\partial k}(z) \\ & \text{for } \mu-\text{a.e. } z \in E \text{ and } \int_E \langle \nabla u, \nabla u\rangle_H \, d\mu < +\infty \Big\}. \end{aligned}\right.$$

Clearly, ∇u is (μ-a.e.) unique and if $K_0 \subset K$ is an orthonormal basis of H and $(\mathcal{E}, D(\mathcal{E}))$ is defined by (1.9) then $S \subset D(\mathcal{E})$ and for $u, v \in S$

$$\mathcal{E}(u,v) = \int \langle \nabla u, \nabla v\rangle_H \, d\mu.$$

The following theorem corresponds to Theorem 1 in [K]. The proof can be found in [A/Röl]:

1.5 Theorem Let H, K, S be as above and $K_0 \subset K$ an orthonormal basis of H. Let $(\mathcal{E}, D(\mathcal{E}))$, $(\tilde{\mathcal{E}}, D(\tilde{\mathcal{E}}))$ be defined as in 1.4 (depending on K_0). Then condition (1.10) holds and hence $D(\tilde{\mathcal{E}}) = \widetilde{\mathcal{F}C_b^\infty}$. Furthermore, $(\mathcal{E}, S)$ is a closed extension of $\left(\tilde{\mathcal{E}}, \widetilde{\mathcal{F}C_b^\infty}\right)$ and $(\mathcal{E}, S)$ is a Dirichlet form.

1.6 Remark It is easy to check that for any $u \in \widetilde{\mathcal{F}C_b^\infty} \subset S$ and $h \in H$

$$\langle \nabla u(z), h\rangle_H = \frac{d}{ds} u(z + sh)\Big|_{s=0} \qquad \text{for } \mu\text{-a.e. } z \in E.$$

It follows by 1.5 that the closure of $\left(\tilde{\mathcal{E}}, \widetilde{\mathcal{F}C_b^\infty}\right)$ only depends on H and K and not on the special orthonormal basis K_0 of H. From now on we fix E, H, K and μ and denote the closure of $\left(\tilde{\mathcal{E}}, \widetilde{\mathcal{F}C_b^\infty}\right)$ by $(\mathcal{E}^0, D(\mathcal{E}^0))$. We have by 1.5

$$(1.13)\qquad \mathcal{E}^0(u,v) = \int_E \langle \nabla u, \nabla v\rangle_H \, d\mu = \sum_{k\in K_0} \int \frac{\partial u}{\partial k}\frac{\partial v}{\partial k}\, d\mu \quad \text{for all } u,v \in D(\mathcal{E}^0)$$

(where ∇ is as in (1.12), $\frac{\partial u}{\partial k}$ as in (1.7) and K_0 is any orthonormal basis of H). From now on we denote the second Dirichlet form $(\mathcal{E}, S)$ in 1.5 by $(\mathcal{E}^+, D(\mathcal{E}^+))$. Of course, if $(\mathcal{E}, D(\mathcal{E}))$ is any Dirichlet form such that $\widetilde{\mathcal{F}C_b^\infty} \subset D(\mathcal{E})$ and $\mathcal{E} = \mathcal{E}^0$ on $\widetilde{\mathcal{F}C_b^\infty}$ then $D(\mathcal{E}^0) \subset D(\mathcal{E})$ and $\mathcal{E} = \mathcal{E}^0$ on $D(\mathcal{E}^0)$, i.e. $(\mathcal{E}, D(\mathcal{E}))$ extends $(\mathcal{E}^0, D(\mathcal{E}^0))$. In this sense $(\mathcal{E}^0, D(\mathcal{E}^0))$ is minimal in the class of all Dirichlet forms which are of the form (1.13) on $\widetilde{\mathcal{F}C_b^\infty}$. By 1.5 $(\mathcal{E}^+, D(\mathcal{E}^+))$ belongs to this class. An interesting problem is in which cases $(\mathcal{E}^+, D(\mathcal{E}^+))$ is the "maximal" extension of $(\mathcal{E}^0, D(\mathcal{E}^0))$ in a certain sense. To explain this more precisely we recall that there is an order "$\prec$" on forms on $L^2(E;\mu)$ defined by

$$(\mathcal{E}^1, D(\mathcal{E}^1)) \prec (\mathcal{E}^2, D(\mathcal{E}^2)) \quad \text{if}$$
$$D(\mathcal{E}^1) \subset D(\mathcal{E}^2) \text{ and } \mathcal{E}^1(u,u) \geq \mathcal{E}^2(u,u) \text{ for all } u \in D(\mathcal{E}^1).$$

Given a form $(\mathcal{E}, D(\mathcal{E}))$ let $L(\mathcal{E})$ with domain $D(L(\mathcal{E}))$ denote the associated generator. Consider the set

$$\underline{\underline{\mathcal{E}}} := \Big\{(\mathcal{E}, D(\mathcal{E})) \,\big|\, (\mathcal{E}, D(\mathcal{E})) \text{ is a Dirichlet form on } L^2(E;\mu)$$
$$\text{extending } (\mathcal{E}^0, D(\mathcal{E}^0)) \text{ such that } \widetilde{\mathcal{F}C_b^\infty(K)} \subset D(L(\mathcal{E}))\Big\}$$

where $\mathcal{F}C_b^\infty(K) := \{u : E \to I\!R \,|\, u(z) = f(l_1(z), \ldots, l_m(z)), z \in E, \text{ for some } f \in C_b^\infty(I\!R^m), l_1, \ldots, l_m \in E' \cap K\}$. We will prove in the subsequent sections that under certain assumptions (i.e. essentially if a partial integration formula holds) that $(\mathcal{E}^0, D(\mathcal{E}^0)) \in \underline{\underline{\mathcal{E}}}$ and that $(\mathcal{E}^+, D(\mathcal{E}^+))$ is the maximal element of $\underline{\underline{\mathcal{E}}}$ (w.r.t. $\prec$). The corresponding theorem (cf. 2.8 below) generalizes a recent result in [A/K]. Note that in [A/K] (and also in [K]) the Dirichlet form corresponding to $(\mathcal{E}^+, D(\mathcal{E}^+))$ are defined in a different way. But it is shown in [A/Rö4] that this definition is equivalent to ours in the special case considered in [A/K], [K].

2. Well-admissibility and maximality

For $k \in E \setminus \{0\}$ let E_k, l_k, π_k, ν_k and $\rho_k(\cdot, ds)$ be as in section 1.

2.1 Definition Let $k \in E$. k is called *well-(μ-)admissible* if $k = 0$ or if for ν_k-a.e. $x \in E_k$, $\rho_k(x, ds) = \rho_k(x,s)\, ds$ for some $\mathcal{B}(I\!R)$-measurable function $\rho_k(x,\cdot)$ such that $\frac{\partial}{\partial s}\rho_k(x,\cdot) \in L^1_{\text{loc}}(I\!R; ds)$ and $\left(\frac{\partial}{\partial s}\rho_k(x,\cdot)/\rho_k(x,\cdot)\right)_{x\in E_k} \in \int^{\oplus} L^2(I\!R; \rho_k(x, ds))\nu_k(dx) = L^2(E;\mu)$ (where the derivative is in the sense of Schwartz distributions on $I\!R^1$ and we set $\frac{a}{0} := (\text{sign } a)(+\infty)$ for $a \in I\!R$ and $(\pm\infty)\cdot 0 = 0$).

2.2 Proposition Each well-(μ-)admissible $k \in E$ is (μ-)admissible.

2.2 follows immediately from 1.2 and the following lemma.

2.3 Lemma Let $\rho \in L^1_{\mathrm{loc}}(\mathbb{R}; \mathrm{d}s)$, $\rho \geq 0$, such that its (distributional) derivative $\frac{\mathrm{d}\rho}{\mathrm{d}s} \in L^1_{\mathrm{loc}}(\mathbb{R}; \mathrm{d}s)$. Then ρ has an absolutely continuous ($\mathrm{d}s$-)version $\tilde{\rho}$, hence in particular ρ satisfies (H).

Proof: If follows by [Mi, Theorem 2.7] that there exists a (non-negative) absolutely continuous ($\mathrm{d}s$-)version $\tilde{\rho}$ of ρ such that its usual derivative $\frac{\mathrm{d}\tilde{\rho}}{\mathrm{d}s}$ which exists $\mathrm{d}s$-a.e. is a ($\mathrm{d}s$-)version of the distributional derivative $\frac{\mathrm{d}\rho}{\mathrm{d}s}$. Clearly, $\mathbb{R} \setminus R(\rho) \subset \{\tilde{\rho} = 0\}$, in particular ρ satisfies (H). □

We will use the following proposition in an essential way below.

2.4 Proposition Let $\rho \in L^1(\mathbb{R}; \mathrm{d}s)$, $\rho \geq 0$, such that $\frac{\mathrm{d}\rho}{\mathrm{d}s} \in L^1_{\mathrm{loc}}(\mathbb{R}; \mathrm{d}s)$ and $\beta := \frac{\mathrm{d}\rho}{\mathrm{d}s}/\rho \in L^2(\mathbb{R}; \rho \mathrm{d}s)$. Let $\tilde{\rho}$ be as in 2.3, then:

(i) $\beta \in L^1_{\mathrm{loc}}(R(\rho); \mathrm{d}s)$ and for any open interval $I \subset R(\rho)$ and $c \in I$

$$\tilde{\rho}(s) = \tilde{\rho}(c) \exp\left[\int_c^s \beta(t)\, \mathrm{d}t\right] \quad \text{for every } s \in I. \tag{2.0}$$

In particular, $R(\rho) = \{\tilde{\rho} > 0\}$.

(ii) For every $u \in L^2(\mathbb{R}; \rho \mathrm{d}s)$ which has an absolutely continuous ($\mathrm{d}s$-)version $\tilde{u}$ such that $\frac{\mathrm{d}\tilde{u}}{\mathrm{d}s} \in L^1(\mathbb{R}; \rho \mathrm{d}s)$ and every $v \in C_b^\infty(\mathbb{R})$

$$\int \frac{\mathrm{d}\tilde{u}}{\mathrm{d}s} v \rho \,\mathrm{d}s = -\int u \frac{\mathrm{d}v}{\mathrm{d}s} \rho \,\mathrm{d}s - \int uv \left(\frac{\mathrm{d}\rho}{\mathrm{d}s}/\rho\right) \rho \,\mathrm{d}s. \tag{2.1}$$

Proof: Let $\frac{\mathrm{d}\tilde{\rho}}{\mathrm{d}s}$ be as in the proof of 2.3. Then, since $\rho > 0$ $\mathrm{d}s$-a.e. on $R(\rho)$ we have that

$$\frac{\mathrm{d}\tilde{\rho}}{\mathrm{d}s} = \beta \tilde{\rho} \quad \mathrm{d}s-\text{a.e. on } R(\rho). \tag{2.2}$$

Since ρ satisfies (H) we have that $L^2(R(\rho); \rho \mathrm{d}s) \subset L^1_{\mathrm{loc}}(R(\rho); \mathrm{d}s)$ (cf. [A/Rö1,2.1]). Therefore, $\beta \in L^1_{\mathrm{loc}}(R(\rho); \mathrm{d}s)$ and consequently by (2.2) for any open interval $I \subset R(\rho)$ and $c \in I$ fixed

$$\frac{\mathrm{d}}{\mathrm{d}s}\left(\exp\left[-\int_c^s \beta(t)\,\mathrm{d}t\right] \tilde{\rho}\right) = 0 \quad \mathrm{d}s-\text{a.e. on } I.$$

Hence

$$\tilde{\rho}(s) = \tilde{\rho}(c) \exp\left[\int_c^s \beta(t)\,\mathrm{d}t\right], \quad s \in I,$$

and therefore $\tilde{\rho} > 0$ on I. Since we already saw in the proof of 2.3 that $\mathbb{R} \setminus R(\rho) \subset \{\tilde{\rho} = 0\}$, it follows that $\{\tilde{\rho} > 0\} = R(\rho)$ and (i) is proven.

(ii): Since $R(\rho)$ as an open subset of $\mathbb{R}$ is a disjoint union of open intervals it is easy to see that the set of isolated points of $\mathbb{R} \setminus R(\rho)$ is countable. Since $\mathbb{R} \setminus R(\rho) \subset \{\tilde{\rho} = 0\}$ it follows therefore that $\frac{\mathrm{d}\tilde{\rho}}{\mathrm{d}s} = 0$ on $\mathbb{R} \setminus R(\rho)$ $\mathrm{d}s$-a.e. Hence by the product rule we conclude that for $\tilde{u}$, v as in the assertion

$$\frac{\mathrm{d}}{\mathrm{d}s}(\tilde{u} v \tilde{\rho}) = \frac{\mathrm{d}\tilde{u}}{\mathrm{d}s} v \tilde{\rho} + u \frac{\mathrm{d}v}{\mathrm{d}s} \tilde{\rho} + \tilde{u} v \frac{\mathrm{d}\tilde{\rho}}{\mathrm{d}s} \cdot 1_{R(\rho)} \quad \mathrm{d}s-\text{a.e. on } \mathbb{R}$$

(where $1_{R(\rho)}$ means indicator function of $R(\rho)$). Since clearly, $\frac{\mathrm{d}\tilde{\rho}}{\mathrm{d}s} \cdot 1_{R(\rho)} = \left(\frac{\mathrm{d}\rho}{\mathrm{d}s}/\rho\right) \cdot \rho$ $\mathrm{d}s$-a.e. on $\mathbb{R}$, it follows that $\frac{\mathrm{d}}{\mathrm{d}s}(\tilde{u} v \tilde{\rho}) \in L^1(\mathbb{R}; \mathrm{d}s)$ and then by assumption that $\int \frac{\mathrm{d}}{\mathrm{d}s}(\tilde{u} v \tilde{\rho}) = 0$. Thus (2.1) is proven. □

We set for a well-(μ-)admissible element $k \in E \setminus \{0\}$

$$\beta(k) := \left(\frac{\partial}{\partial s}\rho_k(x,\cdot)/\rho_k(x,\cdot)\right)_{x \in E_k}. \tag{2.3}$$

Then by (1.6), $\beta(k) \in L^2(E;\mu)$.
The reason for introducing the notion of well-(μ-)admissibility is that we need a partial integration formula. In fact, the validity of such a formula is equivalent with well-(μ-)admissi/-bility. This is the main result of this section.

2.5 Theorem Let $k \in E \setminus \{0\}$ and let $\frac{\tilde{\partial}}{\partial k}$ be as in (1.2). Then the following assertions are equivalent:
(i) k is well-(μ-)admissible.
(ii) There exists $\beta_1(k) \in L^2(E;\mu)$ such that

$$\int \frac{\tilde{\partial} u}{\partial k} v \, d\mu = -\int u \frac{\tilde{\partial} v}{\partial k} \, d\mu - \int u v \beta_1(k) \, d\mu \quad \text{for all } u, v \in \mathcal{F}C_b^\infty. \tag{2.4}$$

(iii) There exists $\beta_2(k) \in L^2(E;\mu)$ such that

$$\int \frac{\tilde{\partial} v}{\partial k} \, d\mu = -\int v \beta_2(k) \, d\mu \quad \text{for all } v \in \mathcal{F}C_b^\infty. \tag{2.5}$$

In this case, $\beta_1(k) = \beta_2(k) = \beta(k)$ (defined as in (2.3)) and if $D(\mathcal{E}_k)$, $\frac{\partial}{\partial k}$ are as in (1.7) then (2.4) extends to

$$\int \frac{\partial u}{\partial k} v \, d\mu = -\int u \frac{\partial v}{\partial k} \, d\mu - \int u v \beta(k) \, d\mu \quad \text{for all } u \in D(\mathcal{E}_k),\ v \in \mathcal{F}C_b^\infty. \tag{2.6}$$

Proof: (i)$\Rightarrow$(ii): (i) implies (2.6) by 2.4(ii) and (1.5).
(ii)$\Leftrightarrow$(iii): Take $u \equiv 1$ in (2.4) respectively replace v in (2.5) by $(u \cdot v)$ and use the product rule for $\frac{\tilde{\partial}}{\partial k}$.
(ii)$\Rightarrow$(i): It follows from (2.4) that (1.3) is satisfied, hence $\frac{\tilde{\partial}}{\partial k}$ can be considered as a linear operator on $L^2(E;\mu)$ with dense domain $\widetilde{\mathcal{F}C_b^\infty}$. It also follows by (2.4) for its adjoint $\left(\frac{\tilde{\partial}}{\partial k}\right)^*$ that $\widetilde{\mathcal{F}C_b^\infty} \subset \text{domain}\left(\left(\frac{\tilde{\partial}}{\partial k}\right)^*\right)$. Hence $\left(\frac{\tilde{\partial}}{\partial k}\right)^*$ is densely defined which is equivalent to $\frac{\tilde{\partial}}{\partial k}$ being closable, i.e. $\left(\widetilde{\mathcal{E}_k}, \widetilde{\mathcal{F}C_b^\infty}\right)$ (defined by (1.4)) is closable. Hence by 1.2(ii) for ν_k-a.e. $x \in E_k$, $\rho_k(x, ds) = \rho_k(x,s)\, ds$ for some $\mathcal{B}(\mathbb{R})$-measurable function $\rho_k(x,\cdot) : \mathbb{R} \to \mathbb{R}^+$ satisfying (H). Furthermore, since E_k is a Souslin space the set $\mathcal{L}$ of all functions $T : E_k \to \mathbb{R}$ of the form

$$T(x) = f(l_1(x), \ldots, l_m(x)), \qquad x \in E_k,\ f \in C_b^\infty(\mathbb{R}^m),\ l_i \in E_k',\ 1 \le i \le m,$$

is dense in $L^2(E_k;\nu_k)$. For all $T \in \mathcal{L}$ and $g \in C_0^\infty(\mathbb{R})$ (i.e. $g \in C_b^\infty(\mathbb{R})$ with compact support) we have by (iii) (applied to $v = (T \circ \pi_k) \cdot (g \circ l_k)$) and (1.5) that

$$\int_{E_k} T(x) \int_{\mathbb{R}} \frac{dg}{ds} \rho_k(x,s)\, ds\, \nu_k(dx) = -\int T(x) \int g(s) \beta_2(k)(x+sk) \rho_k(x,s)\, ds\, \nu_k(dx).$$

We therefore see that there exists a set $\Omega \in \mathcal{B}(E_k)$ with $\nu_k(\Omega) = 1$ such that for all $x \in \Omega$ and all $g \in C_0^\infty(\mathbb{R})$

$$\int_{\mathbb{R}} \frac{dg}{ds} \rho_k(x,s) ds = -\int_{\mathbb{R}} g(s) \beta_2(k)(x+sk) \rho_k(x,s)\, ds,$$

and

$$\beta_2(k)(x+\cdot k)\in L^1(\mathbb{R};\rho_k(x,\cdot)\,\mathrm{d}s)$$

(since $\beta_2\in L^2(E;\mu)\subset L^1(E;\mu)$). Hence in the sense of Schwartz distributions $\frac{\partial}{\partial s}\rho_k(x,\cdot)=\beta_2(k)(x+\cdot k)\rho_k(x,\cdot)\in L^1(\mathbb{R};\mathrm{d}s)$. Since $\rho_k(x,\cdot)>0$ ds-a.e. on $R(\rho_k(x,\cdot))$ and $\rho_k(x,\cdot)=0$ ds-a.e. on $\mathbb{R}\setminus R(\rho)$ it follows that $\frac{\partial}{\partial s}\rho_k(x,\cdot)/\rho_k(x,\cdot)=\beta_2(k)(x+\cdot k)$ $(\rho_k(x,\cdot)\,\mathrm{d}s)$-a.e. on $\mathbb{R}$. Hence, since $\beta_2\in L^2(E;\mu)$, k is well-(μ-)admissible. □

Remark We note that the proof of 2.5 depends on our basic "closability result" stated in Theorem 1.2 in an essential way.

As an immediate consequence of 2.5 we obtain

2.6 Corollary The set W of all well-(μ-)admissible elements of E is a linear space and $k\mapsto\beta(k)$ is linear map from W to $L^2(E;\mu)$.

2.7 Examples (i) (The quasi-invariant case)
Let $k\in E\setminus\{0\}$ such that μ is k-quasi-invariant, i.e. $\tau_{sk}(\mu)$ is absolutely continuous with respect to μ for each $s\in\mathbb{R}$ where, for $z_0\in E$, $\tau_{z_0}(z):=z+z_0$, $z\in E$. We set for $s\in\mathbb{R}$

$$a_{sk}(z):=\frac{\mathrm{d}\tau_{sk}(\mu)}{\mathrm{d}\mu}(z),\quad z\in E.$$

Clearly, (1.3) is satisfied in this case, i.e. $\frac{\tilde\partial}{\partial k}$ (cf. (1.2)) with domain $\mathcal{F}C_b^\infty$ is a linear operator on $L^2(E;\mu)$. Define for $s\in\mathbb{R}$

$$(V(sk)u)(z):=\sqrt{a_{sk}(z)}(u\circ\tau_{sk})(z),\quad z\in E,\ u\in L^2_{\mathbb{C}}(E;\mu)$$

where $L^2_{\mathbb{C}}(E;\mu)$ is the canonical complexification of $L^2(E;\mu)$. We see that $s\mapsto V(sk)$ is a unitary strongly continuous representation of the abelian group $\mathbb{R}$ on $L^2_{\mathbb{C}}(E;\mu)$. Let $\pi(k)$ with domain $D(\pi(k))$ be the generator of $(V(sk))_{s\in\mathbb{R}}$. Obviously, $i\pi(k)$ can be considered as an operator on $L^2(E;\mu)$. Now we assume that

$$1\in D(\pi(k)) \tag{2.7}$$

and we set

$$\beta(k):=2i\pi(k)1\quad(\in L^2(E;\mu)).$$

Then it is easy to check that $\mathcal{F}C_b^\infty\subset D(\pi(k))$ and that for $u\in\mathcal{F}C_b^\infty$

$$i\pi(k)u=\frac{\tilde\partial}{\partial k}u+\frac{1}{2}\beta(k)u.$$

Since $-i\pi(k)=(i\pi(k))^*$ (i.e. the adjoint operator on $L^2(E;\mu)$) it follows that $\mathcal{F}C_b^\infty\subset D\left(\left(\frac{\tilde\partial}{\partial k}\right)^*\right)$ and that for $u\in\mathcal{F}C_b^\infty$

$$\left(\frac{\tilde\partial}{\partial k}\right)^*u=-\frac{\tilde\partial}{\partial k}u-\beta(k)u.$$

Consequently, 2.5(ii) holds and hence k is well-(μ-)admissible.
(ii) For concrete examples of quasi-invariant measures we refer the reader to [A/H-K1-3] and [A/Rö1-3]. They include both Gaussian measures as well as many non-Gaussian measures occuring in Euclidean quantum field theory. Non-quasi-invariant examples are easily constructed from these e.g. by restriction of μ to Borel subsets of E.

Now we can formulate the main result of this paper.

2.8 Theorem Let $(H, <\ ,\ >_H)$ be a real separable Hilbert space such that $H \subset E$ continuously and densely, hence if H' is identified with H and E' is equipped with the strong topology

$$E' \subset H \subset E \quad \text{densely and continuously.}$$

Suppose that there exists a dense linear subspace K of E' consisting of well-(μ-) admissible elements of E. Let $\underline{\underline{\mathcal{E}}}$, $\prec$, $(\mathcal{E}^0, D(\mathcal{E}^0))$, $(\mathcal{E}^+, D(\mathcal{E}^+))$ be defined as in section 1. Then $(\mathcal{E}^0, D(\mathcal{E}^0))$ is the minimal and $(\mathcal{E}^+, D(\mathcal{E}^+))$ is the maximal element of $\underline{\underline{\mathcal{E}}}$ with respect to $\prec$.

2.9 Remark Since we do not assume quasi-invariance or strict-positivity (cf. [A/K, (1.6)] and because of 2.7(i), 2.8 generalizes the main result in [A/K,Sect. 2] which was formulated in the case where E is a separable Banach space. See also [A/K/Rö].

The proof of 2.8 will be given in section 4 below. It is partly based on a lemma proved in [A/K] which still holds in our more general situation. We will present this lemma (cf. 3.4 below) and the main steps for its proof in the next section. Finally, we note that by the following proposition one can replace $\mathcal{F}C_b^\infty$ by $\mathcal{F}C_b^\infty(K)$ throughout the whole discussion, if E' is metrizable.

2.10 Proposition Let $K \subset E'$, dense with respect to the strong topology. If E' is metrizable, then $\widetilde{\mathcal{F}C_b^\infty(K)}$ is dense in $D(\mathcal{E}^0)$ with respect to $\mathcal{E}_1^0$.

Proof: Let $u \in \mathcal{F}C_b^\infty$, $u(z) = f(l_1(z), \ldots, l_m(z))$, $z \in E$, for $f \in C_b^\infty(\mathbb{R}^m)$, $l_1, \ldots, l_m \in E'$. Let $i \in \{1, \ldots, m\}$. Then there exist $l_i^n \in K$, $n \in \mathbb{N}$, such that $l_i^n \to l_i$, $n \to \infty$, in E' hence $l_i^n \to l_i$, $n \to \infty$, in $H' = H$. Define for $n \in \mathbb{N}$

$$u_n(z) := f(l_1^n(z), \ldots, l_m^n(z)), \quad z \in E.$$

Then if K_0 is any orthonormal basis of H and $f_i := \frac{\partial f}{\partial x_i}$ we conclude using 1.6, Minkowski's inequality and Lebesgue's dominated convergence theorem that

$$\begin{aligned}
&\lim_{n\to\infty} \mathcal{E}_1^0(u - u_n, u - u_n)^{\frac{1}{2}} \\
&= \lim_{n\to\infty} \left[\sum_{k\in K_0} \int \left[\sum_{i=1}^m (f_i(l_1(z), \ldots, l_m(z)) l_i(k) - f_i(l_1^n(z), \ldots, l_m^n(z)) l_i^n(k)) \right]^2 \mu(dz) \right]^{\frac{1}{2}} \\
&\le \sum_{i=1}^m \lim_{n\to\infty} \|l_i - l_i^n\|_H \left[\int |f_i(l_1(z), \ldots, l_m(z))|^2 \mu(dz) \right]^{\frac{1}{2}} + \\
&\qquad + \sum_{i=1}^m \lim_{n\to\infty} \|l_i^n\|_H \left[\int |f_i(l_1(z), \ldots, l_m(z)) - f_i(l_1^n(z), \ldots, l_m^n(z))|^2 \mu(dz) \right]^{\frac{1}{2}} \\
&= 0
\end{aligned}$$

Now the assertion follows. □

3. A path space representation formula

Let H, K be as in 2.8 and fix $(\mathcal{E}, D(\mathcal{E})) \in \underline{\underline{\mathcal{E}}}$. Let L with domain $D(L)$ be the associated generator. Clearly, if $v \in \mathcal{F}C_b^\infty(K)$ there exists a finite subspace R of K such that $\nabla v(z) \in R$ for all $z \in E$. It follows by 2.5 that for any linear basis $\{e_j\}_{j=1}^n$ of R which is orthonormal w.r.t. $<\ ,\ >_H$

(3.1)
$$Lv = \sum_{j=1}^n \left(\frac{d^2}{dt^2} v(z+te_j)\Big|_{t=0} + \langle e_j, \nabla v(z)\rangle_H \beta(e_j)(z) \right).$$

For $t \geq 0$, we set $T_t := e^{tL}$. $(T_t)_{(t \geq 0)}$ is a semigroup of operators on $L^2(E;\mu)$ which is Markovian (i.e. $0 \leq T_t u \leq 1$ μ-a.e. if $0 \leq u \leq 1$ μ-a.e.) since $(\mathcal{E}, D(\mathcal{E}))$ is a Dirichlet form.
Define $T_k^{[u]} : \widetilde{\mathcal{F}C_b^\infty(K)} \to I\!R$, $u \in L^2(E;\mu)$ and $k \in K$, by

(3.2)
$$T_k^{[u]}(v) := -\int_E u \frac{\partial v}{\partial k}\, d\mu - \int_E uv\beta(k)\, d\mu, \quad v \in \widetilde{\mathcal{F}C_b^\infty(K)}.$$

Recall that $\frac{\partial v}{\partial k} = \langle \nabla v, k\rangle_H$ where ∇v is as in (1.12). Clearly, $T_k^{[u]}(v)$ is linear in k, u, v. The purpose of this section is to prove the following (cf. [A/K,(2.17)]):

3.1 Proposition (i) $\left|T_k^{[u]}(v)\right| \leq \|k\|_H \|v\|_{L^2(E;\mu)} \mathcal{E}(u,u)^{\frac{1}{2}}$ for any $u \in D(\mathcal{E})$, $k \in K$ and $v \in \widetilde{\mathcal{F}C_b^\infty(K)}$.
(ii) If $T_k^{[u]}$, $u \in D(\mathcal{E})$ and $k \in K$, also denotes the extension of $T_k^{[u]}$ to $L^2(E;\mu)$ then for any $\{k_i\}_{i=1}^\infty \subset K$ an orthonormal basis in $(H, <\ ,\ >_H)$ and $\{g_i\}_{i=1}^\infty \subset L^2(E;\mu)$

$$\sum_{i=1}^\infty \left|T_{k_i}^{[u]}(g_i)\right| \leq \left[\sum_{i=1}^\infty \|g_i\|^2_{L^2(E;\mu)}\right]^{\frac{1}{2}} \mathcal{E}(u,u)^{\frac{1}{2}} \quad \text{for each } u \in D(\mathcal{E}).$$

3.1 follows from a "path space representation formula" for $T_k^{[u]}$ proved in [A/K] which is formulated as lemma 3.4 below. In order to illustrate that the arguments in [A/K] also work in our more general situation we now present a sketch of the proof for this "key lemma". We refer to [A/K] for more details.

Let $Q := \{\frac{m}{2^n} | m$ and n non-negative integers$\}$, $\Omega := E^T$ and $\tilde{\mathcal{A}} := \bigotimes_{t \in T} \mathcal{B}(E)$. By Kolmogorov's extension theorem there exists a unique probability measure on $(\Omega, \tilde{\mathcal{A}})$ such that for any $r_1, \ldots, r_n \in Q$ with $r_1 \leq r_2 \leq \ldots \leq r_n$ and $A_1, \ldots, A_n \in \mathcal{B}(E)$

(3.3)
$$\begin{aligned} &P[\omega(r_1) \in A_1, \ldots, \omega(r_n) \in A_n] = \\ &\qquad = \left\langle 1, 1_{A_n} T_{r_n - r_{n-1}} 1_{A_{n-1}} T_{r_{n-1} - r_{n-2}} \cdots T_{r_2 - r_1} 1_{A_1} \right\rangle_{L^2(E;\mu)} \end{aligned}$$

where 1_A denotes the indicator function of $A \in \mathcal{B}(E)$. Let $(\Omega, \mathcal{A}, P)$ be the completion of $(\Omega, \tilde{\mathcal{A}}, P)$. It is an easy consequence of (3.3) that for any $g \in L^2(E;\mu)$ the process $(g(\omega(r)))_{r \in Q}$ extends to an $L^2(\Omega, \mathcal{A}, P)$-continuous process $(\tilde{g}_t)_{t \in [0,\infty[}$.

Let $\widetilde{\mathcal{F}_r} := \sigma\{\omega(r')|r' \in Q,\ r' < r\} \vee \{A \in \mathcal{A}|P(A) = 0 \text{ or } 1\}$, $r \in Q$, and let $\mathcal{F}_t := \bigcap_{\substack{r \in Q \\ r > t}} \widetilde{\mathcal{F}_r}$, $t \in [0, \infty[$. Then $\{\mathcal{F}_t\}_{t \geq 0}$ is a right continuous increasing family of σ-algebras and $\widetilde{g_t}$ is $\mathcal{F}_t$-measurable for any $g \in L^2(E; \mu)$.
For $u \in D(L)$ let

$$\widetilde{M}_t^{[u]} = \widetilde{u_t} - \widetilde{u_0} - \int_0^t \widetilde{Lu_s}\, ds, \quad t \geq 0.$$

Then it follows by (3.3) that $\left(\widetilde{M}_t^{[u]}\right)_{t \geq 0}$ is an $(\mathcal{F}_t)_{t \geq 0}$-martingale for each $u \in D(L)$. Let $\left(M_t^{[u]}\right)_{t \geq 0}$ be the corresponding right continuous version with left limits (cf. e.g. [I/W, Chap. I, Theorem 6.9]). If $< M >_t$ denotes the quadratic variation process corresponding to a martingale M_t (cf. [I/W, p. 53] or [D/M]) then it follows that for $t \geq 0$

$$E_P\left[M_t^{[u]} M_t^{[v]}\right] = E_P\left[\left\langle M^{[u]}, M^{[v]}\right\rangle_t\right] = t\mathcal{E}(u, v) \tag{3.4}$$

for all $u, v \in D(L)$, where E_P denotes the expectation w.r.t. P. Since $D(L)$ is dense in $D(\mathcal{E})$ w.r.t. $\mathcal{E}_1$ it follows as in [F, Chap. 5] from (3.4) that for any $u \in D(\mathcal{E})$ a martingale $\left(M_t^{[u]}\right)_{t \geq 0}$ is defined as an L^2-limit such that the following holds:

3.2 Proposition For all $u, v \in D(\mathcal{E})$, $t \geq 0$,

$$E_P\left[M_t^{[u]} M_t^{[v]}\right] = E_P\left[\left\langle M^{[u]}, M^{[v]}\right\rangle_t\right] = t\mathcal{E}(u, v).$$

In fact, since $\mathcal{F}\widetilde{C_b^\infty(K)} \subset D(L)$, hence $u^n \in D(L)$ for all $u \in \mathcal{F}\widetilde{C_b^\infty(K)}$ and $n \in I\!N$, it can be shown using Kolmogorov's theorem that $\left(M_t^{[u]}\right)_{t \geq 0}$ is a continuous martingale.
Let $k \in K$ and $\varphi \in C_0^\infty(I\!R)$ with $\varphi(t) = t$, $|t| \leq 1$. Define $g_n^{(k)} \in \mathcal{F}\widetilde{C_b^\infty(K)}$, $n \geq 1$, by $g_n^{(k)}(z) := n \cdot \varphi\left(\frac{1}{n}\, {}_{E'}\!< k, z >_E\right)$, $z \in E$. Then it is easy to check that $\left(g_n^{(k)}\right)_{n \in I\!N}$ is $\mathcal{E}$-Cauchy. Hence using 3.2 and Doob's theorem (see e.g. [I/W, Chap. I, Theorem 6.10]) one can show the following:

3.3 Proposition For each $k \in K$ there exists a continuous martingale $\left(M_t^{[k]}\right)_{t \geq 0}$ such that for each $t \geq 0$

$$\lim_{n \to \infty} E_P\left[\left|M_t^{[k]} - M_t^{[g_n^{(k)}]}\right|^2\right] = 0.$$

In particular, $\left\langle M^{[k]}, M^{[k']}\right\rangle_t = t\langle k, k'\rangle_H$ for all $k, k' \in K$, $t \geq 0$.

Now we can formulate the lemma giving the "path space integration formula" mentioned above. Its proof is based on some basic facts on (sub-)martingales (like e.g. the Doob-Meyer decomposition) and Ito's calculus (cf. [A/K, Section 2]).

3.4 Lemma For any $k \in K$, $u \in D(\mathcal{E})$ and $v \in \mathcal{F}\widetilde{C_b^\infty(K)}$

$$T_k^{[u]}(v) = E_P\left[M_1^{[u]} \int_0^1 \widetilde{v_t}\, dM_t^{[k]}\right].$$

The proof of 3.1 is now an easy consequence of 3.2–3.4.

4. Proof of the main result

Now we prove 2.8. We first remark that $(\mathcal{E}^0, D(\mathcal{E}^0))$, $(\mathcal{E}^+, D(\mathcal{E}^+)) \in \underline{\underline{\mathcal{E}}}$ by (2.6) and that $(\mathcal{E}^0, D(\mathcal{E}^0))$ is clearly minimal in $\underline{\underline{\mathcal{E}}}$ w.r.t. $\prec$. So, it remains to prove the maximality of $(\mathcal{E}^+, D(\mathcal{E}^+))$.

Fix $(\mathcal{E}, D(\mathcal{E})) \in \underline{\underline{\mathcal{E}}}$, $u \in D(\mathcal{E})$ and $\{k_i\}_{i=1}^{\infty} \subset K$ an orthonormal basis in $(H, <\ ,\ >_H)$. Define a linear map $\lambda : L^2(E; H; \mu) \to \mathbb{R}$ by

$$\lambda(G) := \sum_{i=1}^{\infty} T_{k_i}^{[u]}(g_i)$$

(cf. (3.2)), where $G(\cdot) := \sum_{i=1}^{\infty} \langle G(\cdot), k_i\rangle_H k_i \in L^2(E; H; \mu)$ with $g_i := \langle G(\cdot), k_i\rangle_H \in L^2(E; \mu)$. It follows by 3.1(ii) that λ is continuous with norm dominated by $\mathcal{E}(u,u)^{\frac{1}{2}}$, hence there exists $F \in L^2(E; H; \mu)$ such that

$$\lambda(G) = \int_E \langle F(z), G(z)\rangle_H \mu(\mathrm{d}z), \quad G \in L^2(E; H; \mu) \tag{4.1}$$

and

$$\int_E \|F(z)\|_H^2 \mu(\mathrm{d}z) \le \mathcal{E}(u,u). \tag{4.2}$$

Because of 3.1(i) (4.1) in particular implies that

$$T_k^{[u]}(v) = \int_E \langle k, F(z)\rangle_H v(z)\mu(\mathrm{d}z) \quad \text{for all } k \in K,\ v \in L^2(E; \mu). \tag{4.3}$$

Now fix $k \in K$. For $T(x) := f(l_1(x), \dots, l_m(x))$, $x \in E_k$, $f \in C_b^\infty(\mathbb{R}^m)$, $l_1, \dots, l_m \in E_k'$ and $g \in C_0^\infty(\mathbb{R})$ we have by (3.2) and (4.3) (applied to $v := (T\circ\pi_k)\cdot(g\circ l_k)$) that

$$\int_{E_k} T(x) \int_{\mathbb{R}} \frac{\mathrm{d}g}{\mathrm{d}s}(s) u(x+sk)\rho_k(x,s)\mathrm{d}s\ \nu_k(\mathrm{d}x) =$$
$$= -\int_{E_k} T(x) \int_{\mathbb{R}} g(s)\Big(\langle k, F(x+sk)\rangle_H + u(x+sk)\beta(k)(x+sk)\Big)\rho_k(x,s)\ \mathrm{d}s\ \nu_k(\mathrm{d}x).$$

Since such T are dense in $L^2(E;\mu)$ (cf. the proof of 2.5(ii)$\Rightarrow$(i)) we conclude that there exists $\Omega_1 \in \mathcal{B}(E_k)$ with $\nu_k(\Omega_1) = 1$ such that for every $x \in \Omega_1$

$$\frac{\partial}{\partial s}\Big(u(x+sk)\rho_k(x,s)\Big) = \Big(\langle k, F(x+sk)\rangle_H + u(x+sk)\beta(k)(x+sk)\Big)\rho_k(x,s) \tag{4.4}$$

as Schwartz distributions on $\mathbb{R}$. Note that since $\langle k, F\rangle_H$, $\beta(k) \in L^2(E;\mu)$, we may assume that the right hand side of (4.4) is in $L^1(\mathbb{R}; \mathrm{d}s)$ for each $x \in \Omega_1$. Fix $x \in \Omega_1$. By (4.4) and [Mi, Theorem 2.7] there exists an absolutely continuous (ds-)version of $s \mapsto u(x+sk)\rho_k(x,s)$. Hence by 2.4(i) we conclude that there exists an absolutely continuous (ds-)version $\widetilde{u_x}$ of $u(x+\cdot k)$ on $R(\rho_k(x,\cdot))$ and that

$$\frac{\mathrm{d}\widetilde{u_x}}{\mathrm{d}s} = \langle k, F(x+\cdot k)\rangle_H \qquad (\rho_k(x,s)\mathrm{d}s) - \text{a.e. on } \mathbb{R}.$$

Since $\langle k, F\rangle_H \in L^2(E;\mu)$, it follows that $u \in D(\mathcal{E}_k)$ (cf. (1.7)). Since $k \in K$ was arbitrary and $F \in L^2(E;H;\mu)$ we conclude that $u \in D(\mathcal{E}^+)$ with $\nabla u = F$. But by (4.2) we obtain that

$$\mathcal{E}(u,u) \leq \mathcal{E}^+(u,u)$$

and the proof of 2.8 is complete.

4.1 Remark We note that the classical result on $\mathbb{R}^d$ corresponding to 2.8 is usually proved using Weyl's lemma (cf. [F1, Theorem 2.3.1]). Since there is no such regularity result on infinite dimensional spaces available probabilistic methods were used in the proof of 2.8 (cf. 3.4).

For some interesting consequences of 2.8 concerning the connection between "partial and total stochastic differentiability" we refer the reader to [A/K/Rö].

Acknowledgement

We thank the organizers for a very kind invitation. Interesting and stimulating discussions with Shigeo Kusuoka, Jürgen Potthoff and Ludwig Streit are gratefully acknowledged.
The second named author would like to thank Prof. Dr. W. Hansen, Department of Mathematics, University of Bielefeld, and the SFB 237 Bochum-Essen-Düsseldorf for kind invitations which facilitated this work. The skilfull help by Frank Nitzschner in editing the manuscript is also gratefully acknowledged.

References

[A1] Albeverio, S.: *Some points of interaction between stochastic analysis and quantum theory*, pp. 1-26 in **"Stochastic Differential Systems"**, Edts. N. Christopeit, K. Helmes, M. Kohlmann, Lect. Notes Control Inf. Sci. **78**, Springer, Berlin (1986)

[ABR] Albeverio, S., Brasche, J., Röckner, M.: *Dirichlet forms and generalized Schrödinger operators*, to appear in **"Proc. Summer School Schrödinger Operators"**, Sønderborg, Edts. H. Holden, A. Jensen, Lect. Notes Maths., Springer (1989)

[AFH-KL] Albeverio, S., Fenstad, J.E., Høegh-Krohn, R., Lindstrøm, T.: **Nonstandard methods in stochastic analysis and mathematical physics**, Academic Press, Orlando (1986)

[AHPRS] Albeverio, S., Hida, T., Potthoff, J., Röckner, M., Streit, L.: *Dirichlet forms in terms of white noise analysis I+II*, in preparation

[AHPS] Albeverio, S., Hida, T., Potthoff, J., Streit, L.: *The vacuum of the Høegh-Krohn model as a generalized white noise functional*, to appear in Phys. Letts.

[AH-K0] Albeverio, S., Høegh-Krohn, R.: *The Wightman axioms and the mass gap for strong interactions of exponential type in two dimensional space-time.* J. Funct. Anal. **16**, 39-82 (1974).

[AH-K1] Albeverio, S., Høegh-Krohn, R.: *Quasi-invariant measures, symmetric diffusion processes and quantum fields.* In: **Les méthodes mathématiques de la théorie quantique des champs**, Colloques Internationaux du C.N.R.S., no. **248**, Marseille, 23-27 juin 1975, C.N.R.S., 1976.

[AH-K2] Albeverio, S., Høegh-Krohn, R.: *Dirichlet forms and diffusion processes on rigged Hilbert spaces.* Z. Wahrscheinlichkeitstheorie verw. Gebiete **40**, 1-57 (1977).

[AH-K3] Albeverio, S., Høegh-Krohn, R.: *Hunt processes and analytic potential theory on rigged Hilbert spaces.* Ann. Inst. Henri Poincaré, Vol. **XIII**, n°3, 269-291 (1977).

[AH-K4] Albeverio, S., Høegh-Krohn, R.: *Diffusion fields, quantum fields, and fields with values in Lie groups*, pp. 1-98 in **"Stochastic Analysis and Applications"**, Ed. M.A. Pinsky, M. Dekker, New York (1984)

[AH-K5] Albeverio, S., Høegh-Krohn, R.: *Some remarks on Dirichlet forms and their applications to quantum mechanics and statistical mechanics, diffusions, quantum fields and groups of mappings*, pp. 120-145 in **"Functional Analysis in Markov Processes"**, Ed. M. Fukushima, Lect. Notes Maths. **923**, Springer, Berlin (1983)

[AH-KR] Albeverio, S., Høegh-Krohn, R., Röckner, M.: in preparation

[AH-KS] Albeverio, S., Høegh-Krohn, R., Streit, L.: *Energy forms, Hamiltonians and distorted Brownian paths*, J. Math. Phys. **18**, 907-917 (1977)

[AK] Albeverio, S., Kusuoka, S.: *Maximality of infinite dimensional Dirichlet forms and Høegh-Krohn's model of quantum fields*, Kyoto-Bochum Preprint (1988)

[AKRö] Albeverio, S., Kusuoka, S., Röckner, M.: *On partial integration in infinite dimensional space and applications to Dirichlet forms.* Preprint Bochum, in preparation.

[AM] Albeverio, S., Zhiming Ma: these Proceedings

[ARö1] Albeverio, S., Röckner, M.: *Classical Dirichlet forms on topological vector spaces - closability and a Cameron-Martin formula.* To appear in J. Funct. Anal.

[ARö2] Albeverio, S., Röckner, M.: *Dirichlet forms, quantum fields and stochastic quantization.* To appear in Proceedings to **"Summer Stochastic in Warwick 1987"**.

[ARö3] Albeverio, S., Röckner, M.: *Classical Dirichlet forms on topological vector spaces - construction of an associated diffusion process.* Preprint Bochum, August 1988. To appear.

[ARö4] Albeverio, S., Röckner, M.: *New developments in the theory and applications of Dirichlet forms.* To appear in Proceedings **"Ascona July 1988"**.

[BH] Bouleau, N., Hirsch, F.: *Formes de Dirichlet générales et densité des variables aléatoires réelles sur l'espace de Wiener.* J. Funct. Anal. **69**, 229-259 (1986).

[DM] Dellacherie, C., Meyer, P.A.: **Probabilities and potential.** Amsterdam-New York-Oxford: North-Holland 1978.

[Di] Dixmier, J.: **Les algèbres d'opérateurs dans l'espace hilbertien.** Paris: Gauthier-Villars 1969.

[Dy] Dynkin, E.B.: **Green's and Dirichlet spaces for a symmetric Markov transition function**, Lect. Notes of the LMS (1982)

[F1] Fukushima, M.: **Dirichlet forms and Markov processes.** Amsterdam-Oxford-New York: North-Holland 1980.

[F2] Fukushima, M.: *Energy forms and diffusion processes*, pp. 65-97 in **"Mathematics and Physics I"**, ed. L. Streit, World Scientific, Singapore (1985)

[Ha] Hamza, M.M.: *Détermination des formes de Dirichlet sur* $\mathbb{R}^n$. Thèse 3eme cycle, Orsay (1975).

[IWa] Ikeda, N., Watanabe, S.: **Stochastic differential equations and diffusion processes.** Amsterdam-Oxford-New York-Tokyo: North-Holland/Kodansha 1981.

[K] Kusuoka, S.: *Dirichlet forms and diffusion processes on Banach space.* J. Fac. Science Univ. Tokyo, Sec. 1A **29**, 79-95 (1982).

[Ko] Kolsrud, T.: *Gaussian random fields, infinite dimensional Ornstein-Uhlenbeck processes, and symmetric Markov processes*, BiBoS-Preprint, to appear in Acta Appl. Math. (1988)

[Mi] Mizohata, S.: **The theory of partial differential equations.** London: Cambridge University Press 1973.

[O] Oshima, Y.: **Lectures on Dirichlet spaces**, Universität Erlangen-Nürnberg, prepublication, 1988

[Pa] Paclet, Ph.: *Espaces de Dirichlet et capacités fonctionnelles sur triplets de Hilbert-Schmidt*, Sém. Krée, Paris (1978)

[R] Ripley, B.D.: *The disintegration of invariant measures.* Math. Proc. Camb. Phil. Soc. **79**, 337-341 (1976).

[Rö] Röckner, M.: *Traces of harmonic functions and a new path space for the free quantum field*, J. Funct. Anal. **79**, 211-249 (1988)

[RuSp] Rullkötter, K., Spönemann, U.: *Dirichletformen und Diffusionsprozesse.* Diplomarbeit, Bielefeld (1983).

[Sch] Schwartz, L.: **Radon measures on arbitrary topological spaces and cylindrical measures.** London: Oxford University Press 1973.

[Si] Silverstein, M.: **Symmetric Markov processes.** Lect. Notes in Math. **426.** Berlin-Heidelberg-New York: Springer 1974.

[Sp] Spönemann, U.: PhD thesis, Bielefeld, publ. in preparation.

[T1] Takeda, M.: *On the uniqueness of Markovian extensions of diffusion operators on infinite dimensional spaces*, Osaka Math. J. **22**, 733-742 (1985)

[T2] Takeda, M.: *On the uniqueness of the Markovian self-adjoint extension of a diffusion operator on a Wiener space*, Preprint

[W] Wielens, N.: *On the essential self-adjointness of generalized Schrödinger operators*, J. Funct. Anal. **61**, 98-115 (1985)

Nowhere Radon smooth measures, perturbations of Dirichlet forms and singular quadratic forms

by
Sergio Albeverio *
Zhiming Ma ♯

* Fakultät für Mathematik, Ruhr-Universität Bochum, D 4630 Bochum 1 (FRG); BiBoS Bielefeld-Bochum; SFB 237 Bochum-Essen-Düsseldorf; CERFIM (Locarno)
♯ Fakultät für Mathematik, Universität Bielefeld, D 4800 Bielefeld; BiBoS; Institute of Appl. Math. Academia Sinica (Beijing)

Abstract

We expose some new results concerning Dirichlet forms on locally compact spaces, associated Markov processes, and potential theory. In particular to any regular Dirichlet form there exist nowhere Radon smooth measures provided each single-point set is a set of zero capacity. We give examples of such measures of the type of generalized Schrödinger operators. We also present an approximation result of smooth measures by smooth measures in a Kato class considered before in connection with perturbations of Schrödinger operators. We also study perturbations of regular Dirichlet forms by symmetric bilinear forms given by differences of smooth measures, providing in particular new criteria for closability and form cores.

0. Introduction

Dirichlet forms are basic functional analytic objects for a general potential theory associated with L^2-spaces (rather than, as in the classical or the usual axiomatic potential theory, spaces of continuous functions or measures, see e.g. [BlH]). Let us first recall the definition of Dirichlet forms in the case where the basic underlying space is taken to be (as we shall do in this paper) a locally compact space X with countable basis for the topology (i.e. separable). It is well known that X is then a polish space, in particular a metric space. Let m be a positive Radon measure on X having X as support i.e. supp[m]=X. One calls $(\mathcal{E},\mathcal{F})$ a form on $L^2(X;m)$ if $\mathcal{F}$ is a linear subspace of $L^2(X;m)$ and $\mathcal{E}:\mathcal{F}\times\mathcal{F}\to\mathbb{R}$ is a non-negative symmetric bilinear form. $(\mathcal{E},\mathcal{F})$ is a closed form if $\mathcal{F}$ is complete with respect to the metric given by the inner procuct $(f,g)\to<f,g>_{\mathcal{E}}\equiv\mathcal{E}(f,g)+(f,g),\quad f,g\in\mathcal{F}$. A closed form is called a Dirichlet form if $\mathcal{F}$ is dense in $L^2(X;m)$ and given any $T:\mathbb{R}\to\mathbb{R}$ with $T(0)=0$ and $|T(x)-T(y)|\le|x-y|$ for all $x,y\in\mathbb{R}$ then for every $u\in\mathcal{F}\quad,\quad T\circ u\in\mathcal{F}$ and $\mathcal{E}(T\circ u,T\circ u)\le\mathcal{E}(u,u)$. The unique negative definite self-adjoint operator A on $L^2(X;m)$ s.t. the domain $D(\sqrt{-A})$ of $\sqrt{-A}$ is equal $\mathcal{F}$ and $\mathcal{E}(f,g)=(\sqrt{-A}f,\sqrt{-A}g)$ $((\ ,\ )$ being the scalar product in $L^2(X;m)$) is called the generator of the Dirichlet form $(\mathcal{E},\mathcal{F})$. For all concepts relative to Dirichlet forms see [F1]. We recall that the systematic study of Dirichlet forms and their relations to potential theory and symmetric Markov semigroups and processes goes back to Beurling and Deny and was developed particularly by Fukushima and Silverstein, see references in [F1]. Of course a particular case of Dirichlet form is the one where X is a domain of $\mathbb{R}^d$, $\mathcal{F}$ is the Sobolev space of order 1, $H^{2,1}(X),\mathcal{E}(u,v)=\frac{1}{2}\int_X\nabla u\cdot\nabla v dx, u,v\in H^{2,1}(X)$. This is the classical Dirichlet form (arising in Dirichlet principle, e.g.). What makes Dirichlet forms worthwhile studying is their connection with boundary value problems of partial differential equations and variational problems, on one hand, and, on the other hand, with the theory of symmetric Markov semigroups and processes. In fact, if A is the generator of a Dirichlet form, then $T_t\equiv e^{tA}, t>0$ is a Markov semigroup on $L^2(X;m)$, in the sense that each T_t is a (strongly continuous) symmetric contraction in $L^2(X;m)$ and T_t is Markovian in the sense that $0\le T_tf\le 1$ m.a.e. if $0\le f\le 1$ m.a.e. Viceversa, to any such semigroup T_t with generator A there exists an unique Dirichlet form associated with it, namely the one given by $\mathcal{E}(u,v)=(\sqrt{-A}u,\sqrt{-A}v)$, $\mathcal{F}=D(\sqrt{-A})$. A Markov process $(\Omega,(X_t)_{(t\ge0)};(P_x)_{x\in X})$ with state space X is said to be associated to $(\mathcal{E},\mathcal{F})$ if for any $u:X\to\mathbb{R}$, measurable and bounded and every $t\ge0$:

$$(T_tu)(x)=E^x(u(X_t))$$

for m.a.e. $x\in X$, where E^x is the expectation with respect to P_x .
Sufficient conditions on $X,(\mathcal{E},\mathcal{F})$ for having a Markov Hunt process (i.e. a strong Markov process s.t. every sample path is left continuous along any increasing sequence of stopping times) associated with $(\mathcal{E},\mathcal{F})$ are known: e.g. $\mathcal{F}\cap C_\infty(X)$ (with $C_\infty(X)$ denoting the continuous functions vanishing at infinity) dense, both in $\mathcal{F}$ (with the metric given by $<,>_{\mathcal{E}}$) and in $C_\infty(X)$ (with supremum norm). One calls such a Dirichlet form "regular": see also [AR] for recent results on the case where X is not necessarily locally compact. One also shows then that the Hunt process is a diffusion (i.e. has sample

paths continuous up to a death time, P_x a.s., $\forall x \in X$) iff $(\mathcal{E}, \mathcal{F})$ is local (i.e. $\mathcal{E}(u,v) = 0$ whenever $u, v \in \mathcal{F}$, *supp*$[u]$ compact, $v = 1$ in a neighborhood of *supp*$[u]$). One of the main applications of the theory of Dirichlet forms consists in providing the correct setting for studying Markov processes associated with singular differential equations. This has been discussed in connection with quantum mechanics, see e.g. [AHS], [AR], [ABR], [AFHKL], [A]. In the classical theory of Markov processes, see e.g. [BG], one way of constructing new processes from a given known one is to consider additive functionals of the given process. This leads in particular to the study of Feynman-Kac functionals in connection with Schrödinger operators, see e.g. [AS], [BHH], [BM], [DevC], [SV] for recent developments and references. In the setting of the theory of Dirichlet forms additive functionals have also been discussed in general (cfr. Sect. 1 for precise definitions), see e.g. [F1-3], [O] and references therein. The present paper is mainly devoted to give shortly some new properties of the class of smooth measures associated to positive continuous additive functionals to a given regular Dirichlet form. It is based on work by the authors in [AM1]-[AM2], to which we refer for complements and some more detailed proofs. We also examine the relation of smooth measures and measures considered in connection with Schrödinger semigroups [S1] and Feynman-Kac formulae, the so called Kato class measures, see e.g. [AS], [BHH], [BM], [St].

Let us describe shortly the structure of the paper. In Sect. 1, based on [AM1], we start by recalling the concepts of smooth measures and Revuz measures associated with positive continuous additive functionals of a given regular Dirichlet form (an extension of the concept of classical positive continuous additive functionals of Markov processes, see e.g. [BG], [DM]). We give a general theorem affirming the existence of nowhere Radon smooth measures to a given regular Dirichlet form and give examples (§1.1).

In section §1.2 we show that any smooth measure con be approximated by smooth measures in the so called Kato class (cfr. e.g. [AS], [BHH]).

In section 2 based on [AM 2], we report on results concerning perturbations of regular Dirichlet forms by symmetric bilinear forms given by signed measures μ with components being smooth measures in the sense of Sect. 1. We first show that if the negative part of μ is finite, with finite energy integral, and is in Kato class then the perturbed form is again a lower semibounded closed quadratic form. We then provide an extension of this and related results to the general case. We also give general closability results for the perturbed form and results about form cores. These results extend in particular some results of [AHKS].

1. Some remarks on smooth measures

The concept of smooth measures was introduced by M. Fukushima in the description of the class of Revuz measures associated with positive continuous additive functionals in the Dirichlet space setting. We shall start by recalling these concepts, referring to [F1] and [F2] for more details.

Let $(\mathcal{E}, \mathcal{F})$ be a regular Dirichlet form on $L^2(X; m)$, where X is a locally compact separable metric space and m is a positive Radon measure on X with *supp*$[m] = X$. Then there exists a unique (up to equivalence) Hunt process $M = (\Omega, X_t, \zeta, P_x)$ whose

Dirichlet form is $(\mathcal{E}, \mathcal{F})$. A function $A : [0,\infty) \times \Omega \mapsto [-\infty, \infty]$ is said to be an additive functional (AF in abbreviation) if:

(i) $A_t(\cdot)$ is $\mathcal{F}_t$ measurable where $\mathcal{F}_t$ is the smallest complete σ-algebra which contains $\sigma\{X_s : s \leq t\}$.

(ii) There exists a defining set $\Lambda \in \mathcal{F}_\infty$ and an exceptional set $N \subset X$ with $Cap(N) = 0$ such that:

a)
$$P_x(\Lambda) = 1 \quad \text{for all} \quad x \in X - N \tag{1.0.1}$$

b) $\theta_t \Lambda \subset \Lambda$ for all $t > 0$ (θ_t denotes the shift operator on Ω).

c) If $\omega \in \Lambda$, then $A_0(\omega) = 0, |A_t(\omega)| < \infty$ for $t < \zeta(\omega).A_.(\omega)$ is right continuous and has left limit, and $A_{t+s}(\omega) = A_t(\omega) + A_s(\theta_t\omega)$ for $s, t \geq 0$.

An AF A is called a positive continuous additive functional (PCAF in abbreviation) if $A_.(\omega)$ is a non-negative and continuous function for each ω in its defining set Λ. For instance, let $A_t(\omega) = \int_0^t g(X_s(\omega))ds$ for some non-negative $L^1_{loc}(X;m)$ function g, then A is a PCAF.

Given a PCAF A, there exists a unique Borel measure μ on X, which is called the Revuz measure of A, such that

$$\lim_{t \downarrow 0} \frac{1}{t} E_{h \cdot m}\left[\int_0^t f(X_s)dA_s\right] =< f \cdot \mu, h >:= \int_X h(x)(f \cdot \mu)(dx) \tag{1.0.2}$$

for all ν-excessive functions h and $f \in \mathcal{B}^+$($\mathcal{B}^+$ denotes all non-negative Borel functions on X) (ν is any number $\geq$).

A Borel measure is then called a smooth measure if it is the Revuz measure of some PCAF. The above definition of PCAF's is in fact a generalization of the concept of the classical PCAF's (see for example [BG]) of Markov processes. The aim of the generalization, made by M. Fukushima, is to relax the finiteness requirement on PCAF's to get a broader but simpler class of associated Revuz measures. Another motivation is that "it is also desirable to replace the strong duality assumption on M by weaker ones" ([F2] §1). (We should mention here that in addition to the above approach there is another meaningful generalization of PCAF's, namely the one introducing the notion of homogeneous random measures in the weak duality setting. In this report we shall however discuss solely the generalization made by Fukushima in the Dirichlet space setting).

Denote by S the family of all smooth measures, i.e., the class of the associated Revuz measures of PCAF's in the Dirichlet space setting. A simple analytical description of S has been given as follows ([F1] p.72):

A Borel measure μ on X is in S if and only if μ charges no set of zero capacity and there exists an increasing sequence of compact sets $\{F_n\}_{n\geq 1}$, such that

(i)
$$\mu(F_n) < \infty \quad \text{for each n} \quad ; \tag{1.0.3}$$

(ii)
$$\mu(X - \bigcup_{n=1}^{\infty} F_n) = 0; \tag{1.0.4}$$

(iii)
$$\lim_{n\to\infty} Cap(K - F_n) = 0 \quad \text{for any compact set K} . \tag{1.0.5}$$

From the above description it is easy to see that S contains all positive Radon measures

charging no sets of zero capacity. It is also known that any measure in S can be approximated by measures of finite energy integral (c.f. [F1] Th. 3.2.3).
In what follows we shall expose two new observations on further properties of the class S. On the one hand, we observe that there are many smooth measures μ which are "nowhere Radon" in the sense that $\mu(G) = \infty$ for all non-empty open sets $G \subset X$. Thus the class S is much broader than it has been realized up to date. On the other side we observe also that the class S is so nice that each measure in S can be approximated by the measures in Kato class. Recall that in the classical case the measures of Kato class play an important role in connecting the Schrödinger semigroups and Feynman-Kac formula (cf. [AS], [BHH], [S1], [BM]). In the sequel we shall work in a fixed Dirichlet space as described at the beginning of this section.

1.1 On the "nowhere Radon" smooth measures

1.1.1 Theorem Let B be a subset of zero capacity and ν be a smooth measure such that $supp[\nu] \supset \bar{B}$. Then there exists at least one smooth measure μ which is equivalent to ν and $\mu(G) = \infty$ for all open set G such that $G \cap B \neq \emptyset$.

The existence of "nowhere Radon" smooth measures is then simply a corollary of the above theorem.

1.1.2 Corollary Suppose that each single-point set of X is a set of zero capacity, then there are smooth measures μ which are nowhere Radon in the sense tha $\mu(G) = \infty$ for all non-empty open sets G of X.

For example, if $X = I\!R^d, d \geq 2$, and $(\mathcal{E}, \mathcal{F})$ is the classical Dirichlet form associated with the Brownian motion, then each single-set point is a set of zero capacity. Corollary 1.1.2 asserts that there are smooth measures on $I\!R^d$ which are nowhere Radon.
Let us give here some visual examples of nowhere Radon smooth measures and other strange smooth measures in connection to Theorem 1.1.1.

<u>Example 1</u> Let $\{x_j\}, j \geq 1$ be a dense subset of $I\!R^d (d \geq 2)$ and let $\{\alpha_j\}, j \geq 1$ with $\alpha_j \geq d \; \forall j$. Then one can show that there exists a sequence of strictly positive real numbers $\{c_j\}, j \geq 1$ s.t. $f(x)dx$ is a nowhere Radon smooth measure (with respect to the classical Dirichlet form associated with the Laplacian), where $f(x)$ is given by

$$f(x) := \sum_{j \geq 1} c_j |x - x_j|^{-\alpha_j} .$$

Remark In [SV] Stollmann and Voigt constructed a regular potential V satisfying the property $\int_G |V(x)|^p dx = \infty$ for any non-empty open set G and any $p > 0$. It is unknown whether $V(x)dx$ is a smooth measure or not. The above f is similar to the construction of V in [SV], but with different choice of $\{c_j\}, j \geq 1$. Nevertheless, since $W_2^1(\mathbb{R}^d) \cap L^2(\mathbb{R}^d, \mu)$ is dense in $L^2(\mathbb{R}^d, \nu)$ for any smooth measure μ (cfr. [AM2]), we can show that the above function f is regular in the sense of [SV].

Example 2 Let $X = D \cup \partial D$ where D is a bounded domain of $\mathbb{R}^d (d \geq 2)$ with C^3 boundary ∂D. Let m be the Lebesgue measure on X and $(\mathcal{E}, \mathcal{F})$ be the maximal Markovian extension of the form

$$\mathcal{E}(u,u) = \frac{1}{2} \int |\nabla u|^2 m(dx), \quad u \in C_0^\infty(D).$$

Then $(\mathcal{E}, \mathcal{F})$ is a regular Dirichlet form on $L^2(X; m)$ corresponding to the Laplacian operator with Neumann boundary condition on ∂D. Denote by ν the area measure of ∂D. Obviously ν is singular with respect to m. But ν is a smooth measure. We can also prove that each single point of ∂D is of zero capacity. Thus by Theorem 1.1.1 there exists a smooth measure μ concentrated on ∂D (hence singular with respect to m) such that $\mu(G) = \infty$ for all non-empty relatively open subsets G on ∂D.

Example 3 Let $X = \mathbb{R}^{3N}$ and let us write $x \in \mathbb{R}^{3N}$ by $x = \{x_1, \ldots, x_N\}$ with $x_i \in \mathbb{R}^3$. Set

$$\phi(x) = \frac{1}{4\pi} \sum_{i<j}^{N} |x_i - x_j|^{-1} \exp(-\lambda |x_i - x_j|)$$

for some $\lambda \geq 0$. Let $m(dx) = \phi^2(x)dx$ and define

$$E(u,v) = \int_X \nabla u \cdot \nabla v m(dx)$$

for u and v in $C_0^1(\mathbb{R}^{3N})$. Then E is positive and closable and it produces a regular Dirichlet form $(\mathcal{E}, \mathcal{F})$ on $L^2(X, m)$ [AHKS]. (The energy operator H associated with $\mathcal{E}$ is a realization of the Hamiltonian of N particles interacting by δ-interactions.) Notice that in this case each single-point set is of zero capacity.
Denote by

$$D = \{x = \{x_1, \ldots, x_N\} : x_i = x_j \text{ for some } 1 \leq i < j \leq N\}.$$

Applying Theorem 1.1.1 we can construct a smooth measure μ such that $\mu(G) = \infty$ for all open sets G such that $G \cap D \neq \emptyset$. Now consider the positive quadratic form $\mathcal{E}^\mu$:

$$\mathcal{E}^\mu(u,v) = \mathcal{E}(u,v) + \int_X uv\mu(dx), \quad u, v \in \mathcal{F} \cap L^2(\mu).$$

$\mathcal{E}^\mu$ is then a Dirichlet form (cf. Theorem 2.1.2 (iv) below). In this way we obtain a self-adjoint operator $H^\mu := H + \mu$ which describes a Hamiltonian describing N particles interacting by δ-interactions plus Coulomb-like interactions.

1.2 Smooth measures in Kato class

For a given smooth measure μ, we shall denote by A^μ the unique (up to equivalence) PCAF such that μ is the Revuz measure of A^μ.
We denote by $\mathcal{B}(X)$ the family of Borel functions on X. For $f \in \mathcal{B}(X)$, we introduce the norm:

$$\|f\|_q = \inf_{Cap(N)=0} \sup_{x \in X-N} |f(x)| \,. \tag{1.2.1}$$

Now we can state

1.2.1 Definition A smooth measure μ is said to be in Kato class, if

$$\lim_{t \downarrow 0} \|E_\cdot A_t^\mu\|_q = 0. \tag{1.2.2}$$

We shall denote by S_K the set of all smooth measures in Kato class. To justify the name we chose here, we should remark that in the classical case of X_t being Brownian motion on $\mathbb{R}^d$, S_K coincides with the generalized Kato class GK_d introduced in [BM], that is, $\mu \in S_K$ if and only if (we take for example $d \geq 3$),

$$\lim_{\alpha \downarrow 0} \sup_x \int_{|x-y| \leq \alpha} \frac{\mu(dy)}{|x-y|^{d-2}} = 0.$$

Consequently a function f is in Kato class in the sense of [AS], if and only if $f \cdot dx \in S_K$.
The importance of the class S_K is its connection to the Feynman-Kac semigroups as expressed by the following inequality.

1.2.2 Proposition Let $\mu \in S_K$. Then there exist constants c and β such that

$$\left\| E_\cdot e^{A_t^\mu} \right\|_q \leq c e^{\beta t}, \forall t > 0. \tag{1.2.3}$$

Let us denote by S_0 the measure of finite energy integral (c.f. [F1]) §3.2) and introduce the family $S_{K,0}$ as follows:

$$S_{K,0} = \{\mu \in S_K \cap S_0 : \mu(X) < \infty\}.$$

We now claim that any smooth measure μ can be approximated by measures in $S_{K,0}$.

1.2.3 Theorem A positive Borel measure μ on X is smooth if and only if there exists an increasing sequence $\{F_n\}_{n\geq 1}$ of compact sets satisfying the following properties:

(i) $$I_{F_n} \cdot \mu \in S_{K,0}, \forall n \geq 1 \quad , \tag{1.2.4}$$

(ii) $$\mu(X - \bigcup_{n=1}^{\infty} F_n) = 0, \tag{1.2.5}$$

(iii) $$\lim_{n\to\infty} Cap(K - F_n) = 0, \quad \text{for any compact set} \quad K. \tag{1.2.6}$$

The above approximation theorem is a powerful tool for studying the perturbation of Dirichlet forms as well as various properties of PCAF's. As an example we state here a very strong duality of PCAF's which includes [FO] Lemma 3.1 (i) as a special case.

1.2.4 Theorem Let $\mu_1, \mu_2, \nu_1, \nu_2 \in S$ and $f_1, f_2 \in \mathcal{B}^+(X)(\mathcal{B}^+(X)$ denotes nonnegative functions in $\mathcal{B}(X))$. Then for all $0 \leq T \leq \infty$,

$$E_{f_1\cdot\nu_1}\left[\int_0^T e^{A_t^{\mu_1}-A_t^{\mu_2}} f_2(X_t)dA_t^{\nu_2}\right] = E_{f_2\cdot\nu_2}\left[\int_0^T e^{A_t^{\mu_1}-A_t^{\mu_2}} f_1(X_t)dA_t^{\nu_1}\right]. \tag{1.2.7}$$

A simple application of Theorem 1.2.4 yields an analytic description of the class S_K, for which we need some more notations.
Let $(P_t)_{t>0}$ be the Markovian transition function of (X_t) and μ be a Borel measure on X. We set

$$\mu T_t f := \int_X \mu(dx) \int_0^t P_s f(x)ds, \tag{1.2.8}$$

and

$$\mu U^\alpha f := \int_X \mu(dx) \int_0^\infty e^{-\alpha s} P_s f(x)ds, \tag{1.2.9}$$

provided the right hand sides of (1.2.8), (1.2.9) make sense.

1.2.5 Theorem Let μ be a smooth measure, then the following assertions are equivalent to each other.

(i) $\mu \in S_K$;
(ii) $\lim_{\alpha\uparrow\infty} \|E_\cdot \int_0^\infty e^{-\alpha} dA_t^\mu\|_q = 0$;
(iii) μU^α is a bounded functional on $L^1(X;m)$ for each $\alpha > 0$ and $\lim_{\alpha\uparrow\infty} \|\mu U^\alpha\| = 0$;
(iv) μT_t is a bounded functional on $L^1(X;m)$ for each $t > 0$ and $\lim_{t\downarrow 0} \|\mu T_t\| = 0$.
In (iii) and (iv) $\|\cdot\|$ denotes the operator norm of a functional on $L^1(X;m)$.

2. Perturbations of Dirichlet forms

Let $(\mathcal{E},\mathcal{F})$ be a regular Dirichlet form on $L^2(X;m)$ as described in Section 1. We shall freely use the notations introduced previously. For a signed Borel measure $\mu=\mu^+-\mu^-$, we write $\mu\in S-S$ (resp.$\mu\in S-S_{K,0}, \mu\in S_{K,0}-S_{K,0}$, etc.) if $\mu^+\in S$ (resp. $\mu^+\in S_{K,0}$ etc.) and $\mu^-\in S$ (resp. $\mu^-\in S_{K,0}$ etc.) It is evident that $\mu\in S-S$ (resp. $S_{K,0}-S_{K,0}$) if and only $|\mu|\in S$ (resp. $S_{K,0}$). For $\mu\in S-S$ we shall write $A^\mu:=A^{\mu^+}-A^{\mu^-}$, and we call μ the Revuz measure of A^μ. From now on we shall sometimes use the short notation $L^2(\mu)$ for $L^2(X;\mu)$. The following notations will also be alternatively used provided the integral makes sense:

$$\int_X fg\mu(dx):=<f,g\cdot\mu>:=<f,g>_\mu .$$

Recall that a symmetric bilinear form defined on a subset of some Hilbert space is called a quadratic form if its form domain is dense in the Hilbert space. For a signed Borel measure μ on X, we define

$$Q_\mu(f,g):=<f,g>_\mu \quad , \quad f,g\in L^2(|\mu|+m). \tag{2.0.1}$$

Q_μ is naturally a symmetric bilinear form. In general it is not a quadratic form. But if $\mu\in S-S$, then, by virtue of the analytic description of S ((1.0.3)-(1.0.5)), it can be proven that $L^2(|\mu|+m)$ is dense in $L^2(m)$ and hence Q_μ is a quadratic form on $L^2(m)$. For $\mu\in S-S$, we define

$$\mathcal{E}^\mu(f,g):=\mathcal{E}(f,g)+Q_\mu(f,g):=\mathcal{E}(f,g)+<f,g>_\mu \quad , \quad \forall f,g\in\mathcal{F}^\mu \tag{2.1.1}$$

where

$$\mathcal{F}^\mu:=\mathcal{F}\cap L^2(|\mu|+m). \tag{2.1.2}$$

In what follows we shall give the criteria of lower semiboundedness and closability of $(\mathcal{E}^\mu,\mathcal{F}^\mu)$, and discuss the form core and regularity when it is closed.

2.1 Closability and lower semiboundedness

It is well known that there is a one-to-one correspondence between the family of the closed lower semibounded quadratic forms and the family of the lower semibounded self-adjoint operators. Consequently the discussion of the closability and the lower semiboundedness of $(\mathcal{E}^\mu,\mathcal{F}^\mu)$ is of great interest.
We remark first that even for $\mu\in S$ the positive quadratic form Q_μ is in general not closable. A simple example is the Dirac measure $\delta(dx)$ on $\mathbb{R}$. The quadratic form Q_δ is written as

$$Q_\delta(f,g)=f(0)g(0) \quad , \quad f,g\in L^2(\delta+dx).$$

It is easy to see that Q_δ is not closed or even not closable in $L^2(\mathbb{R};dx)$. But we remark that δ is a very nice smooth measure in Kato class with respect to the classical Dirichlet

form on $L^2(\mathbb{R}; dx)$.
We have a simple criterion for the closability of a positive quadratic form Q_μ as follows.

2.1.1 Proposition Let μ be a positive Borel measure on X such that $L^2(\mu + m)$ is dense in $L^2(m)$. Then Q_μ is closable if and only if μ is absolutely continuous with respect to m. Furthermore, if Q_μ is closable, then it is closed.

We shall see that in many cases $(\mathcal{E}^\mu, \mathcal{F}^\mu)$ is a closed quadratic form no matter whether the form Q_μ is closable or not. Let us introduce the following notations:

$$P_t^\mu f(x) := E_x \left[e^{-A_t^\mu} f(X_t) \right] \tag{2.1.3}$$

$$U_\alpha^\mu f(x) := E_x \left[\int_0^\infty e^{-A_t^\mu - \alpha t} f(X_t) dt \right] \tag{2.1.4}$$

provided the right hand sides make sense.
We consider first the easier case of $\mu \in S - S_{K,0}$.

2.1.2 Theorem Let $\mu \in S - S_{K,0}$. Then
(i) $(\mathcal{E}^\mu, \mathcal{F}^\mu)$ is a lower semibounded closed quadratic form;
(ii) $(P_t^\mu)_{t \geq 0}$ is the unique strongly continuous semigroup in $L^2(m)$ corresponding to $(\mathcal{E}^\mu, \mathcal{F}^\mu)$;
(iii) $\mathcal{F}^\mu = \mathcal{F} \cap L^2(\mu^+ + m)$;
(iv) If $\mu \in S$, then $(\mathcal{E}^\mu, \mathcal{F}^\mu)$ is a Dirichlet form.

2.1.3 Corollary
(i) If $\mu \in S$, then $\mathcal{F} \cap L^2(\mu + m)$ is dense in $L^2(m)$.
(ii) Let $\mu \in S - S_{K,0}$. Then there exists $\alpha > 0$ such that

$$U_\alpha^\mu(L^2(m)) \subset L^2(m) \quad . \tag{2.1.5}$$

It is easy to see that if P_t^μ is a strongly continuous semigroup, then (U_α^μ) is its resolvent and hence (2.1.5) is trivial. We isolate (2.1.5) because it becomes a criterion for $\mathcal{E}^\mu$ to be semibounded and for (P_t^μ) to be strongly continuous in the case of $\mu \in S - S$. More specifically we have the following result.

2.1.4 Theorem Let $\mu \in S - S$. Then the following assertions are equivalent to each other.
(i) There exists $\alpha > 0$ such that $U_\alpha^\mu(L^2(m)) \subset L^2(m)$
(ii) $(P_t^\mu)_{t \geq 0}$ is a strongly continuous semigroup on $L^2(m)$.
(iii) Q_{μ^-} is relatively form bounded with respect to $\mathcal{E}^{\mu^+}$ with bound equal or less then one.
(iv) $\mathcal{E}^\mu$ is lower semibounded.

2.1.5 Corollary Let $\mu \in S - S$. Then Q_{μ^-} is relatively form bounded with respect to $\mathcal{E}^{\mu^+}$ with bound less than one if and only if for some $\alpha > 0$ and $\varepsilon > 0$.

$$U_\alpha^{\mu-\varepsilon\mu^-}(L^2(m)) \subset L^2(m) \quad . \tag{2.1.6}$$

For applications of Corollary 2.1.5 see [AM2].
By KLMN Theorem (c.f. [RS2] Th. X. 17), if Q_{μ^-} is relatively form bounded with respect to $\mathcal{E}^{\mu^+}$ with bound less than one, then $\mathcal{E}^\mu$ is closable. Applying KLMN Theorem, monotone convergence theorem for forms (c.f. [RS1], Th. p.16) as well as our Theorems 2.1.2. and 1.2.3 we obtain:

2.1.6 Proposition Let $\mu \in S - S$ such that (2.1.6) holds, then Theorem 2.1.2 (i)-(iii) holds.
According to a general theorem of Simon (c.f. [S2] or [RS1], Th. p.15) for an arbitrary lower semibounded quadratic form Q, there exists a unique largest closable lower semibounded quadratic form Q^r that is smaller than Q. But we notice that in general Q^r has nothing to do with Q, even there is a decreasing sequence of $Q^{(n)}$ such that $Q = \lim_{n\to\infty} Q^{(n)}$ and the associated semigroup $(P_t^{(n)})$ converges to the corresponding semigroup of Q^r in strong resolvent sense (c.f. [S1] p. 374, Example). But in our case we are lucky. Let $\mu \in S - S$ such that $(\mathcal{E}^\mu, \mathcal{F}^\mu)$ is lower semibounded. Denote by $(\mathcal{E}^{\mu r}, \mathcal{F}^{\mu r})$ the largest closed lower bounded quadratic form that is smaller than $(\mathcal{E}^\mu, \mathcal{F}^\mu)$. Then we have enough reason to realize $(\mathcal{E}^{\mu r}, \mathcal{F}^{\mu r})$ as a "weak closure" of $(\mathcal{E}^\mu, \mathcal{F}^\mu)$, as stated in the following theorem:

2.1.7 Theorem Let $\mu \in S - S$ such that $(\mathcal{E}^\mu, \mathcal{F}^\mu)$ is lower semibounded. Let $(\mathcal{E}^{\mu r}, \mathcal{F}^{\mu r})$ be specified as above. Then there exists a subset $\mathcal{F}_0^\mu \subset \mathcal{F}^\mu$ such that

(i) $\mathcal{F}_0^\mu$ is simultaneously a form core for $\mathcal{E}^{\mu^+}$ and $\mathcal{E}^{\mu r}$ respectively, and $\mathcal{F}_0^\mu$ is a dense subset of $L^2(\mu^-)$; (2.1.7)

(ii) $$\mathcal{E}^\mu(f,g) = \mathcal{E}^{\mu r}(f,g) \quad , \quad \forall f, g \in \mathcal{F}_0^\mu. \tag{2.1.8}$$

Furthermore, $(P_t^\mu)_{t\geq 0}$ (defined by (2.1.3)) is exactly the strongly continuous semigroup associated with $(\mathcal{E}^{\mu r}, \mathcal{F}^{\mu r})$.

Let $\mu \in S - S$ such that one of the assertions of Theorem 2.1.4 hold. By virtue of Theorem 2.1.7, and also by noticing that if μ satisfies (2.1.6) then $(\mathcal{E}^{\mu r}, \mathcal{F}^{\mu r})$ coincides with $(\mathcal{E}^\mu, \mathcal{F}^\mu)$, we shall therefore refer to $(\mathcal{E}^{\mu r}, \mathcal{F}^{\mu r})$ as the perturbation of $\mathcal{E}$ by μ.

Remark In general however $\mathcal{E}^\mu$ is not closable even though $\mathcal{E}^\mu$ is lower semibounded and hence $\mathcal{E}^{\mu r}$ is closable. In general we have only $\mathcal{E}^{\mu r} \leq \mathcal{E}^\mu$.
Concerning the closability we have the following result

2.1.8 Theorem Let $\mu \in S - S$. In order that $(\mathcal{E}^\mu, \mathcal{F}^\mu)$ is bounded below and closable, and the corresponding operator domain is contained in $\mathcal{F}^\mu$, a necessary and and sufficient condition is that there exists $\alpha > 0$ such that

$$U^\mu_\alpha(L^2(m)) \subset L^2(m + \mu^-) \quad . \tag{2.1.9}$$

Remark It seems to us that even in the classical case, there is no prior result other than Theorem 2.1.8, concerning necessary and sufficient conditions for $(\mathcal{E}^\mu, \mathcal{F}^\mu)$ to be bounded below and closable and the corresponding operator domain to be contained in $\mathcal{F}^\mu$.

2.2 About the form core and regularity

Recall that a Dirichlet form $(Q, \mathcal{G})$ on $L^2(X; m)$ is called regular if there is a form core $\mathcal{G}_0 \subset C_0(X)$ such that $\mathcal{G}_0$ is dense in $C_0(X)$ in uniform norm, and in this case $\mathcal{G}_0$ is called a core of Q (c.f. [F1] pp 5-6).
Let us set $\mathcal{F}_0 := \mathcal{F} \cap C_0(X)$. Then $\mathcal{F}_0$ is a form core of $(\mathcal{E}, \mathcal{F})$. Furthermore, $\mathcal{F}_0$ satisfies the following property (c.f. [F1] Th. 4.4.2):

2.2.1 Property Let $G \subset X$ be an open set, and let $u \in \mathcal{F}$ such that u is bounded and $supp[u] \subset G$. Then there exists a uniformly bounded sequence $\{u_i\}_{i \geq 1} \subset \mathcal{F}_0 \cap C_0(G)$ such that $u_i \to u$ in $\mathcal{E}_1$ norm.
Our first observation concerning the regularity of $\mathcal{E}^\mu$ is the following.

2.2.2 Theorem Let $\mu \in S$. Then the Dirichlet form $(\mathcal{E}^\mu, \mathcal{F}^\mu)$ is regular if and only if μ is a Radon measure on X. If μ is a Radon measure then any core of $(\mathcal{E}, \mathcal{F})$ satisfying property 2.2.1 is again a core of $(\mathcal{E}^\mu, \mathcal{F}^\mu)$.
In the general case of $\mu \in S - S$, we have the following result.

2.2.3 Theorem Let $\mu \in S - S$ satisfying (2.1.6). Then $\mathcal{F}^\mu \cap C_0(X)$ is a form core of $(\mathcal{E}^\mu, \mathcal{F}^\mu)$ if and only if there is an open set X_0 of X such that $Cap(X - X_0) = 0$ and μ restricted to X_0 is a signed Radon measure on X_0. In the latter case $\mathcal{F}_0 \cap C_0(X)$ is a form core of $(\mathcal{E}^\mu, \mathcal{F}^\mu)$ provided $\mathcal{F}_0$ is a core of $(\mathcal{E}, \mathcal{F})$ and $\mathcal{F}_0$ satisfies property 2.2.1.

Remark This extends some prior results, like e.g. [AHKS] Lemma 2.4.

The above theorem suggests that in general $\mathcal{F}^\mu \cap C_0(X)$ may not be a form core of $(\mathcal{E}^\mu, \mathcal{F}^\mu)$. In fact, if μ is a nowhere Radon smooth measure, then $\mathcal{F}^\mu$ contains even no non-trivial continuous functions. Thus it is desirable to find a relatively nice class of functions which can be used for constituting a form core of a perturbed quadratic form. To this end we define a class of functions $C_q(X)$ as follows.
$C_q(X) := \{f \in \mathcal{B}(X) :$ f is bounded quasi-continuous and with compact support $\}$.
Now we can state the following theorem.

2.2.4 Theorem Let $\mu \in S-S$ such that $\mathcal{E}^{\mu}$ is lower semibounded. Then $\mathcal{F}^{\mu r} \cap C_q(X)$ is a form core of $\mathcal{E}^{\mu r}$. Furthermore, there exists a form core $\mathcal{F}^{\mu}_{0,0} \subset \mathcal{F}^{\mu} \cap C_q(X)$ such that (2.1.7) and (2.1.8) hold.

Acknowledgements We would like to thank the Organizers for a very kind invitation. We are grateful to Prof. W. Karwowski and Prof. M. Röckner for useful discussions. The second named author would like to thank Prof. Ph. Blanchard, Prof. W. Hansen and Prof. L. Streit for the hospitality at the Bielefeld University. He would also like to thank v. Humboldt-Stiftung for financial support.

References

[A] Albeverio, S.: Some points of interaction between stochastic analysis and quantum theory, in *Stochastic Differential Systems*, 1–26, Proc. 3d Bad Honnef conf. 1985, Edts. W. Christopeit, K. Helmes, M. Kohlmann, Lect. Notes Control Inform. Sciences, Springer, Berlin (1986)

[ABR] Albeverio, S.; Brasche, J.; Röckner, M.: Dirichlet forms and generalized Schrödinger operators, to appear in *Lectures on Schrödinger operators*, Søndeborg Nordic School, Edts. H. Holden, A. Jensen, Lect. Notes Maths., Springer, Berlin (1989)

[AFHKL] Albeverio, S.; Fenstad, J. E.; Høegh-Krohn, R.; Lindstrøm, T: Non standard methods in stochastic analysis and mathematical physics, Academic Press, New York (1986)

[AGHKH] Albeverio, S.; Gesztesy, F.; Høegh-Krohn, R.; Holden, H.: Solvable models in quantum mechanics, Springer Verlag, Berlin (1988)

[AHKS] Albeverio, S.; Høegh-Krohn, R; Streit, L.: Energy forms, Hamiltonians, and distorted Brownian paths, J. Math. Phys. **18**, 907–917 (1977)

[AM1] Albeverio, S.; Ma, Zhiming: Additive functionals, nowhere Radon and Kato class smooth measures associated with Dirichlet forms, Bielefeld-Bochum Preprint (1988)

[AM2] Albeverio, S.; Ma, Zhiming: Perturbation of Dirichlet forms - lower semiboundedness, closability and form cores, Bielefeld-Bochum Preprint (1988)

[AR1] Albeverio, S.; Röckner, M.: Classical Dirichlet forms on topological vector spaces - closability and a Cameron-Martin formula, BiBoS-SFB237-Preprint (1988), (to appear in J. Funct. Anal.)

[AR2] Albeverio, S.; Röckner, M.: Classical Dirichlet forms on topological vector spaces - the construction of the associated diffusion process, SFB-237-Preprint (1988)

[AS] Aizenman, M.; Simon, B.: Brownian motion and Harnack's inequality for Schrödinger operators, Comm. Pure Appl. Math. **35**, 209–971 (1982)

[B1] Brasche, J: Perturbation of Schrödinger Hamiltonians by measures - Self-adjointness and lower semiboundedness, J. Math. Phys. **26**, 621–626 (1985)

[B2] Brasche, J: Perturbations of self-adjoint operators supported by null sets, Ph. D. Thesis, Bielefeld (1989)

[BG] Blumenthal, R.M.; Getoor, R.K.: Markov processes and Potential Theory, Academic Press, New York and London, 1968

[BHH] Boukricha, A.; Hansen, W.; Hueber, H.: Continuous solutions of the generalized Schrödinger equation and perturbation of harmonic spaces, Exp. Math. **5**, 97–135 (1987)

[BlH] Bliedtner, J.; Hansen, W.: Potential Theory, Springer, Berlin (1986)

[BM] Blanchard, Ph.; Ma, Zhiming: Semigroup of Schrödinger operator with potentials given by Radon measures, BiBoS preprint No. 262 (1987)

[DevC] Demuth, M.; Van Carsteren, J. A.: On spectral theory for Feller generators, Antwerpen Preprint (1988)

[DM] Dellacherie, C.; Meyer, P. A.: Probabilités et Potentiel, Ch. XII-XVI, Hermann, Paris (1987)

[F1] Fukushima, M.: *Dirichlet forms and Markov processes*, Kodansha and North Holland (1980)

[F2] Fukushima, M.: On two classes of smooth measures for symmetric Markov processes, Osaka preprint

[F3] Fukushima, M.: Energy forms and diffusion processes, in Mathematics and Physics, Lectures on recent results I, Ed. L. Streit, World Scientific, Singapore (1985)

[FO] Fukushima, M.; Oshima, Y.: On skew product of symmetric diffusion processes, preprint

[MT] McKean, H. P.; Tanaka, H.: Additive functionals of the Brownian path, Memoirs Coll. Sci. Univ. Kyoto, A. Math., **33**, 479–506 (1961)

[MD] Dynkin, E. B.: Markov process, Springer Verlag (1965)

[Na] Nakao, S.: Stochastic calculus for continuous additive functionals of zero energy, Z. Wahrscheinlichkeitsth. verw. Geb. **68**, 517–578 (1985)

[O] Oshima, Y.: *Lecture on Dirichlet spaces*, Universität Erlangen-Nürnberg, May-July 1988

[R] Revuz, D.: Mesures associées aux fonctionnelles additives de Markov I, Trans. Amer. Math. **70**, 43–72 (1959)

[RG] Blumenthal, R. M.; Getoor, R. K.: Markov processes and Potential theory, Academic Press, New York and London (1968)

[RS1] Reed, M.; Simon, B.: Methods of Modern Mathematical Physics, Vol. 1, Academic Press, New York (1972)

[RS2] Reed, M.; Simon, B.: Methods of Modern Mathematical Physics Vol. 2, Academic Press, New York (1975)

[S1] Simon, B.: Schrödinger semigroups, Bull. Amer. Math. Soc. (N.S.) **7**, 447–526 (1982)

[S2] Simon, B.: A canonical decomposition for quadratic forms with application to monotone convergence therorems, J. Func. Anal. **28**, 377–385 (1978)

[SV] Stollmann, P.; Voigt, J.: A regular potential which is nowhere in L_1, Lett. Math. Phys. **9**, 227-230 (1988)

[St] Sturm, Th.: Störung von Hunt-Prozessen durch signierte additive Funktionale, Diss., Erlangen (1988)

A GENERALIZATION OF ITO'S FORMULA

A.N. ALHUSSAINI
Department of Statistics and Applied Probability
University of Alberta
Edmonton, Alberta
Canada T6G 2G1

O. SUMMARY

We discuss and further generalize recent extension of Ito's formula to absolutely continuous functions. Some applications are given.

1. INTRODUCTION

X will be a real valued continuous semimartingale $\{X_t : t \geq 0\}$ defined on a filtered probability space $(\Omega, \mathcal{F}, P)$ which satisfies the usual conditions. By considering X^{T_n} where $T_n = \inf(t : |X_t| \geq n)$ we may and do assume that X_t is bounded. $L_t^a =$ local time of X_t up to time t. L_t^a will be the version which is right continuous in $t, a,$ left limited in a and continuous in t, Yor [10]. In the sequel unless stated otherwise, X_t will be a continuous and bounded semimartingale.

For a real a, using Tanak's formula $(X_t - a)^- = (X_0 - a)^- - \int_0^t 1_{X_s \leq a} \, dX_s + \frac{1}{2} L_t^a$, where L_t^a : local time at a upto time t.

Bouleau and Yor [2] have shown:

$$F(X_t) = F(X_0) + \int_0^t \frac{\partial F(X_s)}{\partial x} dX_s - \frac{1}{2} \int_{-\infty}^{+\infty} \frac{\partial F}{\partial x}(a) d_a L_t^a$$

where if $f(u) = \sum_{i=1}^m f_i \, 1_{(a_i, a_{i+1}]}(u)$ then $\sum_{i=1}^n f_i (L_t^{a_{i+1}} - L_t^{a_i})$ is the integral of f relative to $d_a L_t^a$ i.e.

$$\int f(a) d_a L_t^a \stackrel{\text{def}}{=} \sum_{i=1}^n f_i (L_t^{a_{i+1}} - L_t^{a_i})$$

which extends Bouleau and Yor [2] to a vector measure on the Borel sets of the real numbers R

with values in $L^2(\mathcal{F},P)$ so that

$$F(X_t) = F(X_0) + \int_0^t f(X_s)dX_s - \frac{1}{2}\int_{-\infty}^{+\infty} f(a)d_a L_t^a$$

whenever f is locally bounded measurable and $F(x) = \int_0^x f(u)du$.

Yor [9] applied the above to approximate "zero energy" processes discussed previously by T. Yamada [8]. Section (2) is from Alhussaini and Elliott [1]. Section (3) is the same vein as Section (1) but we here use Kunita [6]. A remark about convex functions follows in (4). In Section (5) we discuss briefly similar results for Brownian sheet. Meyer [7] is used freely to justify some of the manipulations below.

2. FIRST GENERALIZATION

A more general than above is the following theorem in Alhussani and Elliott [1].

THEOREM. *If* $F \in C^{1,2}([0,\infty)\times R)$ *then*

$$F(t,X_t) = F(0,X_0) + \int_0^t \frac{\partial F}{\partial t}(s,X_s)ds + \int_0^t \frac{\partial F}{\partial x}(s,X_s)dX_s - \frac{1}{2}\int_{-\infty}^{+\infty}\frac{\partial F}{\partial x}(t,a)d_a L_t^a + \frac{1}{2}\int_0^t\int_{-\infty}^{+\infty}\frac{\partial^2 F}{\partial t\partial x}(s,a)d_a L_s^a ds$$

COROLLARY. *If* $F(t,x)$ *is continuous differential in* t *and absolutely continuous in* x *with locally bounded* $\frac{\partial F}{\partial x}$, $F(t,0) = 0$ *and if* $\frac{\partial F}{\partial t}(t,x) = \int_0^x \frac{\partial^2 F}{\partial t\partial x}(t,y)dy$, *where* $\frac{\partial^2 F}{\partial t\partial x}$ *is locally bounded, then*

$$F(t,x) = F(0,X_0) + \int_0^t \frac{\partial F}{\partial t}(s,X_s)ds + \int_0^t \frac{\partial F}{\partial x}(s,X_s)dX_s - \frac{1}{2}\int_{-\infty}^{+\infty}\frac{\partial F}{\partial x}(t,a)d_a L_t^a + \frac{1}{2}\int_0^t\int_{-\infty}^{+\infty}\frac{\partial^2 F}{\partial t\partial x}(s,a)d_a L_s^a ds$$

This corollary follows from the above theorem by mollifing.

<u>Approximation I</u> Specialize to standard Brownian motion B_t, and suppose $F(t,x)$ is C' in t and $\frac{\partial F}{\partial x} \in L^2_{loc}([0,\infty)\times R)$, then $\frac{\partial^2 F}{\partial x^2}$ exists in the sense of Schwartz distribution.

DEFINITION:

$$A_t^F = \int_0^t \frac{\partial^2 F}{\partial x^2}(s, B_s)ds$$
$$\stackrel{\text{def}}{=} 2\Big(F(t, B_t) - F(0,0) - \int_0^t \frac{\partial F}{\partial x}(s, B_s)dB_s - \int_0^t \frac{\partial F}{\partial t}(s, B_s)dB_s\Big)$$

This definition was introduced by Fukushima ([3]. [4]).

THEOREM [1]. *If* $F(t, x)$ *is continuously differentiable in* t *and twice continuous differentiable in* $x \neq 0$.

Let $f(t, x) = \frac{\partial F}{\partial x}(t, x)$ *and suppose for some* $T > 0$,

$$\sup_{t \leq T} |f(t, x)| \in L^2_{loc}(R),$$
$$\sup_{t \leq T} |\frac{\partial f}{\partial t}(t, x)| \in L^1_{loc}(R).$$

Then

$$\lim_{\epsilon \to 0} E\big[\sup_{t \leq T} |A_t^F - \{\int_0^t \frac{\partial^2 F}{\partial x^2}(s, B_s)1_{|B_s| \geq \epsilon}\, ds + \int_0^t f(s, \epsilon)d_s L_s^\epsilon$$
$$- \int_0^t f(s, -\epsilon)d_s L_s^{-\epsilon}\}|^p\big] = 0, \quad 1 \leq p < +\infty.$$

EXAMPLE:

$$F(t, x) = \phi(t)(x \log x - x) \quad x > 0$$
$$= 0 \qquad x \leq 0.$$

Then

$$A_t^F = \text{Principal value of} \int_0^t \frac{\phi(s)}{(B_s)_+} ds$$
$$= \lim_{\epsilon \to 0}\{\int_0^t \frac{\phi(s)}{B_s} 1_{B_s \geq \epsilon}\, ds + \log \epsilon \int_0^t \phi(s) d_s L_s^\epsilon\}$$

3. SECOND GENERALIZATION

First we state:

THEOREM (KUNITA) [6]. *Let* $F_t(x)$, $t \geq 0$, $x \in R^d$ *be continuous in* (t,x) *a.e. satisfying:*

(i) $F_t(\cdot) \in C^2$.

(ii) For each x, $F_t(x)$ *is a continuous semimartingale represented as:*

$$F_T(x) = F_0(x) + \sum_{j=1}^{m} \int_0^t f_s^j(x)dN_s^j$$

where $N_s^1,\ldots,N_s^m$ *are continuous semimartingale,* $f_s^j(x)$, $s \geq 0$, $x \in R^d$ *are continuous in* (s,x) *such that:*

(a) For each s, $f_s^j(x) \in C^1(R^d)$

(b) For each x, $f_s^j(x)$ *are adapted.*

Let $M_t = (M_t^1,\ldots,M_t^d)$ *be a continuous semimartingale. Then*

$$\begin{aligned} F_t(M_t) = F_0(M_0) &+ \sum_{j=1}^{m} \int_0^t f_s^j(M_s)dN_s^j + \sum_{i=1}^{d} \int_0^t \frac{\partial F_s}{\partial x_i}(M_s)dM_s^i \\ &+ \sum_{i=1}^{d}\sum_{j=1}^{m} \int_0^t \frac{\partial f_s^j}{\partial x_i}(M_s)d\langle N^j, M^i\rangle_s \\ &+ \frac{1}{2}\sum_{i,j=1}^{d} \int_0^t \frac{\partial^2 F_s(M_s)}{\partial x_i \partial x_j}(M_s)d\langle M^i, M^j\rangle_s. \end{aligned}$$

Let $L_t^a =$ *local time at* a *up to time* t *for martingale* M.

THEOREM. *If in Kunita's theorem* $m=1$, $d=1$ *and condition (i) is replaced by* $\frac{\partial}{\partial t}\frac{\partial F}{\partial x}$ *is locally bounded. Then*

$$\begin{aligned} F_t(M_t) = F_0(M_0) &+ \int_0^t f_s(M_s)dN_s + \int_0^t \frac{\partial F_s}{\partial x}(M_s)dM_s \\ &+ \int_0^t \frac{\partial f_s}{\partial x}(M_s)d\langle N, M\rangle_s + \frac{1}{2}\int_0^t \int_{-\infty}^{+\infty} \frac{\partial^2 F_s}{\partial t \partial x}(a)d_a L_s^a ds \\ &- \frac{1}{2}\int_{-\infty}^{+\infty} \frac{\partial F_t(a)}{\partial x} d_a L_t^a \end{aligned}$$

PROOF: The proof is similar to the proofs of the theorem and corollary in the preceding section and as given in [1].

Approximation II. Observe that if $F_t(x) = \int_0^t f_s(x)ds$, then by Kunita [5]:

$$F_t(B_t) = \int_0^t f_s(B_s)ds + \int_0^t \frac{\partial F_s}{\partial x}(B_s)dB_s + \frac{1}{2}\int_0^t \frac{\partial^2 F_s}{\partial x^2}(B_s)ds.$$

Define:

$$A_t^F = \int_0^t \frac{\partial^2 F_s(B_s)}{\partial x^2} ds = 2\big(F_t(B_t) - \int_0^t f_s(B_s)ds - \int_0^t \frac{\partial F_s(B_s)}{\partial x} dB_s\big)$$

THEOREM. *Suppose the conditions of Kunita's theorem are satisfied by* $F_t(x) = \int_0^t f_s(x)ds$ *except that the condition (i) is only to hold for* $x \neq 0$, $\sup_{t\leq T} |\frac{\partial f_t(x)}{\partial x}| \in L^1_{loc}(R)$ *and* $\sup_{t\leq T} |\frac{\partial^2 f_t(x)}{\partial x^2}| \in L^2_{loc}(R)$. *Then*

$$\lim_{\epsilon\to 0} E\big[\sup_{t\leq T} |A_t^F - \Big\{\int_0^t \frac{\partial^2 F_s(B_s)}{\partial x^2} 1_{|B_s|\geq\epsilon}\, ds + \int_0^t \frac{\partial F_s(\epsilon)}{\partial x} d_s L_s^\epsilon - \int_0^t \frac{\partial F_s(-\epsilon)}{\partial x}\, d_s L_s^{-\epsilon}\Big\}|^p\big] = 0,$$
$$1 \leq p < +\infty.$$

PROOF: We shall assume that $f_t(0) = 0$, and $\frac{\partial f_t(0)}{\partial x} = 0$ for all t. Let

$$f_t(\varepsilon, x) = f_t(x) 1_{|x|\geq\epsilon} \quad \text{and}$$
$$F_t(\varepsilon, x) = \int_0^t f_s(\varepsilon, x) ds.$$

By the above theorem

$$F_t(\varepsilon, B_t) = \int_0^t f_s(\varepsilon, B_s)ds + \int_0^t \frac{\partial F_s(\varepsilon, B_s)}{\partial x} dB_s + \frac{1}{2}\int_0^t \int_{-\infty}^{+\infty} \frac{\partial^2 F_s}{\partial t \partial x}(\varepsilon, a) d_a L_s^a ds - \frac{1}{2}\int_{-\infty}^{+\infty} \frac{\partial F_t(\varepsilon, a)}{\partial x} d_a L_t^a.$$

As in [1] let

$$A_t^{F_\epsilon} = \frac{1}{2}\int_0^t \int_{-\infty}^{+\infty} \frac{\partial^2 F_s(\varepsilon, a)}{\partial t \partial x} d_a L_s^a ds - \frac{1}{2}\int_{-\infty}^{+\infty} \frac{\partial F_t(\varepsilon, a)}{\partial x} d_a L_t^a$$

then by manipulations similar to those detailed in [1] again, we have,

$$A_t^{F_\epsilon} = \int_0^t \frac{\partial F_s(\varepsilon)}{\partial x} d_s L_s^\epsilon - \int_0^t \frac{\partial F_s(-\varepsilon)}{\partial x} d_s L_s^{-\epsilon} + \int_0^t \frac{\partial^2 F_s}{\partial x^2}(B_s) 1_{|B_s|\geq\epsilon}\, ds$$

It remains to prove that $A_t^F \to A_t^{F_\epsilon}$ in L^p, $1 \leq p < +\infty$. Now:

$$A_t^F - A_t^{F_\epsilon} = 2(F_t(B_t) - \int_0^t f_s(B_s)ds - \int_0^2 \frac{\partial F_s(B_s)}{\partial x} dB_s) \\ - 2(F_t(\epsilon, B_t) - \int_0^2 f_s(\epsilon, B_s)ds - \int_0^t \frac{\partial F_s(\epsilon, B_s)}{\partial x} dB_s)$$

since

$$F_t(B_t) - F_t(\epsilon, B_t) = \int_0^t \int_0^{B_t} \frac{\partial f_s}{\partial x}(y) 1_{|y| \leq \epsilon}\, dy ds = A$$

$$\int_0^t f_s(B_s)ds - \int_0^t f_s(\epsilon, B_s)ds = \int_0^t f_s(B_s) 1_{|B_s| \leq \epsilon}\, ds =$$

$$\int_0^t \int_0^{B_s} \frac{\partial f_s(y)}{\partial x} 1_{|y| \leq \epsilon}\, dy ds = B$$

$$\int_0^t \frac{\partial F_s}{\partial x}(B_s) dB_s - \int_0^t \frac{\partial F_s(\epsilon, B_s)}{\partial x} dB_s = \int_0^t \int_0^s \frac{\partial f_r(B_s)}{\partial x} 1_{|B_s| \leq \epsilon}\, dr dB_s \\ = \int_0^t \int_0^s \int_0^{B_s} \frac{\partial^2 f_r}{\partial x^2}(y) 1_{|y| \leq \epsilon}\, dy\, dr\, dB_s = C$$

$E \sup_{t \leq T} |A|^p, E \sup_{t \leq T} |B|^p$ and $E \sup_{t \leq T} |C|^p$ converge to zero as $\epsilon \to 0$ using hypothesis and arguments similar to those applied in [1], which completes the proof.

This theorem applied to the example in Section (2) obtains the same results.

Here is a very simple

EXAMPLE:

$$F(x) = x^{\frac{7}{4}} \quad x > 0 \\ = 0 \quad x \leq 0$$

then:

$$A_t^F = \text{ Principal value of } \frac{21}{16} \int_0^t \frac{1}{(B_s^{\frac{1}{4}})_+} ds \\ = \lim_{\epsilon \to 0} \{ \frac{21}{16} \int_0^t \frac{1}{B_s^{\frac{1}{4}}} 1_{B_s \geq \epsilon}\, ds - \frac{7}{4} \epsilon^{\frac{3}{4}} L_t^\epsilon \}$$

4. A REMARK

Based on discussions in Section (1) it follows that if F is a convex function then

$$F(X_t) = F(X_0) + \int_0^t \frac{\partial F(X_s)}{\partial x} dX_s - \frac{1}{2}\int_{-\infty}^{+\infty} \frac{\partial F}{\partial x}(a) d_a L_t^a$$

for semimartingale $\{X_t\}$ and its local time L_t^a at a up to time t.

5. BROWNIAN SHEET

Let us recall Cairoli and Walsh [3].

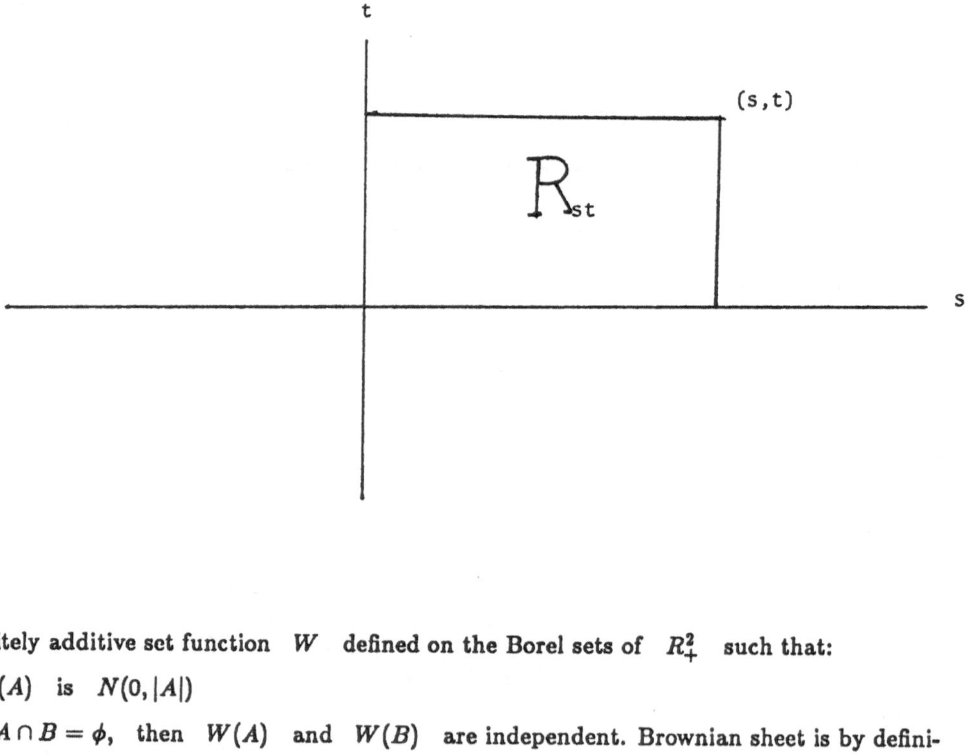

Finitely additive set function W defined on the Borel sets of R_+^2 such that:

(1) $W(A)$ is $N(0, |A|)$

(2) if $A \cap B = \phi$, then $W(A)$ and $W(B)$ are independent. Brownian sheet is by definition,

$$W_{st} \stackrel{\text{def}}{=} W(R_{st}),$$

where R_{st} is as shown in the figure above.

Local time: For any given t, $s \to W_{st}$ is a Brownian motion. Its local time is

$$L_1(x;s,t) = \frac{2}{t}[(W_{st} - x)^+ - x^- - \int_0^s 1_{W_{ut}>x}\, \partial_u W_{ut}]$$

(Tanaka's formula). $L_2(x;s,t)$ is defined similarly.

Local time for W_{st}: Let

$$L(x;s,t) = \int_0^t L_1(x;s,v)dv.$$

It follows that: $\int_0^s \int_0^t f(W_{uv})dudv = \int_0^{+\infty} L(x;s,t)ds$,

$$L(x;s,t) = \int_0^1 L_1(x;s,v)dv = \int_0^t L_2(x,u,t)du.$$

Define $\phi(x;s,t) = \int_0^s \int_0^t uvL(x;du,dv)$.

Ito's formula: Suppose f is four times continuously differential then:

$$\begin{aligned} f(W_{st}) = f(0) &+ \int_{R_{st}} f'(W_{uv})dw + \int_{R_{st}} f(W_{uv})dJ \\ &- \frac{1}{2}\int_{R_{st}} f''(W_{uv})dudv - \frac{1}{4}\int_{R_{st}} uvf''''(W_{uv})dudv \\ &- \frac{1}{2}\int_{\partial R_{st}} f''(udv - vdu). \end{aligned}$$

Here J is as defined in [3].

THEOREM.

$$\begin{aligned} \int_{R_{st}} uvf''''(W_{uv})dudv &= \int_{-\infty}^{+\infty} f''''(x)\phi(x;s,t)dx \\ &= -\int_{-\infty}^{+\infty} f'''(x)d_x\phi(x;s,t). \end{aligned}$$

COROLLARY. *Ito's formula holds when* $f''' = \int g$ *when* g *is locally integral.*

References

1. AlHussaini, A.N. and Elliott, R.J., *An extension of Ito's differentiation formula*, Nagoya Mathematical Journal **105** (1987), 9-118.

2. Bouleau, N. and Yor, M., *Sur la variation quadratic des temps locaux de certaines semimartingales*, C.R. Acad. Sci. Paris **292**.

3. Cairoli, R. and Walsh, J.B., *Stochastic integrals in the plane*, Acta Math. **134** (1975), 434-448.

4. Fukushima, M., *A decomposition of additive functionals of finite energy*, Nagoya Math. Journal . **74** (1979), 137-168.

5. Fukushima, M., "Dirichlet forms and Markov processes," North-Holland, 1981.

6. Kunita, H., *Some extensions of Ito's formula*, Lect. Notes in Math. No. **850**.

7. Meyer, P.A., *Un semimartingales sotochastiques*, Sem. de Probabilités X, Lect. Notes in Math. **511**, 245-400.

8. Yamada, T., *On some representations concerning stochastic integrals.* (preprint).

9. Yor, M., *Sur la transformations de Hilbert des temps locaux Browniens, et une extension de la formula d'Ito*, Sem. de Probabités XVI, Lect. Notes in Math. **920**, 238-247.

10. Yor, M., *Astérique*, Soc. Math. de France (1978), 52-53.

GENERAL FUNCTIONAL LIMIT THEOREMS FOR SEMIMARTINGALES

Svetlana Anulova
Institute of Control Sciences
Profsoyuznaya, 65, Moscow, 117 806 GSP V-342

In this paper sufficient conditions are given for weak convergence of random processes to a semimartingale with given infinitesimal characteristics. For limit processes jump-diffusions in a region with boundary conditions and their generalizations serve.

This problem has been studied by Grigelionis and Mikulevičus for a jump-diffusion with singularities on certain surfaces as a limit process ([1] and other papers). We wanted to obtain results for the most general limit semimartingales, in particular, to admit randomness and dynamics in localization of singularities. This aim is achieved. Besides we have improved the result of Grigelionis and Mikulevičius: we have rejected an assumption of "almost continuity" on limit characteristics and that of uniform integrability of prelimiting local times.

Unfortunately choosing examples to illustrate our results we had to restrict ourselves to the simplest ones, for full value examples of processes with singularities demand too much description (semipermeable hyperplanes etc., cf. [2]).

Fix a positive integer d.

$D = \{X\}$ - the space of càdlàg R^d-valued functions on $[0,\infty)$ with Skorohod topology, $\mathbb{D} = \left\{\mathcal{D}_t = \sigma(X_s,\ s \le t),\ t \in [0,\infty)\right\}$, $\mathcal{D}_\infty = \bigvee_{t \ge 0} \mathcal{D}_t$.

$\mathcal{C}$ - the set of measurable numerical $\mathbb{D}$-adapted processes y such that the set Γ of discontinuity points of y as a function of (t,X) satisfies

$$\int_0^\infty (I_\Gamma(X))_t d\varphi_t = 0$$

for all X and continuous nondecreasing functions $\varphi : [0,\infty) \longrightarrow [0,\infty)$.

$E = \{(X,\varphi)\}$ - the space of R^{d+1}-valued functions on $[0,\infty)$, with first d coordinates X belonging to D and the last φ being a nondecreasing function equal to zero at zero, $\mathbb{E} = \{\mathcal{E}_t = \sigma((X,\varphi)_s, s \le t), t \in [0,\infty)\}$, $\mathcal{E}_\infty = \bigvee_{t \ge 0} \mathcal{E}_t$.

Let Q be a probability measure on $(E, \mathcal{E}_\infty)$. For an $\mathbb{E}$- Markov moment τ $Q_{\wedge\tau}$ denotes the distribution of the random process $((X,\varphi)_{t\wedge\tau}, t \in [0,\infty))$ with respect to the measure Q, for a random variable ξ on $(E, \mathcal{E}_\infty)$ $Q\xi$ denotes $\int_E \xi \, dQ$.

On spaces D and E Skorohod topology is considered.

$\mathcal{T}$ - the set of all $\mathbb{E}$-Markov moments τ with $\tau \vee \varphi_{\tau-} \vee \sup_{[0,\tau)} |X|$ bounded on E.

By convergence of measures on $(D, \mathcal{D}_\infty)$ and $(E, \mathcal{E}_\infty)$ we always mean weak convergence.

$C_b^{1,2}$ (resp. C_b^2) denotes $C_b^{1,2}([0,\infty)\times R^d)$ (resp. $C_b^2(R^d)$).

Suppose we are given some set $\mathcal{R}$ of bounded measurable $\mathbb{D}$-adapted càdlàg processes on $(D, \mathcal{D}_\infty)$ with f(X) belonging to $\mathcal{R}$ for any $f \in C_b^{1,2}$ and a locally bounded operator L, mapping $\mathcal{R}$ into the the set of $\mathbb{D}$-optional numerical processes (locally bounded means that for any $f \in \mathcal{R}$ $Lf_t(X)$ is bounded on sets of the form $\{(t,X): t \le c, \sup_{[0,t]} |X| \le c\}$, $c > 0$.

<u>Definition 1</u>. A probability measure P on $(D, \mathcal{D}_\infty)$ is called an $(\mathcal{R}, L)$-solution, if there exists a $\mathbb{D}^+$-adapted increasing continuous process φ such that for any $f \in \mathcal{R}$

$$f - \int_0^{\cdot} Lf_t \, d\varphi_t$$

is a local martingale with respect to a stochastic basis $(D, \mathcal{D}_\infty, \mathbb{D}, P)$.

<u>Definition 2.</u> A probability measure Q on $(E, \mathcal{E}_\infty)$ is called a weak $(\mathcal{R}, L)$ solution, if φ is continuous Q- a.s. and for every $f \in \mathcal{R}$

$$f - \int_0^{\cdot} Lf_t \, d\varphi_t$$

is an $\mathbb{E}$- local martingale.

<u>Definition 3.</u> A weak $(\mathcal{R}, L)$- solution Q is called special, if

there exists a sequence of $\mathbb{D}$- Markov moments $\tau_n \uparrow \infty$ such that for every $t>0$ $Q\,\varphi_{\tau_n}(X)\wedge t < \infty$, $n = 1, 2, \ldots$

<u>Theorem 1</u>. If a weak $(\mathcal{R},L)$- solution Q is special, then $Q\circ X^{-1}$ is an $(\mathcal{R},L)$- solution.

To prove this theorem one should find the increasing process corresponding to a properly localized measure $\mu(dtdX) = Q(\varphi(dt)dX)$ on $[0,\infty)\times D$.

<u>Definition 4</u>. Let τ be an $\mathbb{E}$- Markov moment. A probability measure Q on $(E, \mathcal{E}_\infty)$ is called a weak $(\mathcal{R},L)$- solution up to τ if $Q = Q_{\wedge\tau}$ and for every $f \in \mathcal{R}$

$$f - \int_0^{\cdot} Lf_t\,d\varphi_t$$

is a Q- local martingale. Such solutions are called local.

<u>Definition 5</u>. We say that an operator L does not admit exploding solutions if for every positive t,R

$$\lim_C \sup_Q Q(\,|X_0| \le R,\quad \sup_{[0,t]} |X| \vee \varphi_t \ge C) = 0,$$

where the supremum is taken over all local weak $(\mathcal{R},L)$- solutions Q.

<u>Theorem 2</u>. Suppose an operator L does not admit exploding weak solutions, Q^n, $n = 1,2,\ldots$, is a sequence of probability measures on $(E, \mathcal{E}_\infty)$. If $Q^n\circ X_0^{-1}$ is relatively compact and for every continuous $\tau\in\mathcal{T}$:

1) $\Delta\varphi_\tau$ is Q^n- a.s. bounded by a constant independent of n, for every $\varepsilon > 0$

$$\lim_n Q^n(\sup_{[0,\tau]} |\Delta\varphi| > \varepsilon\,) = 0,$$

$$\lim_{h\downarrow 0} \overline{\lim_n} \sup_{\sigma} Q^n_{\wedge\tau}(\varphi_{\sigma+h} - \varphi_\sigma) = 0,$$

where the supremum is taken over all $\mathbb{E}$- Markov moments σ ;

2) $$\lim_C \overline{\lim_n}\, Q^n(\,|X_\tau| \ge C) = 0$$

and for every $f \in C_b^2$ there exists a c such that

$$\overline{\lim_n} \sup_{\sigma,\rho} Q^n_{\wedge\tau}(f(X_\sigma) - f(X_\rho) - c(\varphi_\sigma - \varphi_\rho)) \le 0,$$

where the supremum is taken over all $\mathbb{E}$- Markov moments ρ, σ, $\rho \le \sigma$;

3) for every $f \in \mathcal{R}$ one can construct a family of continuous maps $f^{\varepsilon}: D \longrightarrow D$ and of $L^{\varepsilon} f \in \mathcal{C}$, $\varepsilon \in (0,1]$, such that f^{ε} is bounded and $L^{\varepsilon} f$ is bounded on $[0, \tau)$ both uniformly in ε, $f_t^{\varepsilon}(X) \longrightarrow f_t(X)$, $L^{\varepsilon} f_t(X) \longrightarrow Lf_t(X)$ when $\varepsilon \longrightarrow 0$ for all (t,X), and for all $s,t \in [0,\infty)$ and $\mathcal{E}_s$-measurable continuous numerical function Φ with $|\Phi| \leq 1$ holds

$$\lim_n \left| Q^n_{\wedge\tau}\, \Phi(X,\varphi)\left(f_t^{\varepsilon}(X) - f_s^{\varepsilon}(X) - \int_s^t L^{\varepsilon} f_u d\varphi_u\right)\right| \leq \varepsilon .$$

Then:

1) families $\{Q^n\}$ and $\{Q^n \circ X^{-1}\}$ are relatively compact;

2) all limit points of $\{Q^n\}$ are weak $(\mathcal{R}, L)$-solutions;

3) under an additional assumption that all weak $(\mathcal{R}, L)$- solutions are special all limit points of $\{Q^n \circ X^{-1}\}$ are $(\mathcal{R}, L)$- solutions.

In practice condition 1) of Theorem 2 is usually provided by the existence of a Q^n- submartingale F^n for which φ^n is majorated by its increasing process and $Q^n_{\wedge\tau}(F^n_{\sigma+h} - F^n_{\sigma})$ is small uniformly in σ, n.

<u>Corollary</u> (Krylov, §6 ch.II [3]). Let measurable (not continuous) uniformly bounded functions a, a^n, $n = 1,2,\dots$, on $[0,\infty) \times R^d$ with values in some set of symmetric uniformly positive definite matrices be given, $a^n \to a$ when $n \to \infty$ in L_{d+1} on every compact. If P^n is a weak solution of the stochastic differential equation

$$dX_t = \sqrt{a^n}\,(t, X_t)\,dW_t, \quad X_0 = 0,$$

then the family $\{P^n,\ n = 1,2,\dots\}$ is relatively compact and all its limit points are solutions of the stochastic differential equation

$$dX_t = \sqrt{a}\,(t, X_t)\,dW_t, \quad X_0 = 0.$$

The only thing to check up in order to use the theorem is condition 3). Fix $C > 0$, denote

$$\tau = C \wedge \inf\left\{t : \sup_{[0,t]} X + t \geq C\right\} \quad (\inf \emptyset = \infty),$$

$$K = \left\{(t,x) \in [0,\infty) \times R^d :\ t \leq C,\ |x| \leq C\right\}.$$

Approximating a in $L_{d+1}(K)$ by a continuous function $\bar{a}$, with the help of Krylov's estimate (§ 2 ch.II [3]) we derive:

$$P^n \int_0^{\tau} \|a^n - \bar{a}\| (t,X_t)dt \le$$

$$\text{const}\cdot(\|a - \bar{a}\|_{L_{d+1}(K)} + \|a - a^n\|_{L_{d+1}(K)}), \quad n = 1,2,\ldots,$$

where the constant depends neither on n nor $\bar{a}$. It is easy to see that we can take for $L^{\varepsilon} f$ in condition 3)

$$\sum_{i,j=1}^{d} (\bar{a}^{ij}) \frac{\partial^2 f}{\partial x^i \partial x^j}$$

for an appropriately chosen $\bar{a}$.

The following theorem is a corollary of Theorem 2.

Theorem 3. Let for n = 1,2,... the following objects be given: a stochastic basis $(\Omega^n, \mathcal{F}^n, \mathbb{F}^n, P^n)$, an R^d- valued $\mathbb{F}^n$- adapted càdlàg process X^n, an increasing $\mathbb{F}^n$- adapted process φ^n and for every $f \in \mathcal{R}$ an $\mathbb{F}^n$- predictable càdlàg process of bounded variation $A^n f$ such that $f - A^n f$ is an $\mathbb{F}^n$- local martingale. If all f in $\mathcal{R}$ map $D \to D$ continuously, an operator L does not admit exploding weak solutions, the distributions of $\{X_0^n\}$ are relatively compact and for every continuous $\tau \in \mathbb{T}$:

1) (convergence of characteristics) for every $\varepsilon > 0$

$$\lim_n P^n(\sup_{[0,\tau]} |A^n f - \int_0^{\cdot} Lf_t d\varphi_t^n| > \varepsilon) = 0;$$

2) (relative compactness of distributions of $\{\varphi^n\}$) for every $\varepsilon > 0$

$$\lim_n P^n(\sup_{[0,\tau]} |\Delta \varphi^n| > \varepsilon) = 0$$

and there exists a process $f \in \mathcal{R}$ such that $Lf \ge 1$ on $[0, \tau)$ and

$$|f_t(X) - f_s(X)| \le \varepsilon + \text{const}\cdot(t - s);$$

3) (big jumps restriction) for every R, ε one can find a $C > 0$ and an $f \in C_b^2$ such that $f \in [0,1]$, $f\big|_{\{|x| \le R\}} = 0$, $f\big|_{\{|x| > C\}} = 1$ and $|Lf|\big|_{[0,\tau)} \le \varepsilon$;

4) ("almost continuity" of limit characteristics) for every

$f \in \mathcal{R}$, $T > 0$, $\varepsilon > 0$ there exists a ("ε - small") set $\Gamma \subseteq [0,\infty)\times D$ such that changing Lf on Γ one can obtain a process in $\mathcal{C}$ and for some $g \in \mathcal{R}$ $Lg \Big|_{[0,T]} \geq 0$, $g - g_0 \Big|_{[0,T]} \leq \varepsilon$ and

$Lg \Big|_{\Gamma \cap [0,T]} \geq 1$.

Then:

1) the families of distributions $\{(X^n, \varphi^n)\}$ and $\{X^n\}$ are relatively compact;

2) all limit points of distributions of $\{(X^n, \varphi^n)\}$ are weak ($\mathcal{R}$,L)- solutions;

3) under an additional assumption that all weak ($\mathcal{R}$,L)- solutions are special all limit points of distributions of $\{X^n\}$ are ($\mathcal{R}$,L)-solutions.

Example 1. Let d = 1 and continuous functions on R^1 be given: a^1 with values in $[1/2,1]$, a^2 with values in $[0,1]$, b^1 and b^2 with values in $[-1,1]$. Set $\mathcal{R} = \{Y\colon Y = f(X),\ f \in C_b^{1,2}\}$ and denote an operator L:

$$Lf_t(X) = \Big[\frac{\partial}{\partial t} + I_{[0,\tau)}\big(a^1 \frac{\partial^2}{\partial x^2} + b^1 \frac{\partial}{\partial x}\big) +$$

$$+ I_{[\tau,\infty)}\big(a^2 \frac{\partial^2}{\partial x^2} + b^2 \frac{\partial}{\partial x}\big)\Big] f(t,X_t),$$

where $\tau = \inf\{t\colon X_t \geq 1\}$. And let functions $a^{1,n}$, $a^{2,n}$, $b^{1,n}$, $b^{2,n}$ converge to a^1, a^2, b^1, b^2 uniformly. We are interested in the convergence of ($\mathcal{R}$,L^n)- solutions,

$$L^n f_t(X) = \Big[\frac{\partial}{\partial t} + I_{[0,\tau)}\big(a^{1,n} \frac{\partial^2}{\partial x^2} + b^{1,n} \frac{\partial}{\partial x}\big) +$$

$$+ I_{[\tau,\infty)}\big(a^{2,n} \frac{\partial^2}{\partial x^2} + b^{2,n} \frac{\partial}{\partial x}\big)\Big] f(t,X_t)$$

to ($\mathcal{R}$,L)- solutions. Since τ is an essentially discontinuous function of a path this example can't be handled by applying Theorem 1 §2 ch.8 [4]. Let us try to use Theorem 3. The only condition that needs testing is condition 4). Take an $\alpha > 0$. Introduce a continuous $\mathbb{D}$-Markov moment $\tau^\alpha = \inf\{t\colon X_t + \alpha t \geq 1\}$, set $\Gamma = [\tau^\alpha, \tau)$ and define for L an approximation from $\mathcal{C}$ substituting in the formula of L τ^{α} for τ. Take $T > 0$. Employing considerations of the control theory for diffusion processes, we can easily estimate the expectation of $\tau \wedge T - \tau^{\alpha} \wedge T$ for any $(\mathcal{R}, L)$- solution: it does not exceed $u(T\wedge \tau^{\alpha}, X_{T\wedge\tau^{\alpha}})$, where $u = u(t,x)$: $[0,T] \times (-\infty, 1] \longrightarrow [0, \infty)$ is a solution of the problem

$$\frac{\partial u}{\partial t} + \frac{1}{2}\frac{\partial^2 u}{\partial x^2} - \frac{\partial u}{\partial x} + 1 = 0, \quad u\Big|_{t=T} \equiv 0, \quad u\Big|_{x=1} \equiv 0.$$

For this solution $u\ \frac{\partial u}{\partial x} \leq 0$, $\frac{\partial^2 u}{\partial x^2} \leq 0$. The validity of the estimate follows immediately from the definition of an $(\mathcal{R}, L)$- solution. So take for g

$$u(t\wedge \tau^{\alpha}, X_{t\wedge\tau^{\alpha}}) - u(t\wedge\tau, X_{t\wedge\tau}).$$

Observe that $g_0 = 0$, $g - g_0 \leq u(T\wedge\tau^{\alpha}, X_{T\wedge\tau^{\alpha}})$, where the supremum is taken over $t \in [0,T]$, $x \in [1-\alpha, 1]$ and by virtue of properties of u tends to zero when $\alpha \rightarrow 0$.

Example 2. Let a smooth domain G in R^d with a boundary ∂G be given, at every point $x \in G$ a dxd- matrix $a(x)$ defined, at every point $x \in G \cup \partial G$ a d- vector $b(x)$ defined, with a continuous on G and uniformly nondegenerate, b continuous on each of the sets G and ∂G separately, a and b bounded, projection of b on the inner normal to the boundary uniformly positive, for simplicity say not less than 1. Put

$$\mathcal{R} = \left\{Y\colon Y_t = f(t, X_t),\ f \in C_b^{1,2}\right\}$$

and define L:

$$Lf_t(X) = \left[I_G\left(\frac{\partial}{\partial t} + \sum_{i,j=1}^{d} a^{ij}\frac{\partial^2}{\partial x^i \partial x^j}\right) + \sum_{i=1}^{d} b^i \frac{\partial}{\partial x^i}\right] f(t, X_t).$$

Note that for an $(\mathcal{R}, L)$- solution the increasing process φ behaves in the interior of the region as usual time: for $f \equiv t \in \mathcal{R}$ we have $dt = I_G d\varphi_t$.

Let functions a^n, b^n converge uniformly to a, b with $n \rightarrow \infty$. We are interested in the convergence of $(\mathcal{R}, L^n)$- solutions, where

$$L^n = I_G\left(\frac{\partial}{\partial t} + \sum_{i,j=1}^{d} a^{n,ij}\frac{\partial^2}{\partial x^i \partial x^j}\right) + \sum_{i=1}^{d} b^{n,i}\frac{\partial}{\partial x^i},$$

to ($\mathcal{R}$,L)- solutions.

Fix $\varepsilon > 0$ and construct the objects g and f of conditions 2) and 4) of Theorem 4. As in the previous example we exploit the ideas of the control theory for diffusion processes. Define $r(x) = \text{dist}(x, \partial G)$. Under sufficient regularity of the boundary ∂G the assumptions on the coefficients imply that r is a diffusion in $[0,\infty)$ with an instantaneous reflection at zero, bounded coefficients and a uniformly nondegenerate diffusion coefficient. Suppose for simplicity that the diffusion coefficient lays within $[1/2,1]$ and the drift coefficient within $[-1,1]$. Fix $\alpha > 0$, $T > 0$ and set $\Gamma = \{(t,X) \in [0,\infty)\times D: r(X_t) = (0,\alpha)$. It is clear how to change L on Γ in order to obtain a process from $\mathcal{C}$. Roughly speaking, now we have to estimate $\int_0^T I_\Gamma \, dt$. For this purpose a process g suits, $g_t = u(t,X_t)$, where $u = u(t,x): [0,T]\times[0,\infty) \longrightarrow [0,\infty)$ is the solution of the problem

$$\frac{\partial u}{\partial t} + \frac{1}{2}\frac{\partial^2 u}{\partial x^2} - \frac{\partial u}{\partial x} + d, \quad \left.\frac{\partial u}{\partial x}\right|_{x=0} \equiv 0, \quad u\Big|_{t=T} \equiv 0,$$

d being a smooth nonincreasing function equal to 1 on $[0,\alpha]$ and to 0 outside $[0,2\alpha]$. For a sufficiently small α the process g fits condition 4).

Let $\varepsilon > 0$, a smooth function $e = e(x): [0,\infty) \rightarrow [0,\varepsilon]$, $e'(0) = 1$, $e\Big|_{[\varepsilon,\infty)} = \varepsilon$. It is easy to see that for some $c > 0$ the process f, $f_t = (e + ct)(r(X_t))$, fits condition 2).

R E F E R E N C E S

1. Mikulevičius R. On necessary and sufficient conditions for the convergence to singular processes. -In: Statistics and control of stochastic processes. N.Y.: Optimization software, 1985, p.349-369.
2. S.V.Anulova. Diffusion processes: discontinuous coefficients, degenerate diffusion, randomized drift. Soviet Math. Dokl., 1981, vol.24, N°2, p.356-359.
3. N.V.Krylov. Control of diffusion type processes. Moscow, "Nauka", 1977 (in Russian).
4. R.Sh.Liptzer, A.N.Shiryayev. Theory of martingales. Moscow, "Nauka", 1985 (in Russian).

NONLINEAR FILTERING FOR DYNAMIC SYSTEMS WITH SINGULAR PERTURBATIONS

Alain BENSOUSSAN
University of Paris Dauphine and INRIA

Domaine de Voluceau - B.P. 105
Rocquencourt
78153 LE CHESNAY Cedex, FRANCE

INTRODUCTION

Consider a stochastic dynamic system involving slow and fast components

$$dx_t = f(x_t,y_t)dt + \sqrt{2}\, dw_t^1$$

$$\epsilon d\, y_t = g(x_t,y_t)dt + \sqrt{2\epsilon}\, dw_t^2$$

$$x(0) = \xi,\ y(0) = y_o\ .$$

This system is partially observed through the observation process

$$dz_t = h(x_t,y_t)dt + db_t$$

$$z(0) = 0$$

The question at stake is what is the behaviour of the nonlinear filter $E(\psi(x_t)|\ Z^t)$, where ψ is a deterministic test function, as ϵ tends to 0 ?

We show in this article that in some very weak sense, one has convergence towards the nonlinear filter related to the limit of the process x_t.

This supposes of course that the system itself x_t has a nice limit, which requires some properties (of ergodic type) concerning the fast system y_t. The simplest framework to derive good ergodic properties consists in assuming that the dependence in y is periodic. It is the one that we adopt here.

1. SETTING OF THE PROBLEM.

1.1. Assumptions.

Consider

$$f(x,y) : \mathbb{R}^n \times \mathbb{R}^p \to \mathbb{R}^n \qquad (1.1)$$
$$g(x,y) : \mathbb{R}^n \times \mathbb{R}^p \to \mathbb{R}^p$$
$$h(x,y) : \mathbb{R}^n \times \mathbb{R}^p \to \mathbb{R}^p$$

C^2 functions, periodic in y (with period 1 in all components [1]

g and its derivatives (1st and 2nd) are bounded.

(1.2)

The derivatives of f, h (1st and 2nd) are bounded.

Let $(\Omega,\mathcal{A},P)$ be a probability space equipped with a filtration $\mathcal{F}^t$ and let w_t^1, w_t^2, z_t be standard $\mathcal{F}^t$ Wiener processes independent one from each other.

Let also

ξ, y_0 random variables $\mathcal{F}^0$ measurable with values in $\mathbb{R}^n$, Y respectively. (1.3)
The probability law of the pair (ξ, y_0) is denoted by q_0 and the marginal probability law of ξ is π_0.

We define the processes x_t^ϵ, y_t^ϵ by solving the equations :

$$dx_t = f(x_t,y_t)dt + \sqrt{2}\,dw_t^1 \qquad (1.4)$$

$$\epsilon dy_t = g(x_t,y_t)dt + \sqrt{2\epsilon}\,dw_t^2$$

$$x(0) = \xi \,,\; y(0) = y_0 \,.$$

Let next η_t^ϵ be defined by :

$$\eta_t^\epsilon = \exp\left\{\int_0^t h(x_s^\epsilon,y_s^\epsilon)dz_s - \tfrac{1}{2}\int_0^t |h(x_s^\epsilon,y_s^\epsilon)|^2\, ds\right\} \qquad (1.5)$$

i.e. η_t^ϵ has the Ito differential,

---

(1) to fix the ideas. We denote by $Y = (0,1)^p$, identifying the opposite sides.

$$d\eta_t^\epsilon = \eta_t^\epsilon\, h(x_t^\epsilon, y_t^\epsilon) dz_t \tag{1.6}$$

$$\eta_0^\epsilon = 1$$

One can check that

$$E\eta_t^\epsilon = 1$$

and thus we may define a new probability P^ϵ on (Ω, A) by setting :

$$\left.\frac{dP^\epsilon}{dP}\right|_{F^t} = \eta_t^\epsilon \tag{1.7}$$

Consider next the process :

$$b_t^\epsilon = z_t - \int_0^t h(x_s^\epsilon, y_s^\epsilon) ds \tag{1.8}$$

then on the set up $\Omega, A, P^\epsilon, F^t, w_t^1, w_t^2, b_t^\epsilon$ become independent standard F^t Wiener processes and z_t can be written as :

$$dz_t = h(x_t^\epsilon, y_t^\epsilon) dt + db_t^\epsilon \tag{1.9}$$

$$z(0) = 0$$

and appears as the observation process corresponding to the state $(x_t^\epsilon, y_t^\epsilon)$.

We shall be interested by the unormalized conditional probability $p^\epsilon(t)$ given by :

$$p^\epsilon(t)(\psi) = E(\psi(x_t^\epsilon)\eta_t^\epsilon \mid Z^t) \tag{1.10}$$

for any test function ψ (Borel bounded), and the normalized conditional probability $\pi^\epsilon(t)$

$$\pi^\epsilon(t)(\psi) = \frac{p^\epsilon(t)(\psi)}{p^\epsilon(t)(1)} \tag{1.11}$$

$$= E^\epsilon(\psi(x_t^\epsilon) \mid Z^t).$$

The question of interest concerns the behaviour of $p^\epsilon(t)(\psi)$ and $\pi^\epsilon(t)(\psi)$ as ϵ tends to 0.

1.2. A duality argument.

There is a useful characterization of the unnormalized conditional probability, which we describe now. Let $\beta(t) \in L^\infty(0,T;\mathbb{R}^p)$ and $\phi(x) \in C_0^\infty(\mathbb{R}^n)$. Consider the P.D.E. :

$$-\frac{\partial v}{\partial t} - f(x,y)D_x v - \frac{1}{\epsilon}g(x,y) \,.\, D_y\, v - \Delta_x v - \frac{1}{\epsilon}\Delta_y v - iv\; h(x,y).\beta(t) = 0 \qquad (1.12)$$

$$v(x,y,T) = \phi(x).$$

In fact, (1.12) is a system ; if $v = v_1 + iv_2$, then one has :

$$-\frac{\partial v_1}{\partial t} - f(x,y)D_x v_1 - \frac{1}{\epsilon}g(x,y) \,.\, D_y\, v_1 - \Delta_x v_1 - \frac{1}{\epsilon}\Delta_y v_1 + v_2\; h(x,y).\beta(t) = 0$$

$$-\frac{\partial v_2}{\partial t} - f\, D_x v_2 - \frac{1}{\epsilon}g \,.\, D_y\, v_2 - \Delta_x v_2 - \frac{1}{\epsilon}\Delta_y v_2 - v_1\; h \,.\, \beta = 0$$

$$v_1(x,y,T) = \phi(x), \quad v_2(x,y,T) = 0.$$

We shall use the fact (see A. BENSOUSSAN [1], Chap. IV, Proposition 2.1) that there exists a unique solution v_1^ϵ, v_2^ϵ of (1.13) such that :

$$v_1^\epsilon,\ v_2^\epsilon \text{ and their first and second derivatives in } x,y \text{ are bounded} \qquad (1.14)$$

$$|v_{1,t}^\epsilon| \,,\, |v_{2,t}^\epsilon| \le C(1 + |x|).$$

The constants entering in (1.14) depend on ϵ. Considering the process associated to $\beta(t)$,

$$\theta(t) = \exp\{i \int_0^t \beta(s) \,.\, dz + \tfrac{1}{2} \int_0^t |\beta|^2\, ds\} \qquad (1.15)$$

then (see A. BENSOUSSAN [1], Chap. IV, Theorem 2.1)

$$E\, \theta(T)\, p^\epsilon(T)(\phi) = q_0(v_1^\epsilon(0)) \qquad (1.16)$$

where $v_1^\epsilon(0)$ stands for $v_1(x,y,0)$ and emphasizing the dependence in ϵ.

From formula (1.16), it is clear that the convergence of $v_1^\epsilon(x,y,0)$ as $\epsilon \to 0$, plays a fundamental role in the convergence of $p^\epsilon(T)$. In this way we are led to a classical problem of singular perturbations for P.D.E., whose solution is described in the next section.

2. SINGULAR PERTURBATION ANALYSIS.

2.1. Preliminaries.

Let us consider the invariant measure $m(x,y)$ which is the solution of :

$$-\Delta_y m + \mathrm{div}_y(mg) = 0 \qquad (2.1)$$

$$m \in H^1(Y),\ m \text{ periodic}$$

where $Y = (0,1)^p$.

Since g is bounded, there exists for any x a unique solution of (2.1) satisfying :

$$\int_Y m(x,y)dy = 1, \forall x. \tag{2.2}$$

Moreover one has the estimates :

$$0 < \delta_0 \leq m(x,y) \leq \delta_1 \tag{2.3}$$

where δ_0, δ_1 depend only on the bound of g. For details see A. BENSOUSSAN [2].

We can also state the following useful regularity results :

$$\|m(x,.)\|_{W^{2,p}(Y)} \leq C_p , \forall x, \ 2 \leq p < \infty \tag{2.4}$$

$$\|m_\lambda(x,.)\|_{W^{2,p}(Y)} \leq C_p, \forall x$$

where $m_\lambda = \dfrac{\partial m}{\partial x_\lambda}$.

Concerning the system (1.13), it will be useful to introduce the new functions :

$$u_k^\epsilon = v_k^\epsilon \, e^{-G} , \qquad k=1,2 \tag{2.5}$$

where $G(x,t) = e^{Kt}(1 + |x|^2)^{\frac{1}{2}}$, K convenient constant.

2.2. Limit of u_k^ϵ.

We then can state the following estimates :

Lemma 2.1 : One has the estimates

$$\|v_1^\epsilon\|_{L^\infty} , \|v_2^\epsilon\|_{L^\infty} \leq \|\phi\|_{L^\infty} \tag{2.6}$$

$$\|u_k^\epsilon\|_{L^2} , \|D_x \, u_k^\epsilon\|_{L^2} \leq C \tag{2.7}$$

$$\|D_y \, u_k^\epsilon\|_{L^2} \leq C\epsilon \ . \ k=1,2 \ .$$

□

The proof of (2.7) follows from energy inequalities obtained in multiplying the equations for u_1^ϵ, u_2^ϵ by $m\,u_1^\epsilon$, $m\,u_2^\epsilon$. These equations are the following :

$$- \frac{\partial u_1}{\partial t} - (f + 2\, D_x G) \,.\, D_x u_1 - \frac{1}{\epsilon} g \,.\, D_y\, u_1 - \Delta_x u_1 - \frac{1}{\epsilon} \Delta_y u_1 \tag{2.8}$$

$$+ u_1\, \gamma + u_2\, h \,.\, \beta = 0$$

$$- \frac{\partial u_2}{\partial t} - (f + 2\, D_x G) \,.\, D_x u_2 - \frac{1}{\epsilon} g \,.\, D_y\, u_2 - \Delta_x u_2 - \frac{1}{\epsilon} \Delta_y u_2$$

$$+ u_2\, \gamma + u_1\, h \,.\, \beta = 0$$

$$u_1(x,y,T) = \phi(x) e^{-G(x,T)}, \quad u_2(x,y,T) = 0$$

where

$$\gamma(x,y,t) = KG - f \,.\, D_x G - \Delta_x G - |D_x G|^2 \,.$$

The next step is to prove the following result :

<u>*Lemma 2.2*</u> *: We have for* k=1,2

$$u_k^\epsilon \to u_k \,, \quad \text{in } L^2(0,T;H^1(\mathbb{R}^n \times Y)) \text{ } \textit{weakly} \tag{2.9}$$

where u_1, u_2 *are the solutions of* :

$$- \frac{\partial u_1}{\partial t} - (\overline{f} + 2\, D_x G) \,.\, D_x u_1 - \Delta_x u_1 + u_1 \overline{\gamma} + u_2\, \overline{h} \,.\, \beta = 0 \tag{2.10}$$

$$- \frac{\partial u_2}{\partial t} - (\overline{f} + 2\, D_x G) \,.\, D_x u_2 - \Delta_x u_2 + u_2 \overline{\gamma} - u_1\, \overline{h} \,.\, \beta = 0$$

$$u_1(x,T) = \phi(x) e^{-G(x,T)}, \; u_2(x,T) = 0.$$

The functions u_1, u_2 *do not depend on* y, *and* $\overline{f}$, $\overline{\gamma}$, $\overline{h}$ *are defined by :*

$$\overline{f}(x) = \int_Y f(x,y) m(x,y) dy$$

and similarly for $\overline{\gamma}$, $\overline{h}$.

□

One can go a step further and prove the :

<u>*Lemma 2.3*</u> *: We have for* k = 1, 2

$$u_k^\epsilon \to u_k \qquad \textit{in } L^2(0,T;H^1(\mathbb{R}^n \times Y)) \textit{ strongly} \tag{2.11}$$

$$u_k^\epsilon(t) \to u_k(t) \textit{ in } L^2(\mathbb{R}^n \times Y) \qquad \textit{strongly}, \; \forall t \in [0,T].$$

□

The details of the proofs can be found in A. BENSOUSSAN [1].

3. CONVERGENCE.

The convergence is proved in two parts, and requires some technical additional assumptions :

Lemma 3.1 : Assume that $h(x,y)$ *satisfies*

$$|h(x,y)| \le h_0 + H|x| \;, \text{ with } H \text{ not too large} \tag{3.1}$$

and

$$\pi_0(\exp \lambda |x|^2) < \infty, \text{ for } \lambda > 0 \text{ sufficiently large,} \tag{3.2}$$

then one has :

$$E(\eta_T^\epsilon)^2 \le C_T \tag{3.3}$$

□

We can then state the main result of the section :

Theorem 3.1 : We assume (1.1), (1.2), (1.3), (3.1), (3.2), *and*

$$q_0 \text{ has a density } q_0(x,y) \text{ with respect to Lebesgue measure on} \tag{3.4}$$

$$\text{on } \mathbb{R}^n \times Y \text{ and } q_0\, e^{G(x,0)} \in L^2(\mathbb{R}^n \times Y)$$

then one has :

$$p^\epsilon(T)(\psi) \to p(T)(\psi), \ \forall \psi \text{ continuous bounded in, } L^2(\Omega, Z^T, P) \text{ weakly} \tag{3.5}$$

where

$$p(t)(\psi) = E[\psi(x_t)\eta_t \mid Z^t] \tag{3.6}$$

Proof : We first check that

$$E\,|p^\epsilon(T)(\psi)|^2 \le C \tag{3.7}$$

Indeed, since ψ is bounded

$$|p^\epsilon(T)(\psi)| \le \|\psi\|\; E\,(\eta_T^\epsilon \mid Z^T)$$

hence

$$E\,|p^\epsilon(T)(\psi)|^2 \le \|\psi\|^2\; E\,|E(\eta_T^\epsilon \mid Z^T)|^2$$

$$\le \|\psi\|^2\; E(\eta_T^\epsilon)^2 \le C_T$$

by virtue of Lemma 3.1.

The set of linear combinations of varaibles $\Sigma c_i \theta_i(T)$ where $\theta_i(T)$ corresponds to (1.15) with $\beta = \beta_i$, forms a dense subspace of $L^2(\Omega, Z^T, P)$.

Therefore to prove (3.5), it is sufficient to prove that

$$Ep^\epsilon(T)(\psi)\theta(T) \to Ep(T)(\psi)\theta(T), \quad \forall \beta(.).$$

Moreover it is sufficient to prove this result for $\psi = \phi \in C_0^\infty(\mathbb{R}^n)$.
According to the duality argument (1.16), this means

$$q_0(v_1^\epsilon(0)) \to q_0(v_1(0)), \tag{3.8}$$

where the pair $v_1(x,t)$, $v_2(x,t)$ is the solution of

$$\begin{aligned} &-\frac{\partial v_1}{\partial t} - \mathcal{T} . D_x v_1 - \Delta_x v_1 + v_2 \, \bar{h} . \beta = 0 \\ &-\frac{\partial v_2}{\partial t} - \mathcal{T} . D_x v_2 - \Delta_x v_2 + v_1 \, \bar{h} . \beta = 0 \\ &v_1(x,T) = \phi(x), \; v_2(x,T) = 0. \end{aligned} \tag{3.9}$$

Comparing with (2.10) it is clear that

$$\begin{aligned} v_1(x,t) &= u_1(x,t)\, e^{G(x,t)} \\ v_2(x,t) &= u_2(x,t)\, e^{G(x,t)}. \end{aligned} \tag{3.10}$$

Therefore (3.8) amounts to, using (3.4),

$$\iint u_1^\epsilon(x,y,0)\, e^{G(x,0)}\, q_0(x,y)dx\, dy \to \int u_1(x,0)\, e^{G(x,0)}\, p_0(x)dx \tag{3.11}$$

where p_0 is the density of π_0 ; $p_0(x) = \int_Y q_0(x,y)dy$.

Since by (3.4) $q_0(x,y)\, e^{G(x,0)} \in L^2(\mathbb{R}^n \times Y)$, the result (3.11) follows immediately from the 2nd part of (2.11) (Lemma 2.3). The proof has been completed.

□

We can then state a convergence result for the conditional probability itselft as follows :

Theorem 3.2 : Let $\overline{P}$ be the probability on (Ω,A) defined by the Radon-Nikodym derivative :

$$\frac{d\overline{P}}{dP}\Big|_{F^t} = \eta_t$$

then one has :

$$E^\epsilon \pi^\epsilon(T)(\psi)\xi_T \to \overline{E}\,\pi(T)(\psi)\xi_T\,,\ \forall \xi_T \in L^2(\Omega,Z^T,P)$$

where

$$\pi(t)(\psi) = \frac{p(T)(\psi)}{p(T)(1)}$$

Proof : Indeed one has

$$E^\epsilon \pi^\epsilon(T)(\psi)\xi_T = E^\epsilon \psi(x^\epsilon(T))\xi_T$$

$$= E\,\psi(x^\epsilon(T))\,\eta_T^\epsilon\,\xi_T$$

$$= E\,p^\epsilon(T)(\psi)\xi_T$$

$$\to E\,p(T)(\psi)\xi_T = \overline{E}\,\pi(T)(\psi)\,\xi_T$$

and the desired result has been proven.

□

Remark 3.1 : It is worthwhile to notice that the formula :

$$\pi^\epsilon(T)(\psi) = \frac{p^\epsilon(T)(\psi)}{p^\epsilon(T)(1)}$$

with both numerator and denominator converging weakly, does not lead to the convergence of the ratio. Nevertheless, in a weak sense, namely (3.12), there is convergence.

REFERENCES.

[1] A. BENSOUSSAN, Stochastic Control with Partial Observation, to be published.

[2] A. BENSOUSSAN, Perturbation Methods in Optimal Control, Wiley-Gauthier Villars, 1988.

On Recursive Adaptive filtering: Linear Case

T. Bielecki

Institute of Econometrics, The Main College of
Planning and Statistics
Al. Niepodległości 162, 02 - 554 Warsaw, Poland

B. Gołdys

Institute of Mathematics, Polish Academy of Sciences
Śniadeckich 8, P.O. Box 137, 00-950 Warsaw, Poland

0. Introduction

Let $(\Omega, F, (F_t)_{t \geq 0}, P)$ be the underlying probability space on which two independent real-valued Wiener processes (w_t) and (v_t) are given. Let us consider on $(\Omega, F, (F_t)_{t \geq 0}, P)$ the following state-observation system

$$\begin{cases} dx_t = ax_t dt + b_1 dw_t, \\ dy_t = cx_t dt + b_2 dv_t, \\ x_o = x, \; y_o = 0, \; t \geq 0, \end{cases} \tag{1}$$

where x is a Gaussian F_o-measurable random variable with $Ex=0$. The real-valued state process (x_t) is not completely observed and information about its behaviour is given by the real-valued observation process (y_t) depending on unknown parameters c and b_2. Parameters a and b_1 of the state equation are also unknown. In the case of known parameters of the system (1) the well-known Kalman filter $(\hat{x}_t)$ is the best in the mean-square sense estimator of the state x_t given the past observations $(y_s; s \leq t)$.

When all the parameters of the system (1) are unknown then one can hope only for obtaining a "good" in some sense estimator $\hat{\hat{x}}_t$ of the Kalman filter $\hat{x}_t$ multiplied by c. The process $(\hat{\hat{x}}_t)$ is usually called an adaptive filter. The problem of finding an adaptive filter was considered by van Schuppen [5] for continuous-time systems and in the discrete-time case by Kumar and Moore [4].

In the paper the adaptive filtering problem as posed above is solved in the following sense. The results of [1] are used to construct strongly consistent estimators of parameters entering the Kalman filtering equations. In this way recursive formulas for the adaptive filter $\hat{\hat{x}}_t$ are obtained. Moreover, it is shown that the process $(\hat{\hat{x}}_t)$ is close to the process $(c\hat{x}_t)$ in Cesaro sense:

$$\text{a.s.} - \lim_{t\to\infty} \frac{1}{f(t)} \int_0^t (c\hat{x}_s - \hat{\hat{x}}_s)^2 ds = 0,$$

where the weight function f depends on asymptotic properties of the state-process (x_t).

All results that follow can be generalized to the multidimensional case which will be the topic of the forthcoming paper.

1. Formulation of the problem of adaptive filtering and some preliminary results

Consider the state-observation system (1) of the Section 0. Let $\hat{x}_t \overset{\text{def}}{=} E(x_t|F_t^y)$, where $F_t^y = \sigma(y_s, s \le t)$, $t \ge 0$. Then from Kalman filtering theory (see e.g. [3]) we know that $(\hat{x}_t)$ and (y_t) satisfy

$$\begin{cases} d\hat{x}_t = a\hat{x}_t dt + k_t d\nu_t, \\ dy_t = c\hat{x}_t dt + d\nu_t \\ \hat{x}_0 = 0, \ y_0 = 0, \ t \ge 0, \end{cases} \tag{2}$$

where

$$k_t \overset{\text{def}}{=} \frac{p_t c}{b_2^2}$$

is the Kalman gain, and p_t is given by the Ricatti equation

$$\begin{aligned} \dot{p}_t &= b_1^2 + 2ap_t - p_t^2 c^2 / b_2^2 \\ p_0 &= \text{var } x_0, \ t \ge 0. \end{aligned} \tag{3}$$

The process (ν_t) is an innovation process with $E\nu_t^2 = b_2^2 t$.

Let $k \overset{\text{def}}{=} \lim_{t\to\infty} k_t$ be the asymptotic Kalman gain. Our goal is to construct two functionals γ' and γ'', $\gamma'(\gamma'') : R_+ \times C(0,\infty;R) \to R^2(R^1)$, such that processes $\gamma'_t(\omega) = \gamma'(t;y.(\omega))$ and $\gamma''_t(\omega) = \gamma''(t;y.(\omega))$ are non-anticipating, γ'_t and γ''_t can be computed recursively in time and the following strong consistency (SC) and Cesaro convergence (CC) properties hold:

$$\text{a.s.} - \lim_{t\to\infty} \gamma'_t = (a,kc) \tag{SC}$$

$$\text{a.s.} - \lim_{t\to\infty} \frac{1}{f(t)} \int_0^t (\gamma''_s - c\hat{x}_s)^2 ds = 0, \tag{CC}$$

where $f : R_+ \to R_+$ is a deterministic weight function which will be precised later.

The following three lemmas have been proved in [1]. We recall them for the convinience of the reader.

Lemma 1. The observation process satisfies

$$dy_t = ay_t dt + (kc-a)\nu_t dt + \varepsilon_t dt + d\nu_t, \tag{4}$$

$$y_0 = 0, \quad t \geq 0,$$

where

$$\varepsilon_t \stackrel{\text{def}}{=} c \int_0^t (k_s - k) d\nu_t, \quad t \geq 0.$$

Remark 1. Note that by theorem 12.2 [7] we have $kc-a > 0$.

Definition 1. [6]. Let $\kappa > 0$. An operator $K : L^2_{loc}(0,\infty;R) \to L^2_{loc}(0,\infty;R)$ is said to be κ-very strictly passive if for every $T > 0$ its restriction K_T to the space $L^2(0,T;R)$ satisfies the following condition:

For every f $L^2(0,T;R)$

$$\langle K_T f, f \rangle_{L^2} \geq \kappa \|f\|^2_{L^2}. \tag{SP}$$

Lemma 2. Let β be such that $\tilde{c} \stackrel{\text{def}}{=} kc - a \leq \beta$. Put $\delta = \frac{1}{2\beta}$ and define two operators $\tilde{C}$, $D : L^2_{loc}(0,\infty;R) \to L^2_{loc}(0,\infty;R)$ by

$$\tilde{C}f(t) \stackrel{\text{def}}{=} f(t) + \tilde{c} \int_0^t f(s)ds,$$

$$Df(t) \stackrel{\text{def}}{=} f(t) + \delta \int_0^t f(s)ds.$$

Then the operator $U \stackrel{\text{def}}{=} D\tilde{C}^{-1}$ is $\frac{1}{2}$-strictly passive.

Remark 2. Define the integration operator

$$S : L^2_{loc}(0,\infty;R) \to L^2_{loc}(0,\infty;R)$$

$$Sf(t) \overset{def}{=} \int_0^t f(s)ds.$$

Note that the following equality holds

$$MN = NM$$

for $M,N = D, \tilde{C}, D^{-1}, \tilde{C}^{-1}, S$.

Let $(\Theta.)$ be any R^2-valued locally square itegrable function and let $(\phi.)$ be defined by

$$\phi_t^* = (D^{-1}y(t), -\int_0^t \phi_s^* \Theta_s ds), \ t \geq 0.$$

Lemma 3. Put $\Theta^* \overset{def}{=} (kc, kc-a)$, $\tilde{\Theta}_t \overset{def}{=} \Theta - \Theta_t$ and $\xi_t \overset{def}{=} U(\tilde{\Theta}^*.\phi.)(t)$, $t \geq 0$, where the operator U has been introduced in lemma 2. Then we have

$$dy_t = \xi_t dt + D(\Theta^*.\phi.)(t)dt + \tilde{C}^{-1}\varepsilon(t)dt + d\nu_t \qquad (5)$$

$$y_0 = 0, \quad t \geq 0.$$

Remark 3. In sections 3 and 4, $(\Theta_t)_{t\geq 0}$ and $(\phi_t)_{t\geq 0}$ of lemma 3 will be given via systems of SDE's (6)-(9) and processes $(\tilde{\Theta}_t)_{t\geq 0}$ and $(\xi_t)_{t\geq 0}$ will be defined accordingly.

2. Stochastic Gradient Algorithm (SGA) and its Properties

In this section some properties of the stochastic gradient algorithm (6)-(9) are considered. In particular the strong consistency property of the estimator $(\Theta_t)_{t\geq 0}$ generated by (6)-(9) is demonstrated.

The SGA is given via the following equations,

$$d\Theta_t = \frac{\phi_t}{r_t}(dy_t - (\phi_t^*\Theta_t - \delta\phi_t^2)dt, \quad \Theta_0 \text{ given}, \qquad (6)$$

$$d\phi_t^1 = -\delta\phi_t^1 dt + dy_t, \quad \phi_0^1 = 0, \qquad (7)$$

$$\dot{\phi}_t^2 = -\Theta_t^*\phi_t, \quad \phi_0^2 = 0, \qquad (8)$$

$$\dot{r}_t = \|\phi_t\|^2, \quad r_0 = 1, \quad t \geq 0, \qquad (9)$$

where δ R and $\phi_t^* = (\phi_t^1, \phi_t^2)$, $t \geq 0$.

Proposition 1. Let δ be as in lemma 2. Then the system (6)-(9) has a unique global strong solution.

Proof. See proposition 3 of [1].

Define processes (ϕ_t^o) and (r_t^o) by

$$\phi_t^o = \begin{pmatrix} D^{-1}y(t) \\ D^{-1}w(t) - D^{-1}y(t) - D^{-1}\tilde{\eta}(t) \end{pmatrix} \tag{10}$$

$$\dot{r}_t^o = \|\phi_t^o\|^2, \quad r_o^o = 1, \quad t \geq 0, \tag{11}$$

where

$$\tilde{\eta}_t \overset{\text{def}}{=} \int_o^t \eta_s ds, \quad \eta_t \overset{\text{def}}{=} \tilde{c}^{-1} \varepsilon(t), \quad t \geq 0.$$

Remark 4. It follows from lemma 5 in [1] that $\text{as} - \lim_{t\to\infty} r_t = \text{as} - \lim_{t\to\infty} r_t^o = +\infty$

Lemma 4. Let δ be as in lemma 2. Then

$$0 < c_1 \leq \text{as} - \varliminf_{t\to\infty} \frac{r_t}{r_t^o} \leq \text{as} - \varlimsup_{t\to\infty} \frac{r_t}{r_t^o} \leq c_2 < +\infty.$$

Proof. We have

$$\frac{r_t}{r_t^o} = \frac{1 + \int_o^t (\phi_s^1)^2 ds + \int_o^t (\phi_s^2)^2 ds}{1 + \int_o^t (\phi_s^1)^2 ds + \int_o^t (\phi_s^2 + D^{-1} \int_o^s \xi_\lambda d\lambda)^2) ds}$$

$$= \frac{1 + \int_o^t (\phi_s^1)^2 ds + \int_o^t (\phi_s^2)^2 ds}{1 + \int_o^t (\phi_s^1)^2 ds + \int_o^t (2\phi_s^2 - \tilde{c}(h * \phi^2)(s))^2 ds} \tag{12}$$

and $h * \phi^2$ is a convolution of h and ϕ^2, with $h(t) \overset{\text{def}}{=} e^{-\tilde{c}t}$, $t \geq 0$. Now from the Young's inequality and from (12)

$$\frac{r_t}{r_t^o} \leq \frac{1 + \int_o^t (\phi_s^1)^2 ds + \int_o^t (\phi_s^2)^2 ds}{1 + \int_o^t (\phi_s^1)^2 ds + \int_o^t (\phi_s^2)^2 ds (8 + 1/\tilde{c})} \geq c_1 > 0 \tag{13}$$

for all $t \geq 0$. Fix $\rho > 0$. Invoking the Young's inequality again we get, for $t \geq 0$,

$$\frac{r_t}{r_t^o} \le \frac{r_t}{1 + \int_o^t (\phi_s^1)^2 ds + 2\int_\rho^t (\phi_s^2)^2 ds\,(1-e^{-\rho\bar{c}})}$$

which together with (14) implies

$$\frac{r_t}{r_t^o} \le c_2 < +\infty$$

for all $t \ge 0$.

Define processes (Φ_t) and (Φ_t^o) by

$$\dot{\Phi}_t = -\frac{\phi_t \phi_t^*}{r_t} \Phi_t , \tag{14}$$

$$\Phi_o = I, \; t \ge 0,$$

and

$$\dot{\Phi}_t^o = -\frac{\phi_t^o \phi_t^{o*}}{r_t^o} \Phi_t^o , \tag{15}$$

$$\Phi_o^o = I, \quad t \ge 0.$$

Lemma 5. The following two conditions are equivalent,

a) $as - \lim\limits_{t\to\infty} \Phi^o(t) = 0$,

b) $as - \lim\limits_{t\to\infty} \Phi(t) = 0$.

Proof. The proof easily follows from lemma 4, lemma 9 [1], and the proof of Theorem 4 [2] and therefore is ommited.

Theorem 1. Assume δ is as in lemma 2. Suppose also that $as - \lim\limits_{t\to\infty} \Phi^o(t) = 0$. Then

$$as - \lim_{t\to\infty} \Theta_t^* = (kc, kc-a). \tag{16}$$

Proof. It easily follows from lemma 5 and theorem 3 of [1].

Corollary 1. It follows from theorem 1 that for $\gamma_t' = (\Theta_t^1 - \Theta_t^2, \Theta_t^2)$, where $\Theta_t = (\Theta_t^1, \Theta_t^2)$, condition (SC) is satisfied.

Before we close this section we recall an important result from [1].

Lemma 6 ([1], lemma 8). Assume δ is as in lemma 2. Then

$$\int_0^\infty \frac{\xi_s^2}{r_s} ds < +\infty, \text{ a.s.}$$

3. Converegence of adaptive filter

We start with the following result,

Lemma 7 i) If $a < 0$ then we have

$$\sup_{t \geq 0} \frac{r_t^o}{t} < +\infty, \quad \text{a.s.}$$

ii) If $a > 0$ then we have

$$\sup_{t \geq 0} \frac{r_t^o}{e^{2at}} < +\infty, \quad \text{a.s.}$$

Proof.

i) $a < 0$. Let $\tilde{\nu}_t \overset{def}{=} D^{-1}\nu(t)$ and $\tilde{x}_t = D^{-1}\hat{x}(t)$, $t \geq 0$. Then

$$\tilde{x}_t = a \int_0^t \tilde{x}_s ds - k\delta\tilde{\nu}_t + d\nu_t + \frac{1}{c} D^{-1}\varepsilon(t), \quad \tilde{x}_o = 0, \tag{17}$$

$$\tilde{\nu}_t = \mp\delta \int_0^t \nu_s ds + \nu_t, \quad \tilde{\nu}_o = 0, \quad t \geq 0. \tag{18}$$

Define also process $(\bar{x}_t)$ by (18) and

$$d\bar{x}_t = a\bar{x}_t dt - k\delta\tilde{\nu}_t dt + d\nu_t, \quad \bar{x}_o = 0, \quad t \geq 0. \tag{17$^-$}$$

Note that the Markov diffusion $(\bar{x}_t, \tilde{\nu}_t)_{t\geq 0}$ is strongly ergodic and hence

$$\text{as} - \lim_{t\to\infty} \frac{\int_0^t \bar{x}_s^2 ds}{t} = E\bar{x}_1^2 < +\infty,$$

$$\text{as} - \lim_{t\to\infty} \frac{\int_0^t \tilde{\nu}_s^2 ds}{t} = E\tilde{\nu}_1^2 < +\infty,$$

$$\text{as} - \lim_{t\to\infty} \frac{\int_0^t \bar{x}_s\tilde{\nu}_s ds}{t} = E\bar{x}_1\tilde{\nu}_1 < +\infty.$$

It is easy to check that

$$\text{as} - \lim_{t\to\infty} \frac{\int_0^t [(D^{-1}\varepsilon)(s)]^2 ds}{t} < \infty;$$

Hence

$$\text{as}-\lim_{t\to\infty}\frac{\int_0^t \tilde{\tilde{x}}_s^2 ds}{t} \le 2[\text{as}-\lim_{t\to\infty}\frac{\int_0^t \bar{x}_s^2 ds}{t} + \text{as}-\lim_{t\to\infty}\frac{\int_0^t [(D^{-1}\varepsilon)(s)]^2 ds}{t} < +\infty \tag{19}$$

and in consequence

$$\text{as}-\lim_{t\to\infty}\frac{\int_0^t \tilde{\tilde{x}}_s \tilde{\tilde{\nu}}_s ds}{t} < +\infty \tag{20}$$

Now note that (recall $\phi_t^1 = D^{-1}y(t)$)

$$\phi_t^1 = \frac{c}{a}\tilde{\tilde{x}}_t + (1-\frac{ck}{a})\tilde{\nu}_t + \frac{1}{a}D^{-1}\varepsilon(t). \tag{21}$$

From the above discussion and the definition of (r_t^o) one easily sens that

$$\text{as}-\overline{\lim_{t\to\infty}}\frac{r_t^o}{t} < +\infty .$$

The proof of i) is complete.

ii) $a > 0$. It is enough to consider system (17´), (18) and show that

$$\text{as}-\overline{\lim_{t\to\infty}}\frac{1}{e^{2at}}\int_0^t \bar{x}_s^2 ds < +\infty \tag{22}$$

To this end we first note that $\text{as}-\lim_{t\to\infty}\int_0^t \bar{x}_s^2 ds = +\infty$. Since $(\bar{x}_t)$ is a continuous process we can apply de L´Hospital theorem and thus in order to demonstrate (22) it is enough to show

$$\text{as}-\overline{\lim_{t\to\infty}}\frac{\bar{x}_t^2}{e^{2at}} < +\infty \tag{23}$$

To this end let us notice that

$$\frac{1}{e^{at}}\bar{x}_t = -k\delta\int_0^t e^{-as}\tilde{\nu}_s ds + k\int_0^t e^{-as}d\nu_s = -k\delta N_t + kM_t, \quad t \ge 0.$$

By the martingale convergence theorem we get

$$\text{as}-\lim_{t\to\infty} M_t = M < \infty. \tag{24}$$

Now

$$N_t = \int_0^t e^{-\frac{1}{2}as}(e^{-\frac{1}{2}as}\tilde{\nu}_s)ds \le [\int_0^t e^{-as}ds]^{\frac{1}{2}}[\int_0^t e^{-as}\tilde{\nu}_s^2 ds]^{\frac{1}{2}}$$

Due to strong ergodicity of $(\tilde{\nu}_t)$ we have

$$\sup_{t\geq 0} E\int_0^t e^{-as}\nu_s^2 ds < +\infty$$

Hence we get $\int_0^\infty e^{-as}\nu_s^2 ds < +\infty$ almost surely. Hence

$$as-\lim_{t\to\infty} N_t = N_\infty < +\infty. \tag{25}$$

This completes the proof of ii).

Let us define now the adaptive filter $(\hat{\hat{x}}_t)$ by

$$\hat{\hat{x}}_t = D(\phi^*.\Theta.)(t), \quad t\geq 0 \tag{26}$$

where (ϕ_t) and (Θ_t) are generated by the SGA (6)-(9).

Theorem 2. Assume δ is as in lemma 2. Then

i) if $a<0$ we have

$$as-\lim_{t\to\infty} \frac{1}{t}\int_0^t (\hat{\hat{x}}_s - c\hat{x}_s)^2 ds = 0, \tag{27}$$

ii) if $a>0$ we have

$$as-\lim_{t\to\infty} \frac{1}{e^{2at}}\int_0^t (\hat{\hat{x}}_s - c\hat{x}_s)^2 ds = 0. \tag{28}$$

Proof. We first note that by (2) and (5) we have

$$c\hat{x}_t - \hat{\hat{x}}_t = \xi_t + \eta_t, \quad t\geq 0 \tag{29}$$

From lemma 6, lemma A1 of [1] and Kronecker lemma we get

$$as-\lim_{t\to\infty} \frac{1}{r_t}\int_0^t (c\hat{x}_s - \hat{\hat{x}}_s)^2 ds = 0. \tag{30}$$

This together with lemmas 4 and 7 yields the desired results.

Remark 5. Note that for $\gamma''_t \overset{\text{def}}{=} \hat{\hat{x}}_t$ condition (CC) is satisfied.

Remark 6. Note that the strong consistency of parameter estimates generated by (6)-(9) is not required in order to get the Cesaro type convergence of adaptive filter defined by (26).

References

[1] T. Bielecki, B. Gołdys, Recursive Estimation of Parameters for Partially Observed Linear Stochastic Differential Systems, Preprint IM PAN, 1988.

[2] Chen Han Fu, Guo Lei, Strong Consistency of Parameter Estimates for Discrete-Time Stochastic Systems, J. Sys. Sci. and Math. Scis. vol.5 (1985), 81-93.

[3] M.H.A. Davis, Linear Estimation and Stochastic Control, 1977, London, Chapmen and Hall.

[4] R. Kumar, J.B. Moore, Convergence of Adaptive Minimum Variance Algorithms via Weighting Coefficients Selection, IEEE Trans. Automat. Control, vol. AC-27, pp. 146-153, Feb. 1982.

[5] J.H. van Schuppen, Convergence Results for Continuous-Time Adaptive Stochastic Filtering Algorithms, J. Math. Anal. Appl. vol.96 (1983), 209-225.

[6] C.A. Desoer, M. Vidyaçasar, Fedback Systems: Input-Output Properties, Academic Press, New York, 1975.

[7] M. Wohnam, Linear Multivariable Control: a Geometric Approach, Springer-Verlag, New York, 1979, second edition.

On the Smooth Fit Boundary Conditions in the Optimal Stopping Problem for Semimertingales

R.J. Chitashvili
Department of Probability Theory and Mathematical Statistics
Tbilisi Mathematical Institute, Academy of Sciences of the Georgian SSR
150 a, Plekhanov Avenue, Tbilisi 380012, USSR

The well known smooth fit boundary conditions in the optimal stopping problem of Markov diffusion processes are generalized for the optimal stopping problem of semimartingales.

Keywords & Phrases: semimartingales optimal stopping problem.

1. Let a semimartingale $X=M+A$ be given on a probability space $(\Omega, \mathfrak{F}, P)$ with a filtration $(\mathfrak{F}_t)$, $0 \leq t \leq T$, satisfying usual conditions, where M is a martingale and A is a predictable process with integrable variation.

It is well known that in the optimal stopping problem for X the stopping rule (time change) (τ_t^*), $0 \leq t \leq T$, defined as $\tau_t^* = \inf(s \geq t:\ V_s^* = X_s) \wedge T$ is optimal, where $V_t^* = \sup_{t \leq \tau \leq T} E(X_\tau | \mathfrak{F}_t)$ is the value process. In other words, if for each nonnegative semimartingale η, which is right-continuous and admits left-hand limits, we consider the time change $\tau_t^\eta = \inf(s \geq t:\ \eta_s = 0)$ and corresponding expected reward $V_t^\eta = E(X_{\tau_t^\eta} | \mathfrak{F}_t)$, then $V^\eta = V^*$ when $\eta = V^* - X$. This is true, in particular, in the regular case, when process X is supposed to be right-continuous and left-quasicontinuous (see [1] for a surway of the theory and further generalizations).

Thus the construction of the optimal stopping rule is connected with the construction of the value process. The methods of constructing the value process, are in turn, based on the, so called, characterization theorems. The most general *supermartingale characterization,* is given in the following proposition. All the semimartingales considered below are right continuous admitting left-hand limits and having special decompositions.

PROPOSITION 1. *The process V is the value process iff*
a) *V is a supermartingale ($V=M-B$, $m \in \mathfrak{M}$, $dB \geq 0$) $V_T = X_T$;*
b) *$V \geq X$ ($V_t \geq X_t$ a.s. $0 \leq t \leq T$);*
c) *V is the minimal process with the properties a) and b).*
Here $\mathfrak{K}$ denotes the class of martingales and, for convenience, the fact that (predictable) process B is increasing is expressed as $dB \geq 0$.

This characterization is inconvenient because it is often hard to verify the minimality condition c). The following *variational inequalities* allow us to express property c) directly in terms of the process V.

PROPOSITION 2. *The process V is the value process iff*
a) *$V=m-B$, $m \in \mathfrak{M}$, $dB \geq 0$, $V_T = X_T$;*
b) *$V \geq X$;*
c) *$dB = I_{[V_- = X_-]} dB$; i.e. the process B increases only on the set*
$(V_- = X_-) = ((t, \omega):\ V_{t-} = X_{t-})$.

In order to impose additional requirements on this characterization, we represent it now as an optimality test for the given stopping region $S_\eta = (\eta = 0)$ corresponding to some non-negative semimartingale η.

PROPOSITION 3. *The process V^η is the value process if*
a) $V^\eta = m^\eta - B^\eta$, $m^\eta \in \mathfrak{M}$, $dB^\eta \geqslant 0$;
b) $V^\eta \geqslant X$.

Combining this statement with the boundary problem for the process V^η, we can formulate the following *free boundary problem.*

PROPOSITION 4. *The semimartingale $V = m - B$ and the time change (τ_t^η), $0 \leqslant t \leqslant T$, are the value process and optimal stopping rule respectively iff*
a) $dB \geqslant 0$;
b) *$V > X$ on the set $(\eta > 0)$;*
c) *V is a solution of boundary problem*

$$I_{[\eta_- > 0]} dV = I_{[\eta_- > 0]} dm, \quad V = X \text{ on the set } (\eta = 0), \quad V_T = X_T.$$

Obviously condition a) may be written as $I_{[\eta_- = 0]} dB \geqslant 0$.

The aim of reduction of condition a) to the *smooth fit condition* can be explained as follows. The coincidence $V = X$ on the stopping region S_η does not imply of course the coincidence $I_{[\eta_- = 0]} dB = - I_{[\eta_- = 0]} dA$. However, roughly speaking, on the essential part of S_η it might be possible to verify condition a) directly in terms of the process A given apriority, and hence the verification of a) can be reduced to a certain 'small' subset ∂S_η close to the continuation region $\{\eta > 0\}$.

2. To throw more light to the above reasoning, we present now the example of discrete time case, though the natural generalization of the smooth fit conditions on a boundary for diffusion processes ([2], [3]) will be given in Corollary 2 below.

Let the process X be stepwise constant, $X_t = X_{k\Delta}$, $k\Delta \leqslant t < (k+1)\Delta$, $k = 0, \ldots, T/\Delta$; $\Delta > 0$. Then, of course, all processes S^η will be stepwise constant and can be considered only at points $t = 0, \Delta, \ldots, T/\Delta$.

By Proposition 1 the value process V^* is a least solution of the inequality

$$V_{t-1} \geqslant \max(X_{t-1}, E(V_t | F_{t-1})), \quad V_T = X_T, \tag{1}$$

while Proposition 2 determines it as a unique solution of the Wald-Bellman equation

$$V_{t-1} = \max(X_{t-1}, E(V_t | F_{t-1})), \quad V_T = X_T. \tag{2}$$

In fact, conditions a), b) are equivalent to (1), while condition c) with $\Delta B_t = V_{t-1} - E(V_t | F_{t-1})$ leads to (2).

Propositon 4 tells us that the processes V and η can be found by solving the recurrent equation with boundary conditions

$$\begin{aligned} V_{t-1} &= I_{[\eta_{t-1} = 0]} X_{t-1} + I_{[\eta_{t-1} > 0]} E(V_t | F_{t-1}), \quad V_T = X_T; \\ I_{[\eta_{t-1} = 0]} &(V_{t-1} - E(V_t | F_{t-1})) \geqslant 0; \quad (V_t - X_t) I_{[\eta_t > 0]} > 0. \end{aligned} \tag{3}$$

It is convenient now to represent the backward recurrent equation (3) as a stochastic difference equation. Denote by ζ the process determined by the jumps

$$\Delta\zeta_t = E(\Delta(V_t - X_t)I_{[\eta_{t-1}=0,\ \eta_t>0]}|F_t). \tag{4}$$

We have

$$\begin{aligned} \Delta V_t &= I_{[\eta_{t-1}>0]}\Delta M_t + I_{[\eta_{t-1}=0]}\Delta A_t + \Delta\zeta_t, \quad V_T = X_T \\ 0 &\leq I_{[\eta_{t-1}=0]}\Delta\zeta_t = 0 \leq -(1 - I_{[\eta_{t-1}>0,\ V_{t-1}>X_{t-1}]})\Delta A_t. \end{aligned} \tag{5}$$

Thus for the process V we have obtained the stochastic equation with an 'unknown' martingale part M (which obviously is uniquely determined by the boundary condition $V_T = X_T$ at the right end of the time interval), and with a unknown (predictable) process ζ satisfying the boundary conditions, which can be, in turn rewritten separately

$$\begin{aligned} &\Delta\zeta_t \geq 0, \quad I_{[\eta_{t-1}>0,\ V_{t-1}\leq X_{t-1}]}\Delta A_t \leq 0, && \text{(majorising property)}, \\ &\Delta\zeta \leq -I_{[\eta_{t-1}=0]}\Delta A_t, && \text{(supermartingality)}, \\ &\Delta\zeta_t = I_{[\eta_{t-1}=0]}\Delta\zeta_t, && \text{(minimality)}. \end{aligned}$$

It is seen from the expression (4) that the verification of boundary conditions can be restricted to the subset of S_η

$$\partial S_\eta = (\eta_{t-1} = 0,\ P(\eta_t > 0|\eta_{t-1}) > 0).$$

In fact $\Delta\zeta = I_{\partial S_\eta}\Delta\zeta$, and $\Delta B = -\Delta\zeta - I_{[\eta_- = 0]}\Delta A = -\Delta A$ on the set $S_\eta \setminus \partial S_\eta$. In other words it suffieses to verify the conditions

$$(I_{[S_\eta \setminus \partial S_\eta]} + I_{[\eta_- > 0,\ V_{t-1} \leq X_{t-1}]})\Delta A_t \leq 0 \quad \text{a.s.,} \quad 0 \leq t \leq T,$$

and

$$0 \leq \Delta\xi_t \leq -\Delta A_t \quad \text{a.s. relative to the measure} \quad \mu^\eta, \tag{6}$$

where μ^η is Dolean's measure associated with the increasing process

$$\mathfrak{L}_t = \sum_{s \leq t} I_{[\eta_{s-1}=0,\ \eta_s>0]}.$$

Thus, the problem of obtaining the boundary condition for the general case considered below, can be viewed as a generalization of (6).

3. Note that the process V^η does not depend on the choice of the process η within the class of equivalent processes, i.e. the nonnegative semimartingales $\tilde{\eta}$ for which $I_{[\tilde{\eta}_t=0]} = I_{[\eta_t=0]}$, a.s., $0 \leq t \leq T$. Hence we can suppose $|V_t^\eta - X_t| \leq \eta_t$, choosing, if nessesary, the new process $\tilde{\eta} = \eta + |V^\eta - X|$. In the forthcoming we shall assume

$$\sum_{s \leq T} |\Delta X_s| < \infty.$$

THEOREM 1. *The process V^η satisfies the equation*

$$dV^\eta = dM^\eta + I_{[\eta_- = 0]}dA + d\mathfrak{L}(V^\eta - X, \eta), \quad V_T^\eta = X_T, \tag{7}$$

where

$$\mathfrak{L}(V^\eta - X, \eta) = L(V^\eta - X, \eta) + \mathfrak{L}^d(V^\eta - X, \eta),$$

with

$$L_t(V^\eta - X, \eta) = \lim_{\epsilon \to 0} \epsilon^{-1} \int_0^t I_{[\eta_s \leq \epsilon]} d < V^\eta - X, \eta >_s^c,$$

and

$$\mathfrak{L}_t^d(V^\eta = X, \eta) = (\Sigma I_{[\eta_{s-} = 0, \eta_s > 0]} \Delta(V_s^\xi - X_s))_t^P.$$

(Here and elsewhere below $<V^\eta - X, \eta>^c = <V^{\eta c} - X^c, \eta^c>$ *is the square mutual characteristic of continuous martingale parts of* $V^\eta - X$ *and* η*;* $(\ )^P$ *denotes the dual predictable projection).*

PROOF. For $\epsilon > 0$ consider the process $Z_t^\epsilon = \epsilon^{-1}(\epsilon - \eta_t)(V_t^\eta - X_t)$. Evidently $\lim_{\epsilon \to 0} Z_t^\epsilon = I_{[\eta_t = 0]}(V_t^\eta - X_t) = 0$. We shall apply the integral representation formula for the positive part of the semimartingale Y ([4])

$$y_t^+ = y_0^+ + \int_0^t [I_{[y_{s-} > 0]} + \frac{1}{2} I_{[y_{s-} = 0]}) dy_s + \frac{1}{2} L_t^0(y) + \sum_{s \leq t} (\Delta(y)_s^+ - [I_{[y_{s-} > 0]} + \frac{1}{2} I_{[y_{s-} = 0]}] \Delta y_s),$$

where $L^0(y)$ is the local time spent at 0.

We have

$$(\epsilon - \eta_t)^+ = (\epsilon - \eta_0)^+ - \int_0^t (I_{[\eta_{s-} < \epsilon]} + \frac{1}{2} I_{[\eta_{s-} = \epsilon]}) dy_s + \frac{1}{2} L_t^\epsilon(\eta)$$

$$+ \sum_{s \leq t} (\Delta(\epsilon - \eta)_s^+ - (I_{[\eta_{s-} < \epsilon]} + \frac{1}{2} I_{[\eta_{s-} = \epsilon]}) \Delta \eta_s).$$

Applying this we obtain

$$Z_t^\epsilon = Z_0^\epsilon + \epsilon^{-1} \int_0^t (\epsilon - \eta_{s-})^+ d(V_s^\eta - X_s) + \sum_{s \leq t} \epsilon^{-1} \Delta(\epsilon - \eta)_s^+ \Delta(V^\eta - X)_s$$

$$- \epsilon^{-1} \int_0^t (I_{[\eta_s < \epsilon]} + \frac{1}{2} I_{[\eta_s = \epsilon]}) d < V^\eta - X, \eta >^c - \epsilon^{-1} \int_0^t (V_{s-}^\eta - X_{s-})(I_{[\eta_{s-} < \epsilon]}$$

$$+ \frac{1}{2} I_{[\eta_{s-} = \epsilon]}) d\eta_s + (2\epsilon)^{-1} \int_0^t (V_s^\eta - X_s) dL_s^\epsilon(\eta) + \epsilon^{-1} \sum_{s \leq t} (V_{s-}^\eta - X_{s-})[\Delta(\epsilon - \eta_s)^+$$

$$- (I_{[\eta_{s-} \leq \epsilon]} + \frac{1}{2} I_{[\eta_{s-} = \epsilon]}) \Delta \eta_s].$$

Now using the relations

$$\lim_{\epsilon \to 0} \epsilon^{-1} \eta_t - I_{[\eta_{t-} \leq \epsilon]} = 0, \quad \int_0^t I_{[\eta_s = \epsilon]} d < V^\eta - X, \eta >_s^c = 0,$$

it is easily seen that

$$(\int_0^\circ I_{[\eta_{t-} = 0]} d(V^\eta - X)_t)^P = \mathfrak{L}(V^\eta - X, \eta).$$

It remains to show that the process

$$\int_0^t I_{[\eta_{s-}>0]} dV_s^\eta$$

is a martingale. In fact this follows from the general results on balayages ([5]). Indeed

$$V_t^\eta = E(X_{\tau_t^\eta}|F_t) = m_t + E(A_{\tau_t^\eta}|F_t) = m_t + m_t' + A_t^b,$$

where m, m' are martingales, and the predictable projection A^b of the process $A_{\tau_t^\eta}$ has the variation increasing only on the set $((t, \omega): l_t = t) \subset (\eta_- = 0)$, where $l_t = \sup(s < t: \eta_s = 0)$. □

It should be noted also that the process $\tilde{\eta}$ can be chosen so that $I_{[\tilde{\eta}_t = 0]} = I_{[\eta_t = 0]}$ a.s. $0 \leq t \leq T$, and besides

$$(\tilde{\eta}_{t-} = 0) = (l_t = t) \cup (\eta_{t-} = 0,\ \eta_t = 0),$$

by taking

$$\tilde{\eta}_t = E(C_{\tau_t^\eta} - C_t|F_t)$$

with a strongly increasing process C.

4. We shall now present all the above mentioned characterizations of the value process.

THEOREM 2. *The following statements are equivalent*

a) $V_t^* = \sup_{t \leq \tau \leq T} E(X_\tau|F_t)$;

b) V^* *is the minimal solution of the equation*

$$dV = dm - dB,\quad V \geq X,\quad V_T = X_T,\quad m \in \mathfrak{M},\quad dB \geq 0;$$

c) V^* *is the unique solution of the equation*

$$dV = dm - dB,\quad V \geq X,\quad V_T = X_T,\quad m \in \mathfrak{M},\quad 0 \leq dB = I_{[V_- = X_-]} dB;$$

d) V^* *is the maximal solution of the equation*

$$dV = I_{[V_- > X_-]} dm + I_{[V_- = X_-]} dA + d\zeta,\quad V_T = X_T,\quad m \in \mathfrak{M},\quad 0 \leq d\zeta = I_{[V_- = X_-]} d\zeta;$$

e) V^* *is the unique solution of the equation*

$$dV = I_{[V_- > X_-]} dm + I_{[V_- = X_-]} dA + d\zeta,\quad V_T = X_T,\quad m \in \mathfrak{M},$$

$$0 \leq d\zeta = I_{[V_- = X_-]} d\zeta \leq -I_{[V_- = X_-]} dA.$$

PROOF. b) is just a supermartingale characterization. Let us prove c) which expresses the variational inequalities considered for diffusion processes in [6]. As the time change $\tau_t^{V^* - X}$ is optimal we have $V^* = V^{V^* - X}$. By Theorem 1 V^* satisfies the equation c). On the other hand, if V is a solution of c), then V is a supermartingale with $V \geq X$. Hence $V \geq V^\eta$ for each η. But for the time change

$$\tau_t^\epsilon = \inf(s \geq t: V_s \leq X_s + \epsilon) \vee T$$

we have

$$V_t = E(V_{\tau_t^\epsilon}|F_t) \leq E(X_{\tau_t^\epsilon}|F_t) + \epsilon. \tag{8}$$

Thus $V=V^*$, and c) is proved.

To proved d) note that for each solution of equation d) we have the estimation (8). Thus $V \leqslant V^*$. At the same time $V^* = V^{V^*-X}$ is a solution of d) and, besides, by $V^* \geqslant X$ we have

$$d\zeta - = I_{[V^*_- = X_-]} d(V^* - X) \geqslant 0.$$

Finally V^* satisfies the equation e) (the last inequality in the boundary conditions is connected with a supermartingality property). If V is a solution of e), then the process $V-X$ satisfies the relation

$$d(V-X) = I_{[V_- > X_-]} d(V-X) + d\zeta, \quad V_T = X_T,$$

from which it is easily seen that $V \geqslant X$. Thus by d) we have $V \leqslant V^*$ and hence $V = V^*$. □

The decomposition of the increasing process $B = -V^* + m$ in the form $dB = -I_{[\eta_- = 0]} dA - d\zeta$ with $d\zeta \geqslant 0$ was established in [7] for a more general class of processes X.

5. The connection should be mentioned of this theorem with characterization properties of reflecting processes, especially with Skorohod's (direct) equation.

PROPOSITION 5. *The following statements are equivalent*

a) $\overline{V}_t = \sup_{0 \leqslant s \leqslant t} X_s$;

b) $\overline{V}$ *is the minimal solution of the equation*

$$dV = dB, \quad dB \geqslant 0, \quad V_0 = X_0, \quad V \geqslant X;$$

c) $\overline{V}$ *is the unique solution of the equation*

$$dV = dB, \quad V \geqslant X, \quad V_0 = X_0, \quad 0 \leqslant dB = I_{[V_- = X_-]} dB;$$

d) $\overline{V}$ *is the maximal solution of the equation*

$$dV = I_{[V_- = X_-]} DX + d\zeta, \quad V_0 = X_0, \quad 0 \leqslant d\zeta = I_{[V_- = X_-]} d\zeta;$$

e) $\overline{V}$ *is the unique solution of the equation*

$$dV = I_{[V_- = X_-]} dX + d\zeta, \quad V_0 = X_0, \quad 0 \leqslant d\zeta, \quad I_{[V_- = X_-]} dX + d\zeta \geqslant 0.$$

In particular, the process $\overline{X}_t = \overline{V}_t - X_t$ expresses the instantaneous reflection of the process $(-X_t)$ from the zero boundary, and a) is a generalization of Levy's expression for the reflected Wiener process ($X = -W$). For the process $\overline{X}$ c) is Skorokhod's equation ([8], [9])

$$d\overline{X} = dX + dB, \quad \overline{X} \geqslant 0, \quad \overline{X}_0 = 0, \quad 0 \leqslant dB = I_{[\overline{X}_- = 0]} dB. \tag{9}$$

Characterization property d) was noted in [10] for Ito processes.

Using e) we can derive

COROLLARY 1. *For continuous process $X = M + A$, if $varA$ is dominated by $<M>$ ($varA << <M>$), then the process $\overline{X}$ is the unique solution of the boundary problem*

$$I_{[\overline{X} > 0]} d\overline{X} = I_{[\overline{X} > 0]} dm, \quad \overline{X} \geqslant 0, \quad \overline{X}_0 = 0, \quad \int_0^T I_{[\overline{X} = 0]} d<M> = 0. \tag{10}$$

In fact, since $\bar{X} \geqslant 0$, we have $d\bar{X} = I_{[\bar{X}>0]} d\bar{X} + dL^0(\bar{X})$. Comparing this with e) and (9), and having in mind that

$$I_{[\bar{X}=0]} dM + I_{[\bar{X}=0]} dA + d\zeta \geqslant 0$$

implies $I_{[\bar{X}=0]} d<M> = 0$, and hence $I_{[\bar{X}=0]} dA = 0$, we obtain $\zeta = L^0(\bar{X})$. □

Thus the process $\bar{X}$ is characterized by the fact that it spends zero time (measured in terms of $<M>$) on the boundary.

6. Returning to the stopping problem expressed by (in some sense dual to Proposition 5) properties of Theorem 2, we shall derive the assertion which can be viewed as a dual analogy of Corollary 1.

We show meanwhile that there is a nonnegative semimartingale η such that

$$var\, \mathfrak{L}(V^\eta - X, \eta) << \mathfrak{L}(\eta, \eta). \tag{11}$$

In fact it suffices to take

$$\tilde{\eta}_t = E(C_{\tau_t^\eta} - C_t | F_t)$$

with a certain increasing predictable process C such that $var\, A << C$. This is easily seen by relating to increasing process B the process

$$\eta_t^B = E(B_{\tau_t^\eta} - B_t | F_t)$$

with the differential $d\eta^B = I_{[\eta_-^B > 0]} dm^B + d\mathfrak{L}(\eta^B, \eta^B)$, $m^B \in \mathfrak{M}$. For B^1, B^2, with $d(B^1 - B^2) \geqslant 0$, we have $\eta^{B^1} = \eta^{B^2} + \eta^{B^1 - B^2}$ and hence

$$\mathfrak{L}(\eta^{B^1}, \eta^{B^1}) >> \mathfrak{L}(\eta^{B^2}, \eta^{B^2}).$$

The general case $B^2 << B^1$ is treated in the same manner by introducing the decomposition

$$B^2 = \sum_{n=1}^{\infty} I_{[(n+1)^{-1} \leqslant b < n^{-1}]} b \cdot B^1 + I_{[b \geqslant 1]} b \cdot B^1,$$

where $b = dB^2/dB^1$ (i.e. $B^2 = b \cdot B^1$). Obviously, for each n

$$\mathfrak{L}(\eta^{B^n}, \eta^{B^n}) << \mathfrak{L}(\eta^{B^1}, \eta^{B^1}),$$

where $B^n = I_{[(n+1)^{-1} \leqslant b < n^{-1}]} b \cdot B^1$, which leads to (11). Combining the statements e) of Theorems 1 and 2 we obtain

COROLLARY 2. *Let $\eta = \eta^C$ with $var\, A << C$ and μ^η, μ^C be Dolean's measures associated with the processes $\mathfrak{L}(\eta, \eta)$ and C respectively. Furthermore, let $\mu^\eta = \mu_s^\eta + \mu_r^\eta$ be the decomposition of the measure μ^η on the singular and regular parts with respect to the measure μ^C. Then the process V^η is the value process iff*

1) $V^\eta > X$ *on the set* $(\eta > 0)$;

2) $d\mathfrak{L}(V^\eta - X, \eta)/d\mathfrak{L}(\eta, \eta) = 0 \quad \mu_s^\eta$ *a.e.* (12)

$$0 \leqslant d\mathfrak{L}(V^\eta - X, \eta)/dC \leqslant -dA/dC, \quad (\eta_- = 0) - \eta^C \text{ a.e.}$$

Thus if $\mu^\eta = \mu_s^\eta$, the boundary conditions reduce to the smooth fit condition (12) and the condition $dA/dC \leqslant 0 \quad (\eta_- = 0) - \mu^C$ a.e.

For continuous processes with the integral representation property for martingales, the connection of condition (12) with classical Stefan's problem becomes more implicit.

Let X, η be continuous and let there exist the continuous mutually orthogonal martingales M^i, $1 \leqslant i \leqslant n$, such that the martingale parts m^η, M, N in the decomposition of semimartingales V^η, X, η are represented as integrals

$$m^\eta = \sum_i \psi^i \cdot M^i, \quad X = \sum_i \phi^i \cdot M^i, \quad N = \sum_i g^i \cdot M^i.$$

For simplicity suppose $<M^i> = C$, $1 \leqslant i \leqslant n$. We need to require below that the processes ψ, ϕ, g have limits on the boundary within the continuation region $(\eta > 0)$.

For some process Z denote by Z^+ and Z^- the upper and lower limits

$$Z_t^+ = \lim_{\epsilon \to 0} \operatorname{ess\,sup}_{|s-t| \leqslant \epsilon} Z_s, \quad Z_t^- = \lim_{\epsilon \to 0} \operatorname{ess\,inf}_{|s-t| \leqslant \epsilon} Z_s,$$

where ess sup (ess inf) is taken with respect to the measure μ^C.

COROLLARY 3. *Let*

1) $\psi^+ = \psi^- = \hat{\psi}, \ \phi^+ = \phi^- = \hat{\phi}, \ g^+ = g^- = \hat{g} \quad \mu^\eta$ *a.e.*
2) $\sum_i (\hat{g}^i)^2 > 0 \quad \mu^\eta$ *a.e.*

Then the process V^η is the value process if

a) $dA/dC \leqslant 0 \quad (\eta = 0) - \mu^C$ a.e.
b) $V^\eta > X$ *on the set* $(\eta > 0)$ *a.s.*
c) $\sum_i (\hat{\psi}^i - \phi^i)\hat{g}^i = 0 \quad \mu^\eta$ *a.e.*

PROOF. It is easily verified that μ^C a.e. processes $\psi^+, \phi^+, g^+ \quad (\psi^-, \phi^-, g^-)$ are uppersemicontinuous (lowersemicontinuous) in t and hence

$$\lim_{\epsilon \to 0} \epsilon^{-1} \int_0^t I_{[\eta_s < \epsilon]} d < V^\eta - X, \eta >_s \leqslant \lim_{\epsilon \to 0} \epsilon^{-1} \int_0^t I_{[\eta_s < \epsilon]} \Big(\sum_i (\psi^i - \phi^i) g^i / \sum_i (g^i)^\eta\Big)^+ d < \eta >_s$$

$$\leqslant \int_0^t \sum_i ((\psi^i - g^i) g^i / \sum_i (g^i)^\eta)^+ \, dL_s^0(\eta).$$

This and the similar inequality with lower limits, together with assumption 1) implies c). Furthermore, since $I_{[\eta=0]} d < N > = 0$, the measure μ^η is singular w.r.t. the measure $\mu < N >$ and, hence, by assumption 2) w.r.t. the measure μ^C too. □

The smooth fit condition expressed as $\psi_{\tau_\gamma} = \phi_{\tau_\gamma}$ was obtained in [11] for Ito processes X.

7. THE EXAMPLE OF A DIFFUSION PROCESS

Let ξ be a Markov diffusion process with a nonsingular diffusion metrix $B(x)B^*(x)$, $x \in R^{(n)}$, and with a density function $q_x(t,y)$, $x,y \in R^{(n)}$, corresponding to an initial condition $\xi_0 = X$. As for the martingales M^i, $1 \leqslant i \leqslant n$, we can consider the Wiener process $w = B^{-1}(\xi) \cdot \xi$. Moreover, here

$$X_t = \varphi(\xi_t), \quad V_t^\eta = v^\eta(t, \xi_t), \quad \eta_t = G(t, \xi_t),$$

where φ and $G \geqslant 0$ are smooth functions, and

$$v^{\eta}(t,x) = E_x\varphi(\xi^{\eta}_{\tau_t}).$$

The stopping region can be defined in terms of the set $\mathfrak{D}_t = (x: G(t,x)=0)$. In fact $S_\eta = ((t,\omega): \xi_t \in \mathfrak{D}_t)$. Let $C_t = t$. Evidently

$$\psi_t = B^*(\xi_t)\nabla v^{\eta}(t,\xi_t), \quad \phi_t = B^*(\xi_t)\nabla\varphi(\xi_t), \quad g_t = B^*(\xi_t)\nabla G(t,\xi_t).$$

It suffices here to restrict the integration operator corresponding to the measure μ^{η} to the class of processes represented as $F(t,\xi_t)$ for some measurable function $F(t,x)$, $x \in R^{(n)}$, $0 \leqslant t \leqslant T$. It is easily calculated, for fixed initial conditions $\xi_0 = X$, that

$$\iint\limits_{[0,T]\times\Omega} F(t,\xi_t)\mu^{\eta}(dt, d\omega) = E_x\int_0^T F(t,\xi_t)dL_t^0(\eta) = \lim_{\epsilon\to 0}\int_0^T\int_{R^{(n)}} F(t,y).$$

$$\epsilon^{-1}I_{[G(t,y)<\epsilon]}q_x(t,y)(\nabla G(t,y),\ B(y)B^*(y)\nabla G(t,y))dy\,dt =$$

$$= \int_0^T\int_{\partial D_t} F(t,y)(\nabla G(t,s),\ \nabla G(t,y))^{-\frac{1}{2}}(\nabla G(t,s),\ B(y)B^*(y)\nabla G(t,y))q_x(t,y)d\sigma(y)dt,$$

where ∂D_t is a boundary and the integral with respect to $d\sigma(y)$ is understood as the surface integral over ∂D_t.

Now, assume:

1) The functions B, $\nabla\varphi$, ∇G, ∇v^{η} are continuous on ∂D_t,

2) $(\nabla G, \nabla G) > 0$ on D_t.

If, in addition, the density function is also positive and continuous on ∂D_t, then the smooth fit condition c) of Corollary 3 reduces to the pointwise equality condition

$$(\nabla(v^{\eta} - \varphi),\ BB^*\nabla G)(t,x) = 0, \quad x \in \partial D_t, \quad 0 \leqslant t \leqslant T.$$

References

[1] N. El Karoui, Let aspects probabilites du contrôle stochastique, Ecole d'été de probabilités de Saint-Flour, Lect. Notes in Math. no. 876, Springer Verlag, 1980.

[2] A.N. Shiryaev, Statistical sequential analysis: optimal stopping rules. Providence, A.M.S. 1973.

[3] B.I. Grigelionis, A.N. Shiryaev, On Stefan's problem and optimal stopping rules for Markov processes, Theory probab. Applications VXI no. 4 1966 pp. 541-559.

[4] J. Jacod, Calcul stochastique et problems de martingales, Lect. Notes in Math. no. 714, Springer Verlag.

[5] N. El Karoui, Temps local et balayage des semimartingales, Sém. Proba XIII. Lect. Notes in Math. no. 721, Springer Verlag 1979.

[6] A. Bensonssan, J.L. Lions, Problemes de temps d'arret optimal et inequations variationnelles paraboliques, Applicable Analysis, 1973, v. 3, pp. 267-294.

[7] N. El Karoui, Une propriete de domination de domination de l'envelope de Snell des semimartingales fortes, Lect. Notes in Math. Sém. Proba XVI, Springer Verlag, pp. 400-408.

[8] I.I. Gikhman, A.V. Skorokhod, Stochastic differential equations. Berlin, Springer, 1972.

[9] N. El Karoui, M. Chaleyat-Maurel, Un probléme de réflexion et ses applications at temps local et aux équations differentielles stochastiques sur R. Asterisque 52-53, 1978, pp. 117-145.

[10] R.J. Chitashvili, N. L. Lazrieva, Strong solutions of stochastic differential equations with boundary conditions, Stochastics v.5, no. 4, 1981 pp. 255-311.

ORDER DETERMINATION AND ADAPTIVE CONTROL OF ARX MODELS USING THE PLS CRITERION

M.H.A. Davis and E.M. Hemerly[1]
Department of Electrical Engineering
Imperial College, London SW7 2BT, England

Rissanen's "Predictive Least Squares" principle provides a recursive procedure by which the order of a stable autoregression may be consistently estimated. Combining this with a modification of the randomly truncated LQG control with attenuated excitation due to Chen and Guo, we obtain a fully recursive adaptive controller for ARX systems of unknown order which asymptotically minimizes a quadratic performance index while providing consistent estimates of model order and parameters.

1. INTRODUCTION

One of the major achievements of the "prediction error" approach to the identification of linear discrete time stochastic systems is a satisfactory theory of approximate system modelling. This theory – described in Davis and Vinter [1] – states that, under certain conditions, the parameter estimates produced by the prediction error method will converge to values corresponding to the *best predictor* possible within the chosen model set; thus a "best model" is estimated even if – as, of course, will always be the case in practice – the mechanism generating the data is not included in the model set.

No such satisfactory results are available in adaptive control. The classic Goodwin-Ramadge-Caines (GRC) algorithm (see [1], Chapter 7), and most of its successors, assumes that an upper bound is known for the system model order, so that the structure of the controller is at least as complex as that of the system to be controlled. There is a nascent theory of "robust adaptive control" – see for example Middleton *et al* [2] – which shows that certain modified adaptive controllers will stabilize systems in an ϵ-neighbourhood (suitably defined) of the nominal model set, so that small "unmodelled dynamics" can be tolerated. However there is no estimate as to how big ϵ is, and the question of what is the appropriate complexity for the nominal model set is left completely open.

---

[1]On leave from Centro Técnico Aeroespacial, São José dos Campos, SP, Brazil.
Work supported by ITA and CAPES

An alternative approach, which we pursue here, is to estimate model order on-line. This gets away from the limitations of a fixed controller structure, but even so is only a partial answer to the problem of determining a controller of "sufficient complexity". At the moment we assume a known upper bound for the order of the "true" system. This assumption could probably (with some technical difficulty) be dispensed with, but we would still be left with the assumption that the true system is accurately represented by *some* finite-dimensional model. The ultimate algorithm would be one in which the order of the model is increased until some test indicates that the residual unmodelled dynamics are within the ϵ-neighbourhood which robust controllers of that order are able to handle. There are some remarks on this subject in the thesis [3], but basically ideas of this kind are still at the wishful thinking stage.

There is of course a well-developed theory of order determination for linear time series models, the best results being due to Hannan and co-workers — see for example [4], [5]. So far as we are aware, however, there is only one paper dealing with on-line order estimation in adaptive control systems in a rigorous way, namely the paper [6] by Chen and Guo. These authors use a variant of the BIC criterion for order estimation and an LQG-based control algorithm with an additional attenuating dither input signal for persistency of excitation and a truncation of the control input to restrict the rate of growth of the input and output variables. Our purpose here is to summarize the content of the papers [7][8], in which we explore the use of Rissanen's "Predictive Least Squares (PLS)" principle[9], in identification of order and adaptive control of ARX models. The PLS principle is an intuitively appealing way of selecting model order: the order chosen is simply that of the model that has done the best prediction job so far. This avoids the use of the non-data-dependent factors which appear in the BIC and other order determination criteria. But more importantly, as pointed out by Wax [10], use of the PLS criterion has the overriding practical advantage that the models of all orders and the associated prediction errors can be computed simultaneously using the highly efficient "lattice filter" implementation of least squares estimation — see Caines [11], §3.7.

In §2 below we describe the use of the PLS criterion for determining the order of an autoregression. Almost sure convergence of the estimated model order to the true value is obtained. In §3 we discuss adaptive control of ARX models. Model order is estimated by PLS, and the control stategy is similar to that of Chen and Guo [6] with some modifications to the rate of attenuation of the dither signal and the truncation procedure to take account of the somewhat more delicate behaviour of the PLS estimate. As in [6], asymptotically optimal control is obtained. Some concluding remarks are given in §4.

2. ORDER DETERMINATION FOR AUTOREGRESSIONS

Let $(\Omega,\mathcal{F},(\mathcal{F}_t)_{t=0,1,...},P)$ be a filtered probability space and $\{w(t)\}$ a martingale difference process with respect to $\mathcal{F}_t$. The data $\{y(t),\ t\geq 0\}$, with $y(t)=0,\ \forall t<0$, is assumed to be generated by the p_0'th order autoregressive process

$$y(t) = a_1\, y(t-1) + a_2\, y(t-2) +...+ a_{p_0}\, y(t-p_0) + w(t). \tag{2.1}$$

We suppose that an upper bound p^* for the model order p_0 is known. Competing models of order $p = 1,2,...p^*$ are fitted by ordinary least squares. Let

$$\hat{\Theta}^{\mathrm{T}}(p,t) = [\hat{a}_{p,1}(t)\ \hat{a}_{p,2}(t)\ ...\ \hat{a}_{p,p}(t)]$$

denote the estimated predictor coefficients in a p'th order model at time t; then

$$\hat{\Theta}(p,t) = \Big(\sum_{j=1}^{t} \Phi(p,j)\Phi^{\mathrm{T}}(p,j)\Big)^{-1} \sum_{j=1}^{t} \Phi(p,j)y(j)$$

where $\Phi(p,t)$ denotes the regressor vector, $\Phi^{\mathrm{T}}(p,t) = [y(t-1)\ y(t-2)\ ...\ y(t-p)]$. Now define

$$e(p,t+1) = y(t+1) - \hat{y}(p,t+1) = y(t+1) - \Phi^{\mathrm{T}}(p,t+1)\hat{\Theta}(p,t).$$

This is the "honest" prediction error, in the sense that calculation of $\hat{y}(p,t+1)$ involves only the data $y(1)$,..., $y(t)$. The predictive least squares statistic is then

$$\mathrm{PLS}(p,n) = \tfrac{1}{n}\sum_{t=1}^{n} e^2(p,t)$$

and order estimate $\hat{p}(n)$ at time n is

$$\hat{p}(n) = \mathop{\mathrm{Min}}_{p\in \mathrm{M}}\ \mathrm{PLS}(p,n)$$

where $M=\{1,2,...,p^*\}$. Thus $\hat{p}(n)$ is the order of the model which has given the least mean square prediction error up to time n. We have the following result.

Theorem 1 *Suppose that the roots of the characteristic equation* $z^{p_0}-a_1z^{p_0-1}...-a_{p_0}=0$ *all lie inside the unit circle, and that the process* $\{w(t)\}$ *satisfies*

$$\mathbb{E}[w(t)|\mathcal{F}_{t-1}]=0,\ \mathbb{E}[w^2(t)|\mathcal{F}_{t-1}]=\sigma^2 \text{ a.s., } E[|w(t)|^\alpha|\mathcal{F}_{t-1}] < \infty \text{ a.s., } \alpha>2. \tag{2.2}$$

Then $\hat{p}(n) \to p_0$ *a.s. as* $n \to \infty$.

The proof is given in [7]. By definition of the PLS criterion, $\hat{p}(n) \to p_0$ if and only if

$$\mathbb{P}[\mathrm{PLS}(p,n) - \mathrm{PLS}(p_0,n) > 0,\ n\uparrow\infty] = 1 \qquad \text{for all } p \neq p_0 \tag{2.3}$$

We know that

$$\lim_{n\to\infty} \mathrm{PLS}(p,n) = \sigma^2 \qquad \text{a.s. for } p \geq p_0 \tag{2.4}$$

In the *undermodelled case* $p< p_0$ we find that $\liminf_{n\to\infty} \mathrm{PLS}(p,n) > \sigma^2$ and (2.3) follows easily. However the *overmodelled case* $p>p_0$ is much more delicate since, as indicated by (2.4), all models are "asymptotically equivalent". The key result here is a theorem due to C.Z. Wei [12, Theorem 3] describing the "second order" behaviour of the prediction errors $e(p,t)$; this states that under conditions which can be shown to hold here, for $p\geq p_0$

$$\sum_{t=1}^{n}(e(p,t) - w(t))^2 = (1 + o(1))\sigma^2 \log\det \sum_{t=1}^{n} \Phi(p,t)\Phi^T(p,t) \qquad \text{a.s.}$$

It follows easily that for $p\geq p_0$

$$n(\mathrm{PLS}(p,n) - \mathrm{PLS}(p_0,n)) = (1+o(1))\sigma^2(p-p_0)\log n \qquad \text{a.s.,}$$

which implies (2.3).

3. ADAPTIVE CONTROL

The system here is given by the ARX model[2]

$$y(t+1)=a_1y(t)+...+a_{p_0}\, y(t-p_0+1)+b_1u(t)+...+b_{q_0}u(t-q_0+1)+w(t+1) \tag{3.1}$$

where $u(t)$ is the control input and $w(t)$ satisfies (2.2) and

$$\left|\frac{1}{n}\sum_{t=1}^{n} w^2(t)-\sigma^2\right| = O(n^{-\rho}) \qquad \text{for some } \rho > 0, \qquad \text{a.s.} \tag{3.2}$$

Both the coefficients $\theta(p_0,\ q_0) = [a_1 \ldots\ a_{p_0}\ b_1 \ldots\ b_{q_0}]^T$ and the order (p_0,q_0) are unknown, but it is supposed that upper bounds are known, i.e. $(p_0,q_0) \in M := \{(p,q) : p \leq p^*, q \leq q^*\}$. The objective is to choose an $\mathcal{F}_t$-adapted control process $\{u(t)\}$ so as to minimize $J(u) := \limsup_{n\to\infty} J_n(u)$, where

$$J_n(u) = \frac{1}{n}\sum_{t=0}^{n-1} [y^2(t) + \lambda u^2(t)], \qquad \lambda > 0$$

If p_0, q_0 and $\theta(p_0,\ q_0)$ were known, this would, as shown by Chen and Guo[13], be equivalent to a standard LQG problem [1, Chapter 6], for which $\min_u J(u) = J^*$, where J^* is a number which can be computed by solving an algebraic Riccati equation (see Theorem 3 below.) We say that a controller u for the unknown parameter case is *efficient* if $J(u) = J^*$, i.e. if its asymptotic performance is just as good as the best possible if the parameters were known. It is known that efficient controllers exist for various adaptive control problems — for example the GRC algorithm [1, Chapter 7]. There is by now a fairly standard methodology for establishing results of this kind:

a) Give data-dependent conditions on the regressor vector

$$\Phi(p,q,j) := [y(j)\ \ldots\ y(j-p+1)\ u(j)\ \ldots\ u(j-q+1)]^T$$

under which convergence of parameter estimates is obtained.

b) Give conditions on the growth of the input and output variables which ensure satisfaction of the conditions in (a).

---

[2]We restrict ourselves to the single-input, single-output case here for notational simplicity. [8] deals with multivariable systems.

c) Specify a parameter-dependent control policy which is optimal when the parameter take their true values and is such that the resulting input and output variables satisfy the conditions in (b).

With this approach, it may be possible to show that the control algorithm in (c) is both efficient and ensures almost sure convergence of parameter estimates. As regards (a), the standard condition is that of Lai and Wei [14]. Let

$$W(p,q,n) := \sum_{j=0}^{n} \Phi(p,q,j)\ \Phi^{T}(p,q,j),$$

(this is a $(p+q)\times(p+q)$ matrix) and let $\lambda_{min}(p,q,n)$ and $\lambda_{max}(p,q,n)$ denote the minimum and maximum eigenvalues of $W(p,q,n)$. The Lai-Wei condition, under which a.s. convergence of the LS estimate for the parameter vector $\theta(p_0,q_0)$, with known model order (p_0,q_0), is obtained is

$$\lambda_{min}(p_0,q_0,n) \to \infty \qquad \text{a.s.} \quad \text{as} \quad n \to \infty \tag{3.3}$$

$$\log \lambda_{max}(p_0,q_0,n) = O(\lambda_{min}(p_0,q_0,n)) \text{ a.s.} \tag{3.4}$$

In Chen and Guo [6] a slightly stronger condition than (3.4) is used, namely the existence of a deterministic sequence $c_n \uparrow \infty$ such that, as $n \to \infty$, for all $(p,q) \in$ M

$$\frac{\log \lambda_{max}(p,q,n)}{c_n} \to 0 \ a.s., \qquad \frac{c_n}{\lambda_{min}(p,q,n)} \to 0 \ a.s. \tag{3.5}$$

The "CIC" order selection criterion used in [6] uses the sequence c_n, and it is shown that asymptotically correct order selection and LS parameter estimation is obtained under (3.3), (3.5). In our paper [8] a similar conclusion is reached when order selection is done using the PLS criterion, but only if (3.5) is replaced by the much stronger condition

$$\lambda_{max}(p,q,n) = O(\lambda_{min}(p,q,n)(\log \lambda_{min}(p,q,n))^{\gamma}) \quad \text{a.s. , for some } \gamma<1-\tfrac{2}{\alpha} \tag{3.6}$$

($\alpha > 2$ is the constant appearing in (2.2)) , and if the following condition is also satisfied

$$\Phi^{T}(p,q,n)\, W^{-1}(p,q,n)\Phi(p,q,n) \to 0 \text{ a.s. as } n\to\infty \tag{3.7}$$

Moving to part (b) of the program outlined above, we introduce a "dither signal" $\{v(t)\}$, which is a white noise sequence, adapted to $\mathcal{F}_t$ and independent of $\{w(t)\}$, having the following properties:

$$E[v(t)] = 0, \quad v^2(t) \le \sigma_v^2(\log t)^{-\epsilon} \qquad \text{a.s.} \tag{3.8}$$

where σ_v is a constant and

$$\epsilon \in \left(0, \frac{\alpha-2}{\alpha(p^*+q^*)}\right) \tag{3.9}$$

The following result then holds.

Theorem 2 *Consider the system* (3.1) *with input signal*

$$u(t) = u^s(t) + v(t),$$

where $\{u^s(t)\}$ *is an arbitrary* $\mathcal{F}_t$*−predictable process*[3], $\{v(t)\}$ *is a white noise process satisfying* (3.8), *the system noise* $\{w(t)\}$ *satisfies* (2.2), (3.2) *and the polynomial* $z^{p_0} - a_1 z^{p_0-1} - \dots - a_{p_0}$ *is stable. If*

$$\sum_{t=0}^{n-1}(y^2(t) + (u^s(t))^2) = O(n(\log n)^{\delta}), \quad \delta \in \left(0, \frac{\frac{\alpha-2}{\alpha} - (p^*+q^*)\epsilon}{p^*+q^*+1}\right) \tag{3.10}$$

then (3.3) *and* (3.6) *hold. Thus if* (3.7) *also holds, then* $(\hat{p}(n), \hat{q}(n)) \rightarrow (p_0, q_0)$ *and* $\hat{\theta}(\hat{p}(n), \hat{q}(n)) \rightarrow \theta(p_0, q_0)$ *a.s. as* $n \rightarrow \infty$.

The proof that (3.3) and (3.6) hold follows a contradiction argument along the lines of that employed by Chen and Guo [6]. However, in their paper the order $n(\log n)^{\delta}$ in (3.10) is replaced by the faster order $n^{1+\delta'}$ for δ' in some specified range. As mentioned above, we require the slower growth (3.10) to ensure convergence of the PLS order estimate.

It now remains to specify an appropriate control policy, and, again following Chen and

[3] *i.e.* $u^s(t)$ *is* $\mathcal{F}_{t-1}$*−measurable for each* t

Guo, we use an LQG-based control which is truncated in order that (3.7) and (3.10) be satisfied. (Recall that the system (3.1) is assumed to be stable, so (3.10) is certainly satisfied when $u^s \equiv 0$). We represent the system (3.1) in the canonical observable form

$$\begin{aligned} x(t+1) &= Ax(t)+Bu(t)+Cw(t+1) \\ y(t) &= Hx(t)\ ,\ x(0)=[y(0)\ 0\ \cdots\ 0]^{\mathrm{T}} \end{aligned}$$

where

$$x(t+1)=[x_1(t+1)\ x_2(t+1)\ \cdots\ x_s(t+1)]^{\mathrm{T}}\ ,\ s=p_0 \vee q_0$$

and

$$A=\begin{bmatrix} a_1 & 1 & & 0 \\ a_2 & 0 & \ddots & \\ \vdots & \vdots & \ddots & 1 \\ a_s & 0 & \cdots & 0 \end{bmatrix},\ B=\begin{bmatrix} b_1 \\ b_2 \\ \vdots \\ b_s \end{bmatrix},\ C=\begin{bmatrix} 1 \\ 0 \\ \vdots \\ 0 \end{bmatrix},\ H=[\,1\ 0\ \cdots\ 0\,] \tag{3.11}$$

with $a_i=0$, $i>p_0$ and $b_i=0$, $i>q_0$. For all model $(p,q)\in$M, at time t we have the least squares estimate $\hat{\Theta}(p,q,t)$, from which we form $\hat{A}(p,q,t)$ and $\hat{B}(p,q,t)$ and then estimate the state $x(t)$ using the adaptive filter

$$\begin{aligned} \hat{x}(p,q,t+1)=\hat{A}(p,q,t)\hat{x}(p,q,t)+\hat{B}(p,q,t)u(t)+C(p,q)(y(t+1)-H(p,q)\hat{A}(p,q,t)\hat{x}(p,q,t) \\ -H(p,q)\hat{B}(p,q,t)u(t))\ ,\ \hat{x}(p,q,0)=[y(0)\ 0\ \cdots\ 0]^{\mathrm{T}} \end{aligned}$$

By analogy with the LQG control, we set

$$L(p,q,t)=-(\hat{B}^{\mathrm{T}}(p,q,t)S(p,q,t)\hat{B}(p,q,t)+\lambda)^{-1}\hat{B}^{\mathrm{T}}(p,q,t)S(p,q,t)\hat{A}(p,q,t) \tag{3.12}$$

where

$$\begin{aligned} S(p,q,t)=\hat{A}^{\mathrm{T}}(p,q,t)S(p,q,t\text{-}1)\hat{A}(p,q,t)-\hat{A}^{\mathrm{T}}(p,q,t)S(p,q,t\text{-}1)\hat{B}(p,q,t)\times \\ (\hat{B}^{\mathrm{T}}(p,q,t)S(p,q,t\text{-}1)\hat{B}(p,q,t)+\lambda)^{-1}\hat{B}^{\mathrm{T}}(p,q,t)S(p,q,t\text{-}1)\hat{A}(p,q,t)+H^{\mathrm{T}}(p,q)H(p,q), \end{aligned}$$

with $S(p,q,0)\geq 0$. Now, at time t we select the best model order $(\hat{p}(t),\hat{q}(t))$, and as in Chen and Guo [13] we define a randomly truncated adaptive control with attenuating excitation by

$$u(t) = L^0(\hat{p}(t),\hat{q}(t),t)\hat{x}(\hat{p}(t),\hat{q}(t),t)+v(t) \tag{3.13}$$

with $v(t)$ defined as above and

$$L^0(\hat{p}(t),\hat{q}(t),t) = \begin{cases} L(\hat{p}(t),\hat{q}(t),t)\ , \text{ given by (3.12), if } t \text{ belongs to some } [\tau_k,\sigma_k) \\ 0\ , \text{ if } t \text{ belongs to some } [\sigma_k,\tau_{k+1}) \end{cases}$$

where $\{\tau_k\}$ and $\{\sigma_k\}$ are stopping times defined, for any $\iota>0$, as

$$\sigma_k=\sup\Big\{T>\tau_k: \sum_{i=\tau_k}^{j-1} |L(\hat{p}(i),\hat{q}(i),i)\hat{x}(\hat{p}(i),\hat{q}(i),i)|^2 \le (j-1)(\log(j-1))^\delta +$$

$$|L(\hat{p}(\tau_k),\hat{q}(\tau_k),\tau_k)\hat{x}(\hat{p}(\tau_k),\hat{q}(\tau_k),\tau_k)|^2 \text{ and}$$

$$|L(\hat{p}(j-1),\hat{q}(j-1),j-1)\hat{x}(\hat{p}(j-1),\hat{q}(j-1),j-1)|^2 \le \frac{j-1}{(\log(j-1))^{\gamma-\delta+\iota}},\ \forall j\in(\tau_k,T]\Big\}$$

and

$$\tau_{k+1}=\inf\Big\{T>\sigma_k: \sum_{i=\tau_k}^{\sigma_k-1} |L(\hat{p}(i),\hat{q}(i),i)\hat{x}(\hat{p}(i),\hat{q}(i),i)|^2 \le \frac{T(\log T)^\delta}{2^k} \text{ and}$$

$$\sum_{j=1}^{T} |\hat{x}(\hat{p}(j),\hat{q}(j),j)|^2 \le T(\log T)^{\delta/2} \text{ and } \frac{|L(\hat{p}(T),\hat{q}(T),T)\hat{x}(\hat{p}(T),\hat{q}(T),T)|^2}{T/(\log T)^{\gamma-\delta+\iota}} \le 1\Big\}$$

where δ satisfies the condition in (3.10). Qualitatively these sequences of stopping times, which are a slight modification of those in [13], are such as to satisfy the growth rate (3.10), ensure (3.7), and simultaneously minimize the quadratic loss function.

<u>Theorem 3</u> *Suppose the conditions of Theorem 2 hold and that (A,B) is controllable where A, B are given by (3.11). Let the control $\{u(t)\}$ be given by (3.13). Then*

$$(\hat{p}(n),\ \hat{q}(n)) \to (p_0,\ q_0)\ ,\ \hat{\Theta}(\hat{p}(n),\hat{q}(n),n) \to \Theta(p_0,q_0) \quad a.s. \quad as\ n \to \infty.$$

and

$$|\ J_n(u) - J^*| = O((\log\ n)^{-\epsilon})\ ,\ \epsilon\in\left(0,\ \frac{\alpha-2}{\alpha(p^*+q^*)}\right)$$

Here $J^* = tr(SCRC^{\mathsf{T}})$, where S is the solution of the algebraic Riccati equation

$$S = A^{\mathsf{T}}SA - A^{\mathsf{T}}SB(B^{\mathsf{T}}SB+\lambda)^{-1}B^{\mathsf{T}}SA + H^{\mathsf{T}}H$$

This theorem is proved in [8], following Chen and Guo [13]. The truncation stopping times τ_k, σ_k are chosen so that (i) the condition (3.10) of Theorem 2 is met, (ii) condition (3.7) is satisfied, and (iii) $\sigma_k = \infty$ for some $k(\omega) < \infty$, a.s. (i.e. for large t there is no further truncation and $L^0(p_0,q_0,t) = L(p_0,q_0,t)$.) Theorem 3 gives the desired result, in that J^* is the long-run average cost of LQG control when the order and parameter values are known. Thus the control $\{u(t)\}$ of (3.13) is efficient.

4. CONCLUDING REMARKS

It would be very desirable to extend the above results to ARMAX models, i.e. systems of the form

$$\begin{aligned} y(t+1) = a_1\, y(t)+\ldots+a_{p_0}\, y(t-p_0+1) + b_1\, u(t)+\ldots+b_{q_0}\, u(t-q_0+1) \\ + w(t+1) + c_1\, w(t) +\ldots+ c_{r_0}\, w(t-r_0+1) \end{aligned}$$

In this case the parameter vector $[a_1 \ldots a_{p_0}\ b_1 \ldots b_{q_0}\ c_1 \ldots c_{r_0}]$ cannot be estimated by least squares and we have to use one of the quasi-least squares algorithms AML or RML1. However, even in the absence of control we do not know whether the PLS method used with these algorithms provides consistent order estimates. We conjecture that it does not, though there is little convincing evidence either way. The complicated nature of the regressor vector in these algorithms makes analysis extremely difficult, and simulations — even over very long data lengths — are inconclusive. We leave this as a subject for future research.

REFERENCES

[1] M.H.A. DAVIS and R.B. VINTER, *Stochastic Modelling and Control*, Chapman and Hall, London 1985.

[2] R.H. MIDDLETON, G.C. GOODWIN, D.J. HILL and D.Q. MAYNE, Design issues in adaptive control, IEEE Trans Automatic Control AC-33 (1988)50-58.

[3] E.M. HEMERLY, *Model Structure Estimation in Identification and Adaptive Control*, PhD Thesis, University of London 1988.

[4] E.J. HANNAN and B.G. QUINN, The determination of the order of an autoregression, J. Royal Statist. Soc. (B)41(1979)190-195.

[5] H-Z. AN, Z-G. CHEN and E.J. HANNAN, Autocorrelation, autoregression and autoregressive approximation, Ann. Stat. 10(1982)926-936.

[6] H.F. CHEN and L. GUO, Consistent estimation of the order of stochastic control systems, IEEE Trans Automatic Control AC-32(1987)531-535.

[7] E.M. HEMERLY and M.H.A. DAVIS, Strong consistency of the PLS criterion for order determination of autoregressive processes, Ann. Stat. (to appear 1989).

[8] E.M. HEMERLY and M.H.A. DAVIS, Recursive order estimation of stochastic control systems, submitted to Math of Control, Signals and Systems, 1988.

[9] J. RISSANEN, A predictive least-squares principle, IMA J. of Math. Control and Information 3(1986)211-222.

[10] M. WAX, Order selection for AR models by predictive least squares IEEE Trans Acoustics, Speech and Signal Processing 36(1988)581-588.

[11] P. E. CAINES, *Linear Stochastic Systems*, Wiley, New York 1988.

[12] C.Z. WEI, Adaptive prediction by least squares predictors in stochastic regression models with applications to time series, Ann. Stat. 15(1987) 1667-1687.

[13] H.F. CHEN and L. GUO, Optimal adaptive control and consistent parameter estimates for ARMAX model with quadratic cost, SIAM J. Control and Optimization 25(1987)845-867.

[14] T.L. LAI and C.Z. WEI, Least squares estimates in stochastic regression models with applications to identification and control of dynamic systems, Ann. Stat. 10(1982)154-166.

ADAPTIVE CONTROL OF SOME PARTIALLY OBSERVED LINEAR STOCHASTIC SYSTEMS*

T. E. Duncan
Department of Mathematics, University of Kansas
Lawrence, Kansas 66045, U.S.A.

1. Introduction.

The problem of adaptive control of stochastic systems has been studied extensively. While the majority of the work has been for discrete time stochastic systems, the problem that is studied here is for continuous time stochastic systems. The study of continuous time stochastic systems is necessary for some problems that evolve naturally in continuous time and is important in discrete time systems for the case of small sampling times and for the analysis of computational questions. The stochastic systems are modelled by linear stochastic differential equations where the unknown parameters appear affinely in the stochastic differential equation that describes the state. The estimates and the controls are based on the partial observation of the state of the system. While ARMAX models are often assumed in adaptive control, they are not always suitable as mathematical models for some unknown physical systems.

2. Preliminaries.

The partially observed stochastic system for the adaptive control problem is

$$dX(t) = (F_0X(t) + \sum_{i=1}^{q} \alpha_i F_i X(t) + U(t))dt + dW(t) \tag{1}$$

$$dY(t) = HX(t)dt + d\tilde{W}(t) \tag{2}$$

where $X(t) \in \mathbb{R}^n$, $U(t) \in \mathbb{R}^n$, $F_i \in L(\mathbb{R}^n, \mathbb{R}^n)$ for $i = 0,1,\ldots,q$, $(W(t), t \geq 0)$ and $(\tilde{W}(t), t \geq 0)$ are independent $\mathbb{R}^n$-valued and $\mathbb{R}^p$-valued standard Brownian motions respectively, $H \in L(\mathbb{R}^n, \mathbb{R}^p)$, $X(0) \equiv a \in \mathbb{R}^n$ and $Y(0) \equiv 0 \in \mathbb{R}^p$. It is assumed that

* Research partially supported by NSF Grants ECS-8403286-A01 and ECS-8718026.

(A1) $\alpha_i \in I_i \subset \mathbb{R}$ is an unknown parameter where I_i is a bounded, open interval for $i = 1,2,...,q$.

(A2) $(F_0 + \sum_{i=1}^{q} \alpha_i F_i)$ is a stable linear transformation for each $\alpha_i \in I_i$ for $i = 1,2,...,q$.

(A3) The family $(F_i,\ i = 1,2,...,q)$ are linearly independent.

The probability space $(\Omega,\mathcal{F},P)$ can be chosen so that Ω is the Fréchet space of $\mathbb{R}^{n+p}$-valued continuous functions on $\mathbb{R}_+ = [0,\infty)$ with the seminorms of local uniform convergence, P is the Wiener measure on Ω and $\mathcal{F}$ is the P-completion of the Borel σ-algebra of Ω.

The cost functional, C(t), is defined as

$$C(t) = \int_0^t \langle QX(s), X(s)\rangle + \langle PU(s), U(s)\rangle ds \tag{3}$$

where $Q \in L(\mathbb{R}^n, \mathbb{R}^n)$ and $P \in L(\mathbb{R}^n, \mathbb{R}^n)$ are symmetric, positive semidefinite and symmetric, positive definite, respectively.

To provide some perspective for the results on adaptive control of partially observed stochastic systems a brief review is made of some of the results on adaptive control of completely observed stochastic systems. In this latter case the condition (A2) can be replaced by a reachability assumption.

Let $f_{mn}(t)$ be defined by the equation

$$f_{mn}(t) = \int_0^t \langle F_m X(s), F_n X(s)\rangle ds \tag{4}$$

for $m,n \in \{1,2,...,q\}$. Define

$$\tilde{F}(t) = (\tilde{f}_{mn}(t)) \tag{5}$$

where

$$\tilde{f}_{mn}(t) = \frac{f_{mn}(t)}{f_{mm}(t)}$$

If

$$\liminf_{t\to\infty} |\det \tilde{F}(t)| \geq c > 0 \qquad \text{a.s.} \tag{6}$$

where $c \in \mathbb{R}$ then the maximum likelihood estimates based on the observations of the state are strongly consistent [4]. The maximum likelihood estimates can be computed recursively. The condition (6) can be easily verified in many cases. With other assumptions different techniques can be used in verifying strong consistency of the maximum likelihood estimates (e.g. [6]). If the control is $U(t) = K(t)X(t)$ where $K(t)$ is the solution of the algebraic Riccati equation for the deterministic control problem using the maximum likelihood estimates at time t as the true parameters, then the average costs converge almost surely to the optimal average cost [4].

If the likelihood ratio (Radon-Nikodym derivative) is formed for the partially observed stochastic system then it satisfies the equation [3]

$$L_t(\alpha) = \exp\left[\int_0^t \langle H\hat{X}(s), dY(s)\rangle - \frac{1}{2}\int_0^t \langle H\hat{X}(s), H\hat{X}(s)\rangle ds\right]$$

where $\hat{X}(s) = E[X(s)|Y(u), u \leq s]$. Unfortunately the dependence of the likelihood ratio on the parameters appears in a complicated nonlinear fashion so that the maximum likelihood estimate is not computationally practical.

For notational convenience define $A(\alpha)$ as

$$A(\alpha) = F_0 + \sum_{i=1}^{q} \alpha_i F_i \tag{7}$$

where $\alpha = (\alpha_1,\ldots,\alpha_q)'$. To obtain a computationally feasible estimator the following condition is assumed

(C1) There is a fixed vector $\tilde{u} \in \mathbb{R}^n$ such that the map

$$\alpha \mapsto -HA^{-1}(\alpha)\tilde{u} \tag{8}$$

is one to one. Furthermore there is a neighborhood of the range of the map in Range (H) such that the inverse function exists and is continuous.

3. Main Results.

Initially a result is established for a family of estimators under the condition (C1) and then some examples are described of systems satisfying (C1) and (C1) is compared to some other conditions in the literature.

The estimator $\hat{\alpha}(t)$ for $t > 0$ is formed by choosing a vector α that satisfies the equation

$$-HA^{-1}(\alpha)\tilde{u} = \frac{Y(t)}{t} \tag{9}$$

If no value of α satisfies this equation then choose an arbitrary allowable vector of parameters.

Theorem 1. *Assume that (C1) is satisfied for the stochastic system (1-2). Let* $(\hat{\alpha}(t),\ t > 0)$ *be the family of estimators that are described above. Then this family of estimators is strongly consistent, that is*

$$\lim_{t\to\infty} \hat{\alpha}(t) = \alpha_0 \quad \text{a.s.} \tag{10}$$

where α_0 *is the true parameter vector.*

Proof. Let $(X(t),\ t \geq 0)$ be the solution of (1) with the control

$$U(t) = t\tilde{u} = (t\tilde{u}_1,\ldots,t\tilde{u}_n) \tag{11}$$

where $\tilde{u}$ is given in (8). The random vector $X(t)$ can be decomposed as

$$X(t) = X^0(t) + X^W(t) + X^u(t) \tag{12}$$

where $X^0(t)$, $X^W(t)$ and $X^u(t)$ are the responses of the linear stochastic differential equation to the initial condition, the Wiener process and the control (11) respectively. By the stability assumption (A2) it easily follows that

$$\lim_{t\to\infty} \frac{X^0(t)}{t} = 0 \quad \text{a.s.} \tag{13}$$

and

$$\lim_{t\to\infty} \frac{X^W(t)}{t} = 0 \quad \text{a.s.} \tag{14}$$

Let $A = A(\alpha_0)$ where α_0 is the true parameter vector and let $h: (0,\infty) \to \mathbb{R}^n$ be given by

$$h(t) = \frac{1}{t}\int_0^t e^{A(t-s)}U(s)\,ds \tag{15}$$

By differentiation of (15) it follows that

$$\frac{dh}{dt} = -\frac{h}{t} + Ah + \tilde{u} \tag{16}$$

Since $A = A(\alpha_0)$ is stable it easily follows that

$$\lim_{t\to\infty} \frac{h(t)}{t} = 0 \tag{17}$$

and

$$\lim_{t\to\infty} \frac{dh}{dt} = 0 \tag{18}$$

Thus the limit of (16) as $t \to \infty$ exists and is described by the equation

$$Ah(\infty) = -\tilde{u} \tag{19}$$

where

$$h(\infty) = \lim_{t\to\infty} h(t) \tag{20}$$

By the Strong Law of Large Numbers for Brownian motion and (20) it follows that

$$\lim_{t\to\infty} \frac{Y(t)}{t} = Hh(\infty) \qquad \text{a.s.} \tag{21}$$

There is a real number $T(\omega)$ for almost all ω such that if $t \geq T(\omega)$ then

$$\frac{Y(t,\omega)}{t} = -HA^{-1}(\alpha)\tilde{u} \tag{22}$$

can be solved uniquely for a parameter vector. By the continuity assumption in (C1) and the Strong Law of Large Numbers for Brownian motion it follows that

$$\lim_{t\to\infty} \hat{\alpha}(t) = \alpha_0 \qquad \text{a.s.} \tag{23}$$

This completes the proof.

To provide some perspective of the condition (C1) an example is considered that can be generalized to a family of "triangular" systems. For the stochastic system (1-2) let $n = 3$, $q = 2$ and

$$-F_1 = \begin{bmatrix} 1 & 0 & 1 \\ 0 & 0 & 0 \\ 0 & 0 & 0 \end{bmatrix} \tag{24}$$

$$-F_2 = \begin{bmatrix} 0 & 0 & 0 \\ 0 & 1 & 1 \\ 0 & 0 & 1 \end{bmatrix} \tag{25}$$

$$H = \begin{bmatrix} 1 & 0 & 0 \\ 0 & 1 & 0 \\ 0 & 0 & 0 \end{bmatrix} \tag{26}$$

where $\alpha_i \in (a_i, b_i)$ $a_i > 0$, $b_i < \infty$ for $i = 1,2$. $A(\alpha) = \alpha_1 F_1 + \alpha_2 F_2$ is stable for all pairs $\alpha = (\alpha_1, \alpha_2)'$ and

$$A^{-1}(\alpha) = \begin{bmatrix} \alpha_1^{-1} & 0 & -\alpha_2^{-1} \\ 0 & \alpha_2^{-1} & -\alpha_2^{-1} \\ 0 & 0 & \alpha_2^{-1} \end{bmatrix} \tag{27}$$

$$A^{-1}(\alpha)(1,2,3)' = (\alpha_1^{-1} - 3\alpha_2^{-1}, -\alpha_2^{-1}, 3\alpha_2^{-1})' \tag{28}$$

Thus it is clear from (28) that (C1) is satisfied for the unknown system where $\tilde{u} = (1,2,3)'$. Applying the rule (9) for constructing the estimates it is clear that

$$HA^{-1}(\alpha)\tilde{u} = \frac{Y(t)}{t} = (\hat{\alpha}_1^{-1}(t) - 3\hat{\alpha}_2^{-1}(t), -\hat{\alpha}_2^{-1}(t), 0)' \tag{29}$$

If $\hat{\beta} = \hat{\alpha}^{-1} = (\alpha_1^{-1}, \alpha_2^{-1})'$ then $\hat{\beta}$ satisfies the stochastic differential equation

$$d\hat{\beta}(t) = \frac{1}{t} B_1 \hat{\beta}(t)dt + \frac{1}{t} B_2 \, dY(t) \tag{30}$$

where

$$B_1 = \begin{bmatrix} -1 & 6 \\ 0 & -1 \end{bmatrix} \tag{31}$$

and

$$B_2 = \begin{bmatrix} 1 & -3 \\ 0 & 1 \end{bmatrix} \tag{32}$$

Thus for this example there is a recursive estimator of the unknown parameter vector that is strongly consistent. Some additional properties of this example are useful to note. If $\tilde{u} = (1,1,1)'$ then

$$A^{-1}(\alpha)\tilde{u} = (\alpha_1^{-1} - \alpha_2^{-1}, 0, \alpha_2^{-1})' \tag{33}$$

so that there is nonuniqueness of the estimator and (C1) is not satisfied. If $H = E_{11}$, the elementary 3×3 matrix with 1 in the (1,1) position and zeroes elsewhere, then there is also nonuniqueness and (C1) is not satisfied. Furthermore $A(\cdot)$ can be modified so that $(A(\cdot),H)$ is not observable.

Loges [5] has studied a discrete time estimation problem for a partially observed stochastic system with values in a Hilbert space. A major assumption for his result of strong consistency is that

$$\ker H \subset \ker (HF_i) \quad \text{for } i = 1,2,\ldots,q \tag{34}$$

However for the above example it is clear that

$$\ker H \not\subset \ker(HF_i) \quad \text{for } i = 1,2 \tag{35}$$

Bielecki and Goldys [1] provide two recursive estimators for the parameters of partially observed linear stochastic systems. One estimator is a stochastic Newton algorithm. An important assumption for their strong consistency result for this estimator is an eigenvalue condition on a symmetric matrix-valued function that is part of the estimator. While such an eigenvalue condition is common in the literature for such estimators it is usually difficult to verify. The other recursive estimator is a stochastic gradient algorithm. In this case the convergence of a differential equation whose coefficients depend on the estimator is required for strong consistency. In the scalar case, that is $n = p = 1$, the conditions for the strong consistency of this latter estimator simplify as Bielecki and Goldys [2] have shown.

For the control of the unknown stochastic system (1-2) initially it is assumed that there is a strongly consistent family of estimators that do not require use of a control. Subsequently it is briefly described how to translate the cost functional (3) if the control (11) is used to obtain the estimate in Theorem 1.

For the control of (1-2) in addition to (A1-A3) it is assumed that

(A4) $(F_0 + \sum_{i=1}^{q} \alpha_i F_i, H)$ is observable for each $\alpha_i \in I_i$ for $i = 1,2,\ldots,q$.

The separation principle of linear quadratic control implies that the stationary optimal control is

$$U_0(t) = k_0\hat{X}_0(t) \tag{36}$$

where

$$k_0 = -P^{-1}V \tag{37}$$

$$d\hat{X}_0(t) = A(\alpha_0)\hat{X}_0(t)dt + M_0(dY(t) - H\hat{X}_0(t)dt) + U(t)dt \tag{38}$$

$$M_0 = \Sigma H' \tag{39}$$

$$0 = A(\alpha_0)\Sigma + \Sigma A'(\alpha_0) + I - \Sigma H'H\Sigma \tag{40}$$

$$0 = VA(\alpha_0) + A'(\alpha_0)V - VP^{-1}V + Q \tag{41}$$

Since there are unknown parameters in (1-2), the quantities V, Σ and $\hat{X}(t)$ cannot be computed. Thus approximations of these three quantities have to be computed. The approximations on V and Σ, denoted V(t) and $\Sigma(t)$ respectively for $t \geq 0$, are constructed by solving (41) and (40) respectively with the true parameter vector replaced by the estimate $\hat{\alpha}(t)$. An approximation $\hat{X}(t)$ of $\hat{X}_0(t)$ for $t \geq 0$ is formed by solving

$$d\hat{X}(t) = A(\hat{\alpha}(t))\hat{X}(t)dt + M(t)(dY(t) - H\hat{X}(t)dt) + U(t)dt \tag{42}$$

where $M(t) = \Sigma(t)H'$.

The approximation U(t) to the control $U_0(t)$ for $t \geq 0$ is

$$U(t) = K(t)\hat{X}(t) \tag{43}$$

where

$$K(t) = -P^{-1}V(t) \tag{44}$$

Theorem 2. *Let* $(\hat{\alpha}(t), t \geq 0)$ *be a strongly consistent family of estimators for* α_0 *that take their values in the parameter space and are adaptived to* $(\sigma(Y(u), u \leq t - \delta), t \geq 0)$ *where* $\delta > 0$ *is fixed. Let* $(X(t), t \geq 0)$ *be the solution of (1) using the control (43). Assume that there is a* $\gamma \in \mathbb{R}$ *such that*

$$\limsup_{t\to\infty} \frac{1}{t} \int_0^t \langle X(s), X(s)\rangle ds \leq \gamma \qquad \text{a.s.} \tag{45}$$

Then the following equation is satisfied

$$\lim_{t\to\infty} \frac{1}{t} E[C(t)] = tr(V) + tr(VP^{-1}V\Sigma) \tag{46}$$

where C(t) *is given by (3),* V *and* Σ *are the solutions of (41) and (40) respectively and* tr *is the trace. The optimal average cost is (46).*

Proof. Apply the change of variables formula of Itô to $\langle VX(t), X(t)\rangle$ to obtain

$$\begin{aligned} &\langle VX(t), X(t)\rangle - \langle VX(0), X(0)\rangle \\ &= 2\int_0^t \langle VX(s), A(\alpha_0)X(s) + K(s)\hat{X}(s)\rangle ds \\ &+ 2\int_0^t \langle VX(s), dW(s)\rangle + \int_0^t tr\ V\ ds \end{aligned} \tag{47}$$

Let N(t) be defined by rewriting (46) as

$$\begin{aligned} N(t) &= \langle VX(t), X(t)\rangle - \langle VX(0), X(0)\rangle \\ &- 2\int_0^t \langle VX(s), A(\alpha_0)X(s) + K(s)(X(s) - X(s) + \hat{X}(s))\rangle ds \\ &- \int_0^t tr\ V\ ds = 2\int_0^t \langle VX(s), dW(s)\rangle \end{aligned} \tag{48}$$

Since

$$2\langle VX, AX\rangle + \langle X, QX\rangle - \langle k_0X, Pk_0X\rangle = 0 \tag{49}$$

$$\langle VX, KX\rangle = -\langle k_0X, PKX\rangle \tag{50}$$

from (37, 41), the equation (48) can be rewritten as

$$\begin{aligned} N(t) &= 2\int_0^t \langle QX(s), X(s)\rangle ds + \int_0^t \langle K(s)X(s), PK(s)X(s)\rangle ds \\ &- t\, tr\ V + \langle VX(t), X(t)\rangle - \langle VX(0), X(0)\rangle \\ &- \int_0^t \langle P(K(s) - k_0)X(s), (K(s) - k_0)X(s)\rangle ds \end{aligned} \tag{51}$$

$$+ 2 \int_0^t \langle VX(s), K(s)(X(s) - \hat{X}(s))\rangle ds$$

Since the algebraic Riccati equation (41) is a continuous function of the parameters in the equation and the family of estimates $(\hat{\alpha}(t), t > 0)$ are strongly consistent, it follows that

$$\lim_{t\to\infty} K(t) = k_0 \qquad \text{a.s.} \tag{52}$$

Since the family of feedback gains $(K(t), t \geq 0)$ are uniformly bounded and (52) is satisfied it follows by a simple estimate from (45) that

$$\lim_{t\to\infty} \frac{1}{t} E \int_0^t \langle P(k_0 - K(s))X(s), (k_0 - K(s))X(s)\rangle ds = 0 \tag{53}$$

It is useful to compare the solutions of (1) with the control (43) and

$$dX_0(t) = (A(\alpha_0)X_0(t) + k_0\hat{X}_0(t))dt + dW(t) \tag{54}$$

If $Z(t) = X_0(t) - X(t)$ for $t \geq 0$ then $(Z(t), t \geq 0)$ satisfies the differential equation

$$\begin{aligned} dZ(t) = {} & (A(\alpha_0)Z(t) + (A(\alpha_0) - A(\hat{\alpha}(t))\hat{X}(t))dt \\ & + (k\hat{X}_0(t) - K(t)\hat{X}(t))dt \end{aligned} \tag{55}$$

Since the homogeneous linear differential equation associated with (55) is uniformly asymptotically stable, the fundamental matrix Φ for this differential equation satisfies

$$\|\Phi(t)\| \leq \beta e^{-at} \tag{56}$$

where $a > 0$. From this inequality and the variation of parameters formula it easily follows that

$$\lim_{t\to\infty} \frac{1}{t} E \int_0^t \langle Z(t), Z(t)\rangle dt = 0 \tag{57}$$

and

$$\lim_{t\to\infty} \frac{1}{t} E\, [\langle VX(t), X(t)\rangle] = 0 \tag{58}$$

Similarly it is useful to compare the solutions of (38) and (42). Let $\hat{Z}(t) = \hat{X}_0(t) - \hat{X}(t)$. Then $(\hat{Z}(t), t \geq 0)$ satisfies the stochastic differential equation

$$d\hat{Z}(t) = [(A(\alpha_0) - M_0H)\hat{Z}(t) + (A(\hat{\alpha}(t)) - M(t)H - A(\alpha_0) + M_0H)\hat{X}(t)]dt + (M(t) - M_0)dY(t) \tag{59}$$

Since $(A(\alpha_0) - M_0H)$ is stable by the observability assumption we can proceed as in the comparison of (1) and (54) to verify that

$$\lim_{t\to\infty} \frac{1}{t} E \int_0^t \langle \hat{Z}(t), \hat{Z}(t)\rangle dt = 0 \tag{60}$$

Using the Schwarz inequality, (57, 60) and the ergodic property it follows that

$$\lim_{t\to\infty} \frac{1}{t} E\, 2 \int_0^t \langle VX(s), K(s)(X(s) - \hat{X}(s))\rangle ds$$

$$= \lim_{t\to\infty} \frac{1}{t} E\, 2 \int_0^t \langle VX_0(s), k_0(s)(X_0(s) - \hat{X}_0(s))\rangle ds \tag{61}$$

$$= -2 \operatorname{tr}(VP^{-1} V \Sigma)$$

where V is the solution of (41) and Σ is the solution of (40).

In addition the solutions of the following two stochastic differential equations are compared. The first equation is the error of the estimate (42) and the second equation is the error of the estimate (38).

$$d\tilde{X}(t) = [A(\alpha_0)X(t) - (A(\hat{\alpha}(t)) - M(t)H)\hat{X}(t)]dt + M(t)dY(t) + dW(t) \tag{62}$$

$$d\tilde{X}_0(t) = [A(\alpha_0)\tilde{X}_0(t) - M_0H\hat{X}_0(t)]dt + M_0dY(t) + dW(t) \tag{63}$$

Let $\tilde{Z}(t) = \tilde{X}(t) - \tilde{X}_0(t)$. From (62-63) it follows that $(\tilde{Z}(t), t \geq 0)$ satisfies the stochastic differential equation

$$d\tilde{Z}(t) = [A(\alpha_0)\tilde{Z}(t) + (A(\alpha_0) - A(\hat{\alpha}(t))\hat{X}(t) \tag{64}$$

$$+ M(t)H\hat{X}(t) - M_0 H\hat{X}_0(t)\}dt + (M(t) - M_0)dY(t)$$

Since the homogeneous linear ordinary differential equation associated with (64) is uniformly asymptotically stable it follows in analogy with the comparison of solutions above that

$$\lim_{t\to\infty} \frac{1}{t} E \int_0^t \langle \tilde{Z}(t), \tilde{Z}(t)\rangle dt = 0 \tag{65}$$

Now

$$\begin{aligned}
&\frac{1}{t} E \int_0^t \langle PK(s)X(s),\ K(s)X(s)\rangle ds \\
&\quad = \frac{1}{t} E \int_0^t \langle PK(s)(\hat{X}(s) + \tilde{X}(s)),\ K(s)(\hat{X}(s) + \tilde{X}(s))\rangle ds \\
&\quad = \frac{1}{t} E \int_0^t \langle PK(s)\hat{X}(s),\ K(s)\hat{X}(s)\rangle ds \\
&\qquad + \frac{1}{t} E\, 2 \int_0^t \langle PK(s)\hat{X}(s),\ K(s)\tilde{X}(s)\rangle ds \\
&\qquad + \frac{1}{t} E \int_0^t \langle PK(s)\tilde{X}(s),\ K(s)\tilde{X}(s)\rangle ds
\end{aligned} \tag{66}$$

Using (57, 60, 65) and the ergodic property it follows that

$$\begin{aligned}
&\lim_{t\to\infty} \frac{1}{t} E\, 2 \int_0^t \langle PK(s)\hat{X}(s),\ K(s)\tilde{X}(s)\rangle ds \\
&\quad = \lim_{t\to\infty} \frac{1}{t} E\, 2 \int_0^t \langle PK(s)\hat{X}_0(s),\ K(s)\tilde{X}_0(s)\rangle ds = 0
\end{aligned} \tag{67}$$

It follows by (65) and the ergodic property that

$$\lim_{t\to\infty} \frac{1}{t} E \int_0^t \langle PK(s)\tilde{X}(s),\ K(s)\tilde{X}(s)\rangle ds \tag{68}$$

$$= \lim_{t\to\infty} \frac{1}{t} \int_0^t \langle PK(s)\tilde{X}_0(s), K(s)\tilde{X}_0(s)\rangle ds$$

$$= tr(VP^{-1}V\Sigma)$$

Taking the expectation of (51), dividing by t and using (53, 58, 61, 66-68) verifies (46). This completes the proof.

To use the estimator in Theorem 1 for the adaptive control problem the cost must be modified to take into account the control (11) introduced by the estimator. This is accomplished by replacing the state and the estimate of the state in the computation of the cost by their translations by the responses of the equations for the state and the estimator due to the control input (11).

References

[1] T. Bielecki and B. Goldys, Recursive estimation of parameters for partially observed linear stochastic differential systems, preprint.

[2] T. Bielecki and B. Goldys, On recursive adaptive filtering: Linear case, this proceedings.

[3] T. E. Duncan, Evaluation of likelihood functions, Information and Control 13 (1968), 62-74.

[4] T. E. Duncan and B. Pasik-Duncan, Adaptive control of continuous time linear stochastic systems, to appear in Mathematics of Control, Signals and Systems.

[5] W. Loges, Estimation of parameters for Hilbert space-valued partially observable stochastic processes, J. Multivariate Anal. 20 (1986), 161-174.

[6] P. Mandl, T. E. Duncan, and B. Pasik-Duncan, On the consistency of a least squares identification procedure, Kybernetika 24 (1988).

The Adjoint Process in Stochastic Optimal Control

ROBERT J. ELLIOTT
DEPARTMENT OF STATISTICS AND APPLIED PROBABILITY
UNIVERSITY OF ALBERTA, EDMONTON, ALBERTA
CANADA T6G 2G1

MICHAEL KOHLMANN
FAKULTÄT FÜR WIRSTSCHAFTSWISSENSCHAFTEN UND STATISTIK
UNIVERSITÄT KONSTANZ, POSTFACH 5560
D-7750 F.R. GERMANY

Abstract. Using stochastic flows a minimum principle is obtained when a diffusion is controlled using stochastic open loop controls. An equation for the adjoint process is then derived using an explicit formula for the integrand in a certain stochastic integral.

1. Introduction.

There have been many proofs of minimum principles in stochastic control. For a small sample see the works of Kushner [15], Bismut [2], Haussmann [10], [11], [12], Davis and Varaiya [6], and the book by Elliott [8]. In this paper we consider a diffusion and stochastic open loop controls, that is, controls which are adapted to the filtration of the driving Brownian motion process. For such controls the dynamical equations have strong solutions, and the results on the differentiability of the solution, due originally to Blagovescenskii and Freidlin [1], can be applied. The work of Kunita [14] and Bismut [2] on stochastic flows enables the variation in the expected cost, due to a perturbation of the optimal control, to be calculated explicitly. The minimum principle follows by differentiating this quantity.

If the optimal control is Markov the stochastic integral representation result of [9] is applied to give an expression for a quantity associated with the adjoint process. Stochastic calculus is then used to derive the equation satisfied by the adjoint process.

ACKNOWLEDGEMENTS. Dr. Kohlmann wishes to thank the Department of Statistics and Applied Probability of the University of Alberta for hospitality and support during the spring of 1987, when this work was carried out. Both authors gratefully acknowledge the support of the Natural Sciences and Engineering Research Council of Canada under grant A-7964. The work of the first author was also partially supported by the Air Force Office of Scientific Research, United States Air Force, under grant AFOSR-86-0332 and European Office of Aerospace Research and Development, London, England.

2. Dynamics.

Suppose the state of a system is described by a stochastic differential equation:

$$d\xi_t = f(t, \xi_t, u)dt + g(t, \xi_t)dw_t$$
$$\xi_t \in R^d, \quad \xi_0 = x_0, \quad 0 \le t \le T. \tag{2.1}$$

The control parameter u will take values in a compact subset U of some Euclidean space R^k. We shall make the following assumptions.

A_1: $f : [0,T] \times R^d \times U \to R^d$ is Borel measurable, continuous in u for each (t,x), continuously differentiable in x and for some constant K

$$(1+|x|)^{-1}|f(t,x,u)| + |f_x(t,x,u)| \le K_1.$$

A_2: $g : [0,T] \times R^d \to R^d \otimes R^n$ is a matrix valued, Borel measurable function, continuously differentiable in x, and for some constant K_2

$$|g(t,x)| + |g_x(t,x)| \le K_2.$$

The columns of g will be denoted by $g^{(k)}$ for $k = 1, \dots, n$.

A_3: $w = (w^1, \dots, w^n)$ is an n-dimensional Brownian motion on a probability space (Ω, F, P) with a right continuous, complete filtration $\{F_t\}$, $0 \le t \le T$.

DEFINITION 2.1. *The set of admissible controls $\underline{U}$ will be the F_t-predictable functions on $[0,T] \times \Omega$ with values in U. These are sometimes called 'stochastic open loop' controls, [3].*

REMARKS 2.2. For each $u \in \underline{U}$ there is, therefore, a strong solution of (2.1), and we shall write $\xi^u_{s,t}(x)$ for the solution trajectory given by

$$\xi^u_{s,t}(x) = x + \int_s^t f(r, \xi^u_{s,r}(x), u_r)dr + \int_s^t g(r, \xi^u_{s,r}(x))dw_r. \tag{2.2}$$

Then, because u is a (predictable) parameter, the result of Blagovenscenskii and Freidlin [1] extends to this situation, so the Jacobian $\frac{\partial \xi^u_{s,t}}{\partial x}(x) = D^u_{s,t}$ exists and is the solution of

$$D^u_{s,t} = I + \int_s^t f_\xi(r, \xi^u_{s,r}(x), u_r)D^u_{s,r}\,dr + \sum_{k=1}^n \int_s^t g^{(k)}_\xi(r, \xi^u_{s,r}(x))D^u_{s,r}\,dw^k_r. \tag{2.3}$$

Here I is the $d \times d$ identity matrix. In fact, if the coefficients f and g are C^k the map $x \to \xi^u_{s,t}(x)$ is C^{k-1}.

Consider the matrix valued process H defined by:

$$H^u_{s,t} = I - \int_s^t H^u_{s,r}\big(f_\xi(r, \xi^u_{s,r}(x), u_r) - \sum_{k=1}^n g^{(k)}_\xi(r, \xi^u_{s,r}(x))^2\big)dr$$
$$- \sum_{k=1}^n \int_s^t H^u_{s,r} g^{(k)}_\xi(r, \xi^u_{s,r}(x))dw^k_r. \tag{2.4}$$

Then using the Ito rule we see $d(H^u_{s,t} D^u_{s,t}) = 0$ and $H^u_{ss} D^u_{ss} = I$, so $H^u_{s,t} = (D^u_{s,t})^{-1}$.

Write $\|\xi^u(x_0)\|_t = \sup_{0 \le s \le t} |\xi^u_{0,s}(x_0)|$. Then, as in Lemma 2.1 of [12], for any p, $1 \le p < \infty$, using Gronwall's and Jensen's inequalities

$$\|\xi^u(x_0)\|^p_T \le C\Big(1 + |x_0|^p + \Big|\int_0^t g(r, \xi^u_{0,r}(x_0))dw_r\Big|^p\Big)$$

almost surely for some constant C. Therefore, using Burkholder's inequality and hypothesis A_2, $\|\xi^u(x_0)\|_T$ is in L^p for all p, $1 \le p < \infty$. Write

$$\|D^u\|_T = \sup_{0 \le s \le T} |D^u_{0,s}|$$
$$\|H^u\|_T = \sup_{0 \le s \le T} |H^u_{0,s}|.$$

Then, because f_ξ and g_ξ are bounded, an application of Gronwall's, Jensen's and Burkholder's inequalities again implies

$$\|D^u\|_T \text{ and } \|H^u\|_T \quad \text{are in } L^p \text{ for all } p, \quad 1 \le p < \infty.$$

COST 2.3. Suppose for simplicity that the cost associated with the process is purely terminal and given by a bounded C^2 function

$$c(\xi^u_{0,T}(x_0)).$$

A_4: We suppose $|c(x)| + |c_x(x)| + |c_{xx}(x)| \le K_3(1 + |x|^q)$ for some $q < \infty$.

The expected cost if a control $u \in \underline{U}$ is used is, therefore,

$$J(u) = E[c(\xi^u_{0,T}(x_0))].$$

We shall suppose there is an optimal control $u^* \in \underline{U}$ so

$$J(u^*) \leq J(u) \quad \text{for all} \quad u \in \underline{U}.$$

NOTATION 2.4. *If u^* is an optimal control write ξ^* for ξ^{u^*}, D^* for D^{u^*} etc.*

REMARKS 2.5. Consider a d-dimensional semimartingale of the form

$$z_t = z_s + A_t$$

where A is a predictable bounded variation process. Then Kunita's formula [14] for the composition of processes can be applied, (see also Bismut [5]), and we have

$$\begin{aligned} \xi^*_{s,t}(z_t) &= z_s + \int_s^t f(r, \xi^*_{s,r}(z_r), u^*_r)dr \\ &\quad + \int_s^t \frac{\partial \xi^*_{s,r}}{\partial x}(z_r)dA_r + \sum_{k=1}^n \int_s^t g^{(k)}(r, \xi^*_{s,r}(z_r))dw^k_r. \end{aligned} \tag{2.5}$$

DEFINITION 2.6. *Consider perturbations of the optimal control u^* of the following kind: For $s \in [0,T]$, $h > 0$ such that $0 \leq s < s+h < T$, and $A \in F_s$ define, for any other admissible control $\tilde{u} \in U$,*

$$u(t,w) = \begin{cases} u^*(t,w) & \text{if } (t,w) \notin [s,s+h] \times A \\ \tilde{u}(t,w) & \text{if } (t,w) \in [s,s+h] \times A. \end{cases}$$

Applying (2.5) we have, similarly to Theorem 5.1 of [4], the following result.

THEOREM 2.7. *For the perturbation u of u^* consider the process*

$$z_t = x + \int_s^t \left(\frac{\partial \xi^*_{s,r}(z_r)}{\partial x}\right)^{-1} (f(r, \xi^*_{s,r}(z_r), u_r) - f(r, \xi^*_{s,r}(z_r), u^*_r))dr. \tag{2.6}$$

Then the process $\xi^*_{s,t}(z_t)$ is indistinguishable from $\xi^u_{s,t}(x)$.

PROOF. Substituting (2.6) in (2.5) we see

$$\begin{aligned}
\xi^*_{s,t}(z_t) &= x + \int_s^t f(r, \xi^*_{s,r}(z_r), u^*_r)dr \\
&\quad + \int_s^t \Big(\frac{\partial \xi^*_{s,r}(z_r)}{\partial x}\Big) \Big(\frac{\partial \xi^*_{s,r}(z_r)}{\partial x}\Big)^{-1} \Big(f(r, \xi^*_{s,r}(z_r), u_r) - f(r, \xi^*_{s,r}(z_r), u^*_r)\Big) dr \\
&\quad + \int_s^t g(r, \xi^*_{s,r}(z_r))dw_r \\
&= x + \int_s^t f(r, \xi^*_{s,r}(z_r), u_r)dr + \int_s^t g(r, \xi^*_{s,r}(z_r))dw_r.
\end{aligned}$$

However, the solution to (2.2) is unique so $\xi^*_{s,t}(z_r) = \xi^u_{s,t}(x)$.

REMARKS 2.8. Note that $u(t) = u^*(t)$ if $t > s+h$ so $z_t = z_{s+h}$ if $t > s+h$. Therefore

$$\xi^*_{s,t}(z_t) = \xi^*_{s,t}(z_{s+h}) = \xi^*_{s+h,t}(\xi^u_{s,s+h}(x))$$

if $t > s+h$.

3. A Minimum Principle.

Now

$$J(u^*) = E[c(\xi^*_{0,T}(x_0))]$$
$$= E[c(\xi^*_{s,T}(x))] \quad \text{where } x = \xi_{0,s}(x_0),$$

because, by uniqueness, $\xi^*_{0,T}(x_0) = \xi^*_{s,T}(x)$. Similarly,

$$J(u) = E[c(\xi^u_{0,T}(x_0))]$$
$$= E[c(\xi^u_{s,T}(x))]$$
$$= E[c(\xi^*_{s,T}(z_{s+h}))].$$

Therefore,

$$J(u) - J(u^*) = E[c(\xi^*_{s,T}(z_{s+h})) - c(\xi^*_{s,T}(x))].$$

Because $\xi^*_{s,T}(\cdot)$ is differentiable this is

$$= E\Big[\int_s^{s+h} c_\xi(\xi^*_{s,T}(z_r))\frac{\partial \xi^*_{s,T}(z_r)}{\partial x}\cdot\Big(\frac{\partial \xi^*_{s,r}(z_r)}{\partial x}\Big)^{-1}(f(r,\xi^*_{s,r}(z_r),u_r) - f(r,\xi^*_{s,r}(z_r),u^*_r))dr\Big]. \tag{3.1}$$

This gives an explicit formula for the change in the cost resulting from a 'strong' variation in the optimal control. It involves only a time integration. The only remaining problem is to justify the differentiation of the right hand side of (3.1).

Write $\Gamma(s,r,z_r) = c_\xi(\xi^*_{s,T}(z_r))\dfrac{\partial \xi^*_{s,T}(z_r)}{\partial x}\Big(\dfrac{\partial \xi^*_{s,r}(z_r)}{\partial x}\Big)^{-1}$.

Then

$$J(u) - J(u^*) = \int_s^{s+h} E\Big[(\Gamma(s,r,z_r) - \Gamma(s,r,x))(f(r,\ \xi^*_{s,r}(z_r),\ u_r) - f(r,\ \xi^*_{s,r}(z_r),\ u^*_r))\Big]dr$$
$$+ \int_s^{s+h} E\Big[(\Gamma(s,r,x) - \Gamma(r,r,x))(f(r,\ \xi^*_{s,r}(z_r),\ u_r) - f(r,\ \xi^*_{s,r}(z_r),\ u^*_r))\Big]dr$$
$$+ \int_s^{s+h} E\Big[\Gamma(r,r,x)(f(r,\ \xi^*_{s,r}(z_r),\ u_r) - f(r,\ \xi^*_{s,r}(z_r),\ u^*_r)$$
$$- f(r,\ \xi^*_{s,r}(x),\ u_r) + f(r,\ \xi^*_{s,r}(x),\ u^*_r))\Big]dr$$
$$+ \int_s^{s+h} E\Big[\Gamma(r,r,x)(f(r,\ \xi^*_{0,r}(x_0),\ u_r) - f(r,\xi^*_{0,r}(x_0),\ u^*_r))\Big]dr$$
$$= I_1(h) + I_2(h) + I_3(h) + I_4(h), \quad \text{say.}$$

Now,

$$
\begin{aligned}
|I_1(h)| &\le K_4 \int_s^{s+h} E\Big[|\Gamma(s,r,z_r) - \Gamma(s,r,x)|(1 + \|\xi^u(x_0)\|_{s+h})\Big]\,dr \\
&\le K_4 h \sup_{s\le r\le s+h} E\Big[|\Gamma(s,r,z_r) - \Gamma(s,r,x)|(1 + \|\xi^u(x_0)\|_{s+h})\Big] \\
|I_2(h)| &\le K_5 \int_s^{s+h} E\Big[|\Gamma(s,r,x) - \Gamma(r,r,x)|(1 + \|\xi^u(x_0)\|_{s+h})\Big]\,dr \\
&\le K_5 h \sup_{s\le r\le s+h} E\Big[|\Gamma(s,r,z_r) - \Gamma(r,r,x)|(1 + \|\xi^u(x_0)\|_{s+h}\Big] \\
|I_3(h)| &\le K_6 \int_s^{s+h} E\Big[|\Gamma(r,r,x)|\;\|x - z_r\|\Big]\,dr \\
&\le K_6 h \sup_{s\le r\le s+h} E\Big[|\Gamma(r,r,x)|\;\|x - z.\|_{s+h}\Big].
\end{aligned}
$$

The differences $|\Gamma(s,r,z_r)-\Gamma(s,r,x)|$, $|\Gamma(s,r,x)-\Gamma(r,r,x)|$ and $\|x-z\|_{s+h}$ are all uniformly bounded in some L^p, $p>1$, and

$$
\begin{aligned}
&\lim_{r\to s} |\Gamma(s,r,z_r) - \Gamma(s,r,x)| = 0 \quad \text{a.s.} \\
&\lim_{r\to s} |\Gamma(s,r,x) - \Gamma(r,r,x)| = 0 \quad \text{a.s.} \\
&\lim_{h\to 0} \|x - z.\|_{s+h} = 0.
\end{aligned}
$$

Therefore,

$$
\begin{aligned}
&\lim_{r\to s} \|\Gamma(s,r,z_r) - \Gamma(s,r,x)\|_p = 0 \\
&\lim_{r\to s} \|\Gamma(s,r,x) - \Gamma(r,r,x)\|_p = 0 \\
\text{and}\quad &\lim_{h\to 0} \|(\|x - z\|_{s+h})\|_p = 0 \quad \text{for some } p.
\end{aligned}
$$

Consequently, $\lim_{h\to 0} h^{-1} I_k(h) = 0$, for $k = 1,2,3$.

The only remaining problem concerns the differentiability of

$$
I_4(h) = \int_s^{s+h} E\Big[\Gamma(r,r,x)(f(r,\ \xi^*_{0,r}(x_0),\ u_r) - f(r,\ \xi^*_{0,r}(x_0),\ u^*_r))\Big]\,dr.
$$

The integrand is almost surely in $L^1([0,T])$ so $\lim_{h\to 0} h^{-1} I_4(h)$ exists for almost every $s \in [0,T]$. However, the set of times $\{s\}$ where the limit may not exist might depend on the control u. Consequently we must restrict the perturbations u of the optimal control u^* to perturbations from a countable dense set of controls. In fact:

1) Because the trajectories are, almost surely, continuous, F_ρ is countably generated by sets $\{A_{i\rho}\}$, $i = 1, 2, \dots$ for any rational number $\rho \in [0, T]$. Consequently F_t is countably generated by the sets $\{A_{i\rho}\}$, $r \leq t$.

2) Let G_t denote the set of measurable functions from (Ω, F_t) to $U \subset R^k$. (If $u \in \underline{U}$ then $u(t, w) \in G_t$.) Using the L^1-norm, as in [7], there is a countable dense subset $H_\rho = \{u_{j\rho}\}$ of G_ρ, for rational $\rho \in [0, T]$. If $H_t = \bigcup_{\rho \leq t} H_\rho$ then H_t is a countable dense subset of G_t. If $u_{j\rho} \in H_\rho$ then, as a function constant in time, $u_{j\rho}$ can be considered as an admissible control over any time interval $[t, T]$ for $t \geq \rho$.

3) The countable family of perturbations is obtained by considering sets $A_{i\rho} \in F_t$, functions $u_{j\rho} \in H_t$, where $\rho \leq t$, and defining as in 3.1

$$u^*_{j\rho}(s, w) = \begin{cases} u^*(s, w) & \text{if } (s, w) \notin [t, T] \times A_{i\rho} \\ u_{j\rho}(s, w) & \text{if } (s, w) \in [t, T] \times A_{i\rho}. \end{cases}$$

Then for each i, j, ρ

$$\lim_{h \to 0} h^{-1} \int_s^{s+h} E\Big[\Gamma(r, r, x)(f(r,\ \xi^*_{0,r}(x_0),\ u^*_{j\rho}) - f(r,\ \xi^*_{0,r}(x_0),\ u^*))\Big] dr \tag{3.2}$$

exists and equals

$$E\Big[\Gamma(s, s, x)(f(s,\ \xi^*_{0,s}(x_0),\ u_{j\rho}) - f(s,\ \xi^*_{0,s}(x_0),\ u^*))I_{A_{i\rho}}\Big]$$

for almost all $s \in [0, T]$.

Therefore, considering this perturbation we have

$$\lim_{h \to 0} h^{-1}(J(u^*_{j\rho}) - J(u^*)) = E\Big[\Gamma(s, s, x)(f(s,\ \xi^*_{0,s}(x_0),\ u_{j\rho}) - f(s,\ \xi^*_{0,s}(x_0),\ u^*))I_{A_{i\rho}}\Big]$$
$$\geq 0 \quad \text{for almost all } s \in [0, T].$$

Consequently there is a set $S \subset [0, T]$ of zero Lebesgue measure such that, if $s \notin S$, the limit in (3.2) exists for all i, j, ρ, and gives

$$E\Big[\Gamma(s, s, x)(f(s,\ \xi^*_{0,s}(x_0),\ u_{j\rho}) - f(s,\ \xi^*_{0,s}(x_0),\ u^*))I_{A_{i\rho}}\Big] \geq 0.$$

Using the monotone class theorem, and approximating an arbitrary admissible control $u \in \underline{U}$ we can deduce that if $s \notin S$

$$E\Big[\Gamma(s,s,x)\big(f(s,\ \xi^*_{0,s}(x_0),\ u) - f(s,\ \xi^*_{0,s}(x_0),\ u^*)\big)I_A\Big] \geq 0 \quad \text{for any } u \in U \text{ and } A \in F_s. \tag{3.3}$$

Write

$$p_s(x) = E\Big[c_\xi(\xi^*_{0,T}(x_0))\frac{\partial \xi^*_{s,T}(x)}{\partial x} \mid F_s\Big] = E[\Gamma(s,s,x) \mid F_s] \tag{3.4}$$

where, as before, $x = \xi^*_{0,s}(x_0)$. Then $p_s(x)$ is the adjoint variable and we have in (3.3) proved the following minimum principle:

THEOREM 5.1. *If $u^* \in \underline{U}$ is an optimal control there is a set $S \subset [0,T]$ of zero Lebesgue measure such that if $s \notin S$*

$$p_s(x)f(s,x,u^*) \leq p_s(x)f(s,x,u) \quad \text{a.s.}$$

That is, the optimal control u^* almost surely minimizes the Hamiltonian and the adjoint variable is $p_s(x)$.

REMARKS 3.2. Under certain conditions the minimum cost attainable under the stochastic open loop controls is equal to the minimum cost attainable under the Markov, feedback controls of the form $u(s,\ \xi^u_{0,s}(x_0))$. See for example [2], [10]. If u_M is a Markov control, with a corresponding, possibly weak, solution trajectory ξ^{u_M}, then u_M can be considered as a stochastic open loop control $u_M(w)$ by putting

$$u_M(w) = u_M(s,\ \xi^{u_M}_{0,s}(x_0, w)).$$

This means the control in effect 'follows' its original trajectory ξ^{u_M} than any new trajectory. That is the control is similar to the adjoint strategies considered by Krylov [13]. The significance of this is that when we consider variations in the state trajectory ξ, and derivatives of the map $x \to \xi_{s,t}(x)$, the control does not react, and so we do not introduce derivatives in the u variable.

If the optimal control u^* is Markov the process ξ^* is Markov and

$$\begin{aligned} p_s(x) &= E[\Gamma(s,s,x) \mid F_s] \\ &= E[\Gamma(s,s,x) \mid x]. \end{aligned} \tag{3.5}$$

4. The Adjoint Process.

Suppose the optimal control u^* is Markov. As noted above, u^* can and will be considered as an open loop control. The Jacobian $\frac{\partial \xi^*_{s,T}}{\partial x}$ exists, as does $\left(\frac{\partial \xi^*_{s,T}}{\partial x}\right)^{-1}$ and higher derivatives.

THEOREM 4.1. *Suppose the optimal control u^* is Markov. Then*

$$\begin{aligned} p_s(x) = E[c_\xi(\xi^*_{0,T}(x_0))D_{0,T}] &- \int_0^s p_r(\xi^*_{0,r}(x_0)) f_\xi(r, \xi^*_{0,r}(x_0), u^*_r)dr \\ &+ \int_0^s p_x(r, \xi^*_{0,r}(x_0)) g(r, \xi^*_{0,r}(x_0))dw_r \\ &- \int_0^s p_x(r, \xi^*_{0,r}(x_0)) g(r, \xi^*_{0,r}(x_0)) g_\xi(r, \xi^*_{0,r}(x_0))dr. \end{aligned}$$

PROOF. Write $f_\xi(r)$ for $f_\xi(r, \xi^*_{0,r}(x_0), u^*_r)$ and $g(r)$ for $g(r, \xi^*_{0,r}(x_0))$, etc. By uniqueness of the solutions to (2.1)

$$\xi^*_{0,T}(x_0) = \xi^*_{s,T}(\xi^*_{0,s}(x_0)) \tag{4.1}$$

so, differentiating,

$$D_{0,T} = D_{s,T}\, D_{0,s} \tag{4.2}$$

where $D_{0,T} = D^*_{0,T}$ etc. (without the *).

From (3.4) and (3.5)

$$p_s(x) = E[c_\xi(\xi^*_{0,T}(x_0))D_{s,T} \mid F_s]$$

so from (4.2)

$$p_s(x)D_{0,s} = E[c_\xi(\xi^*_{0,T}(x_0))D_{0,T} \mid F_s] \tag{4.3}$$

and this is a $(P, \{F_t\})$ martingale. Write $x = \xi^*_{0,s}(x_0)$, $D = D_{0,s}$. From the martingale representation result [9], the integrand in the representation of $p_s(x)D$ as a stochastic integral is obtained by the Ito rule, noting that only the stochastic integral terms will appear. These involve the derivatives in x and D. Therefore

$$p_s(x)D = E[c_\xi(\xi_{0,T}(x_0))D_{0,T}] + \int_0^s p_x(r, \xi^*_{0,r}(x_0))g(r)dw_r D_{0,r}$$
$$+ \sum_{k=1}^n \int_0^s p_r(\xi^*_{0,r}(x_0))g^{(k)}_\xi(r)D_{0,r}\, dw^k_r. \tag{4.4}$$

Recall from (2.4) that $H_{0,s} = D^{-1}$ so forming the product of (2.4) and (4.4), using the Ito rule:

$$\begin{aligned}
p_s(x) &= (p_s(x)D)H_{0,s} \\
&= E[c_\xi(\xi^*_{0,T}(x_0))D_{0,T}] - \int_0^s p_r(\xi^*_{0,r}(x_0))f_\xi(r)dr \\
&\quad - \sum_{k=1}^n \int_0^s p_r(\xi^*_{0,r}(x_0))g^{(k)}_\xi(r)dw^k_r + \sum_{k=1}^n \int_0^s p_r(\xi^*_{0,r}(x_0))(g^{(k)}_\xi(r))^2 dr \\
&\quad + \int_0^s p_x(r, \xi^*_{0,r}(x_0))g(r)dw_r + \sum_{k=1}^n \int_0^s p_r(\xi^*_{0,r}(x_0))g^{(k)}_\xi(r)dw^k_r \\
&\quad - \sum_{k=1}^n \int_0^s p_x(r, \xi^*_{0,r}(x_0))g(r)g^{(k)}_\xi(r)dr - \sum_{k=1}^n \int_0^s p_r(\xi^*_{0,r}(x_0))(g^{(k)}_\xi(r))^2 dr \\
&= E[c_\xi(\xi^*_{0,T}(x_0))D_{0,T}] - \int_0^s p_r(\xi^*_{0,T}(x_0))f_\xi(r)dr \\
&\quad + \int_0^2 p_x(r, \xi^*_{0,r}(x_0))g(r)dw_r - \sum_{k=1}^n \int_0^s p_x(r, \xi^*_{0,r}(x_0))g(r)g^{(k)}_\xi(r)dr
\end{aligned}$$

so establishing the result.

This verifies by a simple, direct method the formula of Haussman [10].

References

[1] J.N. Blagovescenskii and M.I. Freidlin, *Some properties of diffusion processes depending on a parameter.* Dokl. Akad. Nauk. 138(1961), Soviet Math. 2(1961), 633–636.

[2] J.M. Bismut, *Theorie probabiliste du contrôle des diffusions.* Mem. Amer. Math. Soc. 167(1976).

[3] J.M. Bismut, *An introductory approach to duality in optimal stochastic control.* SIAM Review 20(1978), 62–78.

[4] J.M. Bismut, *Mecanique Aléatoire.* In Ecole d'Eté de Probabilités de Saint-Flour X. Lecture Notes in Mathematics 929, Springer-Verlag 1982, 1–100.

[5] J.M. Bismut, *A generalized formula of Ito and some other properties of stochastic flows.* Zeits. für Wahrs. 55(1981), 331–350.

[6] M.H.A. Davis and P.P. Varaiya, *Dynamic programming conditions for partially observable stochastic systems.* SIAM Jour. Control 11(1973), 226–261.

[7] R.J. Elliott, *The optimal control of a stochastic system.* SIAM Jour. Control and Opt. 15(1977), 756–778.

[8] R.J. Elliott, *Stochastic Calculus and Applications.* Applications of Math. Vol. 18, Springer-Verlag, 1982.

[9] R.J. Elliott and M. Kohlmann, *A short proof of a martingale representation result.* University of Alberta, pre-print.

[10] U.G. Haussmann, *On the adjoint process for optimal control of diffusion processes.* SIAM Jour. Control and Opt. 19(1981), 221–243, and 710.

[11] U.G. Haussmann, *Extremal controls for completely observable diffusions.* Lecture Notes in Control and Information Sciences 42, Springer-Verlag 1982, 149–160.

[12] U.G. Haussmann, *A Stochastic Minimum Principle for Optimal Control of Diffusions.* Pitman/Longman Research Notes in Mathematics 151, Longman U.K., 1986.

[13] N.V. Krylov, *Controlled Diffusion Processes.* Applications of Math. Vol. 14. Springer-Verlag, 1980.

[14] H. Kunita, *Some extensions of Ito's formula.* Lecture Notes in Math. 850, Springer-Verlag 1980, 118–141.

[15] H. Kushner, *Necessary conditions for continuous parameter stochastic optimization problems*, SIAM Jour. Control 10(1972), 550–565.

Integration by parts and the Malliavin calculus

Robert J. Elliott
Department of Statistics
and Applied Probability
University of Alberta
Edmonton, Alberta
Canada T6G 2G1

Michael Kohlmann
Fakultät für Wirtschafts-
wissenschaften und Statistik
Universität Konstanz
D7750 Konstanz
F.R. Germany

1. Introduction.

From a very simple representation of the integrand in the integral representation of a martingale, we derive an integration by parts formula. This is used to give a new proof of the existence of a density of a diffusion process under the hypothesis that the inverse of the Malliavin matrix is in some L^p-space, a result implied by Hörmander's condition H1.

Following Malliavin's original proof of this result there have been other approaches to what is now known as Malliavin's calculus, including those of Stroock [17], Shigekawa [16], Bismut [4], Bichteler and Fonken [2], and Norris [15]. The main simplification in this paper is the observation that no infinite dimensional calculus of variations is required. This calculus can be replaced by ordinary differentiation in finite dimensional spaces.

Acknowledgment: This paper was written during a visit by Professor Elliott to the University of Konstanz in Spring 1988. Professor Elliott would like to thank the Deutsche Forschungsgemeinschaft for its support and the University of Konstanz for its hospitality.

2. Some history and the H1 condition.

Let us consider the unique solution $\xi_{s,t}(x)$ of the stochastic differential equation

$$d\xi_{s,t}(x) = X_0(t, \xi_{s,t}(x))dt + X_i(t, \xi_{s,t}(x))dw_t^i$$

$$\xi_{s,s}(x) = x \in \mathbb{R}^d \tag{1}$$

where $(w_t) = (w_t^1, \ldots, w_t^m)$ is an m-dimensional Brownian motion on $(\Omega, \mathcal{F}, \mathcal{F}_t, P)$ and $X_0, X_1, \ldots, X_m$ are smooth vector fields on $[0, \infty) \times \mathbb{R}^d$, all of whose derivatives are bounded.

It is a well known fact from harmonic analysis that $\xi_{0,T}(x)$ has a density if

$$|Ec_\xi(\xi_{0,T}(x_0))| \le K \sup_{x \in \mathbb{R}^d} |c(x)|, \tag{2}$$

where c is any bounded, smooth function with bounded derivatives [14,17,4,20]. Using different methods Malliavin [14], Stroock [17], Shigekawa [16], and Bismut [4] showed that (2) is true if the inverse of the Malliavin matrix M_{0T} is in some $L^p(\Omega)$, and they linked this result with Hörmander's famous result to conclude that M_{0T}^{-1} is in all $L^p(\Omega)$, $p < \infty$, if Hörmander's condition H1 is satisfied:

Condition H1: $X_1, \ldots, X_m$, $[X_i, X_j]$, $[X_i, [X_j, X_k]], \ldots$, $i, j, k = 0, \ldots, m$ at x_0 span $\mathbb{R}^d$.

Malliavin's approach is based on a function space martingale calculus which comes from the Ornstein–Uhlenbeck process on Wiener space [14]; this is now known as Malliavin's calculus of variations. Shigekawa [16] provided an alternative formulation which relies on a Sobolev-type extension of Fréchet derivatives with Wiener measure replacing the Lebesgue measure in the finite dimensional situation, and he makes no use at all of the Ornstein–Uhlenbeck process. Stroock [17,18] also avoids this process in his entirely functional-analytic reformulation of the Malliavin calculus. So far the approaches of Shigekawa and Stroock (also cf. Ikeda and Watanabe's contribution [12]) are reformulations of Malliavin's approach.

Roughly speaking, these approaches rely on the analysis of a differential operator $\mathcal{L}$, which may be seen on the one hand as an operation on the Wiener chaos decomposition of a Brownian functional $F(w)$

$$\mathcal{L}F(w) = \sum_{m=1}^{\infty} m \int_0^T \cdots \int_0^{t_2} f_m(t_1,\ldots,t_m) dw_{t_1} \ldots dw_{t_m},$$

or as the generator of a time changed Brownian sheet $\{S_\tau(t) \mid (\tau,t) \in [0,\infty)^2\}$, namely

$$V_\tau(t) = e^{-\frac{1}{2}\tau} S_{e^\tau}(t)$$

seen as a process on $C(0,\infty)$. For a "good" function c, we then find

$$c'(F) = \frac{1}{2}(-\mathcal{L}(Fc(F)) + c(F)\mathcal{L}F + F\mathcal{L}F) \cdot A^{-1}, \tag{3}$$

where A is the inverse of the Malliavin matrix $A = (DF, DF) = \sum_i (D_{h^i}F)^2$ and D_{h^i} is the directional derivative in the direction of the integrated element h^i of a complete orthonormal system on $[0,T]$. The analysis of the right hand side of (3) then leads to a bound on $E|c'(F)|$ as required in (2).

Zakai pointed out that $\mathcal{L}F$ may also be seen as the L^2-limit of

$$\frac{F(w) - E[F(\sqrt{1-\varepsilon}\, w + \sqrt{\varepsilon}\tilde{w}) \mid \mathcal{F}^w]}{\varepsilon},$$

where the relation to the generator of the infinite dimensional Ornstein–Uhlenbeck process becomes apparent [19,20], as this non-coherent derivative may be interpreted as

$$\begin{aligned} &\frac{\partial F(\zeta w)}{\partial \zeta}\Big|_{\zeta=1} - \sum \frac{\partial^2}{\partial \varepsilon^2} F(w + \varepsilon \int h_s^i ds)\Big|_{\varepsilon=0} \\ &= D^\zeta F - \text{trace } D^2 F. \end{aligned} \tag{4}$$

Bismut however gives a different approach which expresses the Wiener space derivatives as function space derivatives in a Girsanov functional. The basic idea here is a perturbation of Brownian motion by a small drift $\varepsilon \cdot \int u_s ds$ (u_s a predictable function). Then

$$D_u F(w) = \frac{\partial}{\partial \varepsilon} F(w + \varepsilon \int uds)\Big|_{\varepsilon=0}.$$

However,

$$E[F(w)] = E[F(w + \varepsilon \int uds) \cdot \gamma_T]$$

where γ_T is the Girsanov functional, the solution of

$$\gamma_T = 1 - \varepsilon \int \gamma_s u_s dw_s.$$

With

$$\begin{aligned} E[F(w)] &= E[F(w + \int \varepsilon ud_s) \cdot \gamma_T] \\ &\approx E[F(w)] + \varepsilon E[D_u F(w)] - \varepsilon E[F(w) \int uds] \end{aligned}$$

we find the Bismut integration by parts formula

$$E[F(w) \int u_s dw_s] = E[D_u F(w)].$$

Applying this to "nice" functions $c(F) \cdot g$, formally we find

$$E[c(F)(D_u F)^{-1} \int udw] = E\left[c'(F)(D_u F)^{-1} D_u F + c(F) D_u((D_u F)^{-1})\right]$$

and

$$Ec'(F) = E\big(c(F)(D_u F)^{-1} \int udw\big) - E\big(c(F) D_u (D_u F)^{-1}\big). \tag{5}$$

$(D_u F)$ now plays the role of the Malliavin matrix, and the assumption that $D_u F > 0$ for a suitable predictable (u_s) leads to a bound on $|Ec'(F)|$ in (4) as required in (2).

In the survey article, [20], Zakai points out that the Malliavin and Bismut approaches are not equivalent.

We follow here, more-or-less, the Bismut approach, but where Bismut considers variations in a function space our formulation reduces the Malliavin calculus to differentiation in a finite dimensional space for the situation where the Wiener functional is just a solution of a diffusion equation as in (1), $F(w) = \xi_{0T}(x_0)$. The key observation which leads to our result is a martingale representation formula which might be seen as coming from the folklore of mathematics, but it provides us with a new formulation of the integration by parts formula, which – as is well known – always plays the fundamental role in Malliavin's calculus.

3. Representation of martingales.

Consider the solution $\xi_{0,t}(x_0)$ of (1) and let c be a twice continuously differentiable function for which $c(\xi_{0,T}(x_0))$ and the components of $c_\xi(\xi_{0,T}(x_0))$ are integrable. We then have the following representation for the right continuous version of $E[c(\xi_{0,T}(x_0)) \mid \mathcal{F}_t] =: m_t$.

THEOREM 3.1. *The martingale m_t, $0 \le t \le T$, has a representation as*

$$m_t = E[c(\xi_{0,T}(x_0))] + \int_0^t \gamma_i(s) dw_s^i$$

with

$$\gamma_i(s) = E[c(\xi_{0,T}(x_0)) D_{0,T} \mid \mathcal{F}_s] D_{0,s}^{-1} X_i(s, \xi_{0,s}(x_0)), \tag{6}$$

where $D_{s,t}$ is the Jacobian of the stochastic flow,

$$D_{s,t} = \frac{\partial \xi_{s,t}}{\partial x}.$$

Note that from the following theorem cited from [3,8] $D_{s,t}$ exists as a solution of a stochastic differential equation.

THEOREM 3.2. *There is a map $\xi : \Omega \times [0,\infty) \times [0,\infty) \times \mathbb{R}^d \to \mathbb{R}^d$ such that*

(i) for $0 \le s \le t \le T$, $x \in \mathbb{R}^d$, $\xi_{s,t}(x)$ is the essentially unique solution of (1);

(ii) for each ω, s, t the map $\xi_{s,t}(\omega,\cdot)$ is $C^\infty(\mathbb{R}^d,\mathbb{R}^d)$ with a Jacobian which satisfies

$$dD_{s,t} = \frac{\partial X_0}{\partial \xi}(t,\xi_{s,t}(x))D_{s,t}dt + \frac{\partial X_i}{\partial \xi}(t,\xi_{s,t}(x))D_{s,t}dw_t^i$$

$$D_{s,s} = I, \quad \text{the identity matrix;}$$

(iii) the second derivative $W_{s,t} = \dfrac{\partial^2 \xi_{s,t}}{\partial x^2}$ satisfies

$$dW_{s,t} = \frac{\partial X_0}{\partial \xi}(t,\xi_{s,t}(x))W_{s,t}dt + \frac{\partial X_i}{\partial \xi}(t,\xi_{s,t}(x))W_{s,t}dw_t^i$$
$$+ \frac{\partial^2 X_0}{\partial \xi^2}(t,\xi_{s,t}(x))D_{s,t}\otimes D_{s,t}dt + \frac{\partial^2 X_i}{\partial \xi^2}(t,\xi_{s,t}(x))D_{s,t}\otimes D_{s,t}dw_t^i$$

$$W_{s,s} = 0 \in \mathbb{R}^d \otimes \mathbb{R}^d \otimes \mathbb{R}^d.$$

Proof of 3.1: Any $\mathcal{F}_t$-martingale (m_t) may be represented as

$$m_t = m_0 + \int_0^t \gamma_i(s)dw_s^i$$

for a predictable integrand γ_i. As $\xi_{0,t}(x_0)$ is Markov

$$\begin{aligned} m_t &= E[c(\xi_{0,T}(x_0)) \mid F_t] \\ &= E[c(\xi_{t,T}(x)) \mid F_t] \\ &= E_{t,x}[c(\xi_{t,T}(x))] \\ &=: V(t,x), \qquad \text{where} \quad x = \xi_{0,t}(x_0). \end{aligned}$$

Then applying Itô's rule to $V(t,x)$, $x = \xi_{0,t}(x_0)$ gives

$$V(t,\xi_{0,t}(x_0)) = V(0,x_0) + \int_0^t \Big(\frac{\partial V}{\partial s} + LV\Big)ds$$
$$+ \int_0^t \frac{\partial V}{\partial x}(s,\xi_{0,s}(x_0))X_i(s,\xi_{0,s}(x_0))dw_s^i$$

with

$$L = X_0^i \frac{\partial}{\partial x_i} + \frac{1}{2} \sum X_k^i X_k^j \frac{\partial^2}{\partial x_i \partial x_j}.$$

As (m_t) is a martingale, from the uniqueness of semimartingale decomposition we must have

$$\left(\frac{\partial V}{\partial s} + LV\right) = 0$$

and

$$\gamma_i(s) = \frac{\partial V}{\partial x}(s, \xi_{0,s}(x_0)) X_i(s, \xi_{0,s}(x_0)).$$

Differentiating V we thus arrive at

$$\gamma_i(s) = E[c_\xi(\xi_{0,T}(x_0)) D_{0,T} \mid F_s] D_{0,s}^{-1} X_i(s, \xi_{0,s}(x_0)). \tag{7}$$

□

Now let $u(s) = (u_1(s), \ldots, u_m(s))$ be a square integrable predictable process. Applying the above representation we find the desired integration by parts formula.

THEOREM 3.3. *Under the above assumptions the following equality holds*

$$E\left[c(\xi_{0,T}(x_0) \int_0^T u_i(s) dw_s^i\right] = E \int_0^T E[c_\xi(\xi_{0,T}(x_0)) D_{0,T} \mid F_s] D_{0,s}^{-1} X_i(s) u_i(s) ds$$

$$= E\left[c_\xi(\xi_{0,T}(x_0)) D_{0,T} \int_0^T D_{0,s}^{-1} x_i(s) u_i(s) ds\right]$$

by Fubini's theorem.

In particular, putting $u_i(s) = (D_{0,s}^{-1} X_i(s))^*$ and considering the product function $h(\xi_{0,T}(x_0)) = c(\xi_{0,T}(x_0)) g(\xi_{0,T}(x_0)$ we have

THEOREM 3.4.

$$E[c(\xi_{0,T}(x_0)) g(\xi_{0,T}(x_0))] \int_0^T (D_{0,s}^{-1} x_i(s))^* dw_s^i = E[(c_\xi g + c g_\xi) D_{0T} M_{0T}],$$

where

$$M_{s,t} = \sum_{i=1}^m \int_s^t D_{s,u}^{-1} X_i(u) X_i^*(u) D_{s,u}^{*-1} du.$$

$M_{s,t}$ is the Malliavin matrix. □

In order to obtain a bound on c_ξ we now would like to take

$$g = M_{0,T}^{-1} D_{0,T}^{-1},$$

but this function not only depends on $\xi_{0,T}$. To get around this difficulty we have to enlarge the system in the following way.

4. Existence of a density for $\xi_{0,T}(x_0)$.

When enlarging the system the results of 3.2 might no longer hold for $\xi_{s,t}$ replaced by the new system as the coefficients are no longer bounded.

We consider the flow defined by (1), its Jacobian $D_{s,t}$, the martingale $R_{s,t}(x) = \int_s^t (D_{s,u}^{-1} X_i(u))^* dw_u^i$, and the inverse of the Malliavin matrix $M_{s,t} = \sum \int_s^t D_{s,u}^{-1} X_i(u) X_i^*(u) D_{s,u}^{*-1} du$. Then for

$$\phi^{(0)}(w,s,t,x) = \xi_{s,t}(x), \qquad x = \xi_{0,s}(x_0)$$

$$D_{s,t}^{(0)}(x) = D_{s,t}(x), \qquad D = D_{0,s}(x_0)$$

$$D_{0,t}^{(0)}(x_0) = D_{s,t} D$$

$$R_{s,t}^{(0)}(x) = \int_s^t (D_{s,u}^{-1} X_i(u))^* dw^i u \tag{8}$$

$$R_{0,t}^{(0)} = R + D^{-1} R_{s,t}^0(x), \qquad R = R_{0,s}^0$$

$$M_{s,t}^{(0)} = M_{s,t}(x)$$

$$M_{0,t}^{(0)} = M + D^{-1} M_{s,t}(x) D^{*-1}, \qquad M = M_{0,s}^0$$

the enlarged system $\phi^{(1)} = (\phi^0, D^0, R^0, M^0)$ is Markov. We now would like to apply a result similar to 3.2 to this enlarged system. Introduce the set $S_\alpha(d_1,\ldots,d_k)$, $\alpha, d, d_1, \ldots, d_k$

positive integers, of C^∞ functions $X : \mathbb{R}^d \to \mathbb{R}^d$ of the triangular form

$$X(x) = \begin{pmatrix} X^{(1)}(x^1) \\ X^{(2)}(x^1, x^2) \\ \vdots \\ X^k(x^1, x^2, \ldots, x^k) \end{pmatrix} \qquad \text{for} \qquad x = \begin{pmatrix} x^1 \\ \vdots \\ x^k \end{pmatrix}, \qquad x^i \in \mathbb{R}^{d_i}$$

and $\mathbb{R}^d = \mathbb{R}^{d_1} \times \cdots \times \mathbb{R}^{d_k}$, which satisfy

$$\|X\|_{S(\alpha,N)} = \sup_{x \in \mathbb{R}^d} \Big(\sup_{0 \le n \le N} \frac{|DX(x)|}{1+|x|^\alpha} \vee \sup_{1 \le j \le k} |D_j X^{(j)}(x)| \Big) < \infty$$

for all N.

Note that ϕ^1 is Markov with coefficients in $S(d,\ d+d^2,\ 2d+d^2,\ 2d+d^2)$, and following Norris [15] we may state the extension of 3.2.

THEOREM 4.1. *Let $x_0, x_1, \ldots, x_m \in S_\alpha(d_1, \ldots, d_k)$. Then there is a map $\phi : \Omega \times [0,\infty)^2 \times \mathbb{R}^d \to \mathbb{R}^d$ such that*

(i) for $0 \le s \le t$, $x \in \mathbb{R}^d$, ϕ is the essentially unique solution of

$$dx_t = X_0(x)dt + X_i(x)dw_t^i, \quad x_s = x;$$

(ii) for all (ω, s, t), ϕ is C^∞ with derivatives of all orders satisfying stochastic differential equations;

(iii)

$$\sup_{|x| \le R} E\Big[\sup_{s \le u \le t} |D^n \phi(\omega, s, u, x)|^p \Big]$$
$$\le C(p, R, N, d_1, \ldots, d_k,\ \alpha,\ \|X_0\|_{S_{\alpha,N}}, \ldots, \|X_m\|_{S_{\alpha,N}}).$$

□

Furthermore, we can consider the Jacobian of $\phi^{(1)}$, say $D^{(1)}$, and construct $R^{(1)}_{s,t} = \int_s^t D^{(1)-1}_{s,u} X_i^{(1)}(u)dw_u^i$, and let $M^{(1)}_{s,t} = \langle R^{(1)}_{s,t} \otimes R^{(0)*}_{s,t} \rangle$ be the predictable quadratic variation of R^1 and R^{0*}.

This 4-tuple defines $\phi^{(2)} = (\phi^{(1)}, D^{(1)}, R^{(1)}, M^{(1)})$ and inductively we can proceed to define $\phi^{(n)}$ for all n, and Norris' result holds for all $\phi^{(n)}$.

Now apply 3.4 to $c(\phi^{(0)}) \cdot g(\phi^{(1)})$ to obtain:

COROLLARY 4.2.

$$E[c(\phi^{(0)})g(\phi^{(1)}) \otimes R^{(0)}] = E[(\nabla_0 c)(\phi^0)g(\phi^1)D_{0,T}M_{0,T}]$$
$$+ E[c(\phi^0)(\nabla_1 g)(\phi_1)D^1 M^1],$$

and for

$$g(\phi^{(1)}) = M_{0,T}^{-1}D_{0,T}^{-1}$$

we find

$$E[c_\xi(\xi_{0,T}(x_0))] = E[c(\xi_{0,T}(x_0))M_{0,T}^{-1}D_{0,T}^{-1} \otimes R_{0,T}]$$
$$- E[c(\xi_{0,T}(x_0))(\nabla_1 g)D_{0,T}M_{0,T}D_{0,T}^{(1)}M_{0,T}^{(1)}].$$

□

An application of Jensen's, Burkholder's, and Gronwall's inequalities with Norris' result implies that all terms, except possibly $M_{0,T}^{-1}$, are in all L^p, $p < \infty$. If now we assume that $M_{0,T}^{-1}$ is in some L^p, e.g., if we assume H1 to hold, then we have the desired result.

THEOREM 4.3. *Let $\xi_{0,T}(x_0)$ be the solution of (1) and c a bounded C^∞ function with bounded derivatives. Then if $M_{0,T}^{-1}$ is in some L^p*

$$|E[c_\xi(\xi_{0,T}(x_0))]| \le K \sup_{x \in \mathbb{R}^d} |c(x)|.$$

With this result, D. Williams' 'ridiculous' example on the existence of a density for the Brownian motion really becomes trivial:

$$|E[c'(w_1)]| \le |E[c(w_1) \cdot w_1]| \le \sup_{x \in \mathbb{R}^d} c(x) \cdot \text{const.}$$

5. Application.

The Malliavin calculus could not have attracted so much attention if there were not many important applications, together with the remarkable fact that it links the Hörmander partial differential equation methods with probabilistic aspects. Within stochastic analysis it provides many helpful tools, such as, for example, the integration by parts formula which is equivalent to a martingale representation theorem. In filtering theory, D. Michel and J.M. Bismut [5] used the calculus of variations to prove the existence of densities for optimal filters, and Jacod and Bichteler [1] extended these results to diffusion processes with jumps.

Many of these results can be simplified by using the finite dimensional calculus developed above. The full details are found in [9,10,11].

J.M. Bismut [7] (also cf. [13]) applied the results from Malliavin's calculus to the theory of index theorems in algebraic topology and to large deviations problems [6].

Recently, there have been several attempts to develop a notion of anticipative stochastic integrals. This would allow one to consider functions $u(s)$ above which might not be predictable and, in turn, this would then allow the development of Bismut's Malliavin calculus to its full strength.

References

[1] Bichteler, K.; Jacod, J.: Calcul de Malliavin pour les diffusions avec sauts: existence d'une densité dans le cas unidimensionel. Sém. Probab. XVII, LN Math 986, Springer (1983).

[2] Bichteler, K.; Fonken, D.: A simple version of the Malliavin calculus in dimension one. LN Math 939, Springer (1982).

[3] Bismut, J.M.: Mécanique aléatoire, LN Math 866, Springer (1981).

[4] Bismut, J.M.: Martingales, the Malliavin calculus and hypoellipticity under general Hörmander's conditions, *Zeitschrift f. Wahrscheinlichkeitstheorie* 56(1981), 469–505.

[5] Bismut, J.M.; Michel, D.: Diffusions conditionnelles 1, Hypoellipticité partielle, *J. Funct. Anal.* 44(1981), 174–211.

[6] Bismut, J.M.: *Large deviations and the Malliavin calculus*, Birkhäuser Verlag (1984).

[7] Bismut, J.M.: The Atiyah–Singer theorems: a probabilistic approach, *J. Funct. Anal.* 57(1984), 56–99, 329–348.

[8] Carverhill, A.P.; Elworthy, K.D.: Flows of stochastic dynamical systems, the functional analytic approach, *Zeitschrift f. Wahrscheinlichkeitstheorie* 65(1983), 245–267.

[9] Elliott, R.J.; Kohlmann, M.: Integration by parts, homogeneous chaos expansions, and smooth densities, preprint Univ. Konstanz 1987, to appear in *Ann. Probab.*

[10] Elliott, R.J.; Kohlmann, M.: The existence of smooth densities for the prediction, filtering, and smoothing problems, preprint Univ. Konstanz (1987).

[11] Elliott, R.J.; Kohlmann, M.: Integration by parts and densities for jump processes, preprint Univ. Konstanz (1988).

[12] Ikeda, N.; Watanabe, S.: An introduction to Malliavin's calculus, *Taniguchi Symp.* SA Katata 1983, North Holland Publ. Co. (1984).

[13] Ikeda, N.; Watanabe, S.: Malliavin calculus and its applications, in: *From local times to global geometry, control and physics*, Pitman RN Math 150, Longman (1986).

[14] Malliavin, P.: Stochastic calculus of variations and hypoelliptic operators, *Proc. Int. Conf.*, Kyoto 1976, Wiley (1978), 195–263.

[15] Norris, J.: Simplified Malliavin calculus, LN Math 1204, Springer (1986).

[16] Shigekawa, I.: Derivatives of Wiener functionals and absolute continuity of induced measures, *J. Math. Kyoto Univ.* 202(1980), 263–289.

[17] Stroock, D.W.: The Malliavin calculus, a functional analytic approach, *J. Funct. Anal.* 44(1981), 212–257.

[18] Stroock, D.W.: Some applications of stochastic calculus in partial differential equations, LN Math 976, Springer (1983).

[19] Williams, D.: To begin at the beginning ..., LN Math 851, Springer (1981).

[20] Zakai, M.: The Malliavin calculus, *Acta Appl. Math.* 3(1985), 175–207.

PATHWISE STABILITY OF RANDOM DIFFERENTIAL EQUATIONS AND THE SOLUTION OF AN ADAPTIVE CONTROL RELATED PROBLEM

László Gerencsér

Department of Electrical Engineering, Research Centre for Intelligent Machines
3480 University Street, McGill University, Montréal, Quebec, H3A 2A7

In this paper we present the solution of a long-standing open problem related to adaptive control: a method and its analysis for the recursive estimation of time-varying parameters. The method of analysis is the use of an earlier result in Gerencsér (1988b) on the pathwise stability of random differential equations, which we present (in an improved form) in Section 1.

The paper has been written during a visit to the Department of Electrical Engineering, McGill University, Montreal, Quebec, while being on leave from the Computer and Automation Institute of the Hungarian Academy of Sciences, Budapest. The invitation and support of the organizers of the present conference is also gratefully acknowledged.

1. Introduction and the first theorem

In this paper we present the solution of a long-standing open problem related to adaptive control: a method and its analysis for the recursive estimation of time-varying parameters. The method is similar to the scheme presented by Djereveckii and Fradko (1981) and Ljung (1977), however the analysis is very different. The first partially successful result about recursive identification of time-varying parameters is due to Caines and Meyn (1986), however they assume a basically stationary (but nonlinear) model. In our paper we work with unmodelled parameter process.

The main Theorem (Theorem 2.1) can be extended without diffculty to controlled systems. To ensure closed loop identifiability a suitable modification of the methods in Caines and Lafortune (1984) proved to be useful

The main contribution of the paper is that we give a characterization of the tracking error as a random process, without which the analysis of the closed loop behaviour of an adaptive control system is hopeless.

The method of analysis is the use of an earlier result in Gerencsér (1988b) on the pathwise stability of random differential equations, which we present (in an improved form) in Section 1.

The central notion of L-mixing and a few more notations are summarized in the Appendix.

Let us consider the random differential equation

$$\dot{x}(t) = \epsilon H^{\epsilon}(t, x(t), \omega) \qquad x(s) = \xi \tag{1.1}$$

where $H^{\epsilon}(t, x, \omega)$ is a random field defined on some probability space (Ω, F, P) for $t \geq 0$ and $x \in D$, where D is a compact domain in R^p. We assume

Condition 1.1. $H^{\epsilon}(t, x, \omega)$ is continuous and bounded in (t, x, ω) say

$$|H^{\epsilon}(t, x, \omega)| < K \quad \text{and} \quad |\Delta H^{\epsilon}/\Delta x(t, x, x+h, \omega)| < L.$$

This condition ensures that (1.1) has a unique solution in for all $\xi \in \mathrm{int} D$ in some finite or infinite interval for all ω.

To describe the probabilistic structure of $H^{\epsilon}(t, x, \omega)$ we need the concept of L-mixing defined in the Appendix. With this we have:

Condition 1.2. H^{ϵ} and $\Delta H^{\epsilon}/\Delta x$ are uniformly L-mixing in x for $x \in D$ and in $x, x+h \in D$ respectively. with respect to a pair of families of σ-algebras $(\mathcal{F}_t, \mathcal{F}_t^+)$.

Let

$$G^{\epsilon}(t, x) = \mathrm{E} H^{\epsilon}(t, x, \omega)$$

and let us consider the ordinary differential equation

$$\dot{y}(t) = G^{\epsilon}(t, y(t)) \qquad y(s) = \xi \tag{1.2}$$

We impose on G^{ϵ} the following conditions

Condition 1.3. $G^{\epsilon}(t, y)$ is defined on $R^+ \times D$, it is continuous and bounded in (t, y) together with its first and second partial derivatives as indicated:

$$|G^{\epsilon}(t, y)| < K, \quad \|\partial G^{\epsilon}(t, y)/\partial y\| < L \quad \|\partial^2 G^{\epsilon}(t, y)/\partial y^2\| < L.$$

We also set

$$b = \sup_{t \geq 0, x \in D} |H^{\epsilon}(t, x, \omega) - G^{\epsilon}(t, x)|.$$

Here $|\cdot|$ denotes the Euclidean norm of a vector and $\|\cdot\|$ denotes the operator norm of a matrix.

This condition ensures the existence and uniqueness of the solution of (1.1) in some finite or infinite time interval for all $\xi \in \mathrm{int}\, D$, with int D denoting the interior of D. Let $g^{\epsilon}(t, s, \xi)$ denote the solution of (1.1).

Let $D' \subset D$ be any subset of D and consider the set of points which are reachable from D' along the trajectories of (1.2), i.e. we consider the set

$$\{y : y = g^\varepsilon(t,s,\xi) \quad \text{for some} \quad t \geq s, \xi \in D'\}.$$

The closure of this set will be denoted by $g^\varepsilon(D')$.

The ϵ neighborhood of the set D will be denoted by $S(D',\epsilon)$ i.e.

$$S(D',\epsilon) = \{x : |x - z| < \epsilon \quad \text{for some} \quad z \in D'\}.$$

Our next condition can now be formulated as follows:

Condition 1.4 There exist compact domains $D_\xi \subset D_y \subset D_x \subset D_0 \subset \text{int} D$ (int D denoting the interion of D) such that we have for some $\beta > 0$

$$g^\varepsilon(D_\xi) \subset D_y, \qquad S(D_y,\beta) \subset D_x \quad \text{and} \quad g^\varepsilon(D_x) \subset D_0.$$

It is well-known (c.f. Pontryagin (1970), Ch.24 Th 17) or Hartman (1964 Ch. V. Th. 1.1), that $g^\varepsilon(t,s,\xi)$ is a continuously differentiable function of (t,s,ξ). To ensure exponential asymptotical stability of (1.2) we impose the following

Condition 1.5 For some $\gamma > 1$ $\alpha > 0$ we have for all $0 < s < t, x \in D_x$

$$\|\frac{\partial g^\varepsilon}{\partial x}(t,s,x)\| < \gamma^{-\varepsilon\alpha(t-s)}.$$

Theorem 1.1 Assume that Conditions 1.1-1.5 are satisfied. Then we have for $\varepsilon \in D_\varepsilon$ and any initial times

$$|x(t) - y(t)| < D(t)$$

where $(D(t))$ is an L-mixing process with respect to $(\mathcal{F}_t, \mathcal{F}_t^+)$ and it is independent of the initial time s. Furthermore for $q > 2$ and any $r > p$

$$M_q(D) < C\varepsilon^{\frac{1}{2}} M'^{\frac{1}{2}}_{qr2}(\overline{H}^\varepsilon)\Gamma'^{\frac{1}{2}}_{qr2}(\overline{H}^\varepsilon)$$

where C depends on $p,q,r,K,L,\gamma,\alpha,D_x$ and D.

2. Recursive estimation of time-varying parameters

Let us consider the time-invariant linear system

$$\overline{\overline{y}}(\theta^*) = H(\theta^*)e \tag{2.1}$$

where $H(\theta) = H(\theta,s)$ denotes an $m \times m$ rational matrix-valued transfer function such that for each system parameter θ under consideration we have $H(\theta,\infty) = 0$ and $H(\theta,0) = I$ where I is an $m \times m$ unit matrix.

We assume that we have fixed a unique time-domain description of (2.1) (e.g. state-space or ARMA) so that there is a one-to-one correspondence between the set of matrices appearing in this description and $H(\theta)$. Thus these matrices are parameterized by the same parameter θ.

We assume that the following conditions are satisfied.

Condition 2.1. $H(\theta)$ is three-times continuously differentiable in a neighborhood of D.

Remark. This condition means that the matrices in the time-domain description are three-time continuously differentiable functions of θ in a neighborhood of D.

Condition 2.2. $H(\theta)$ and $H^{-1}(\theta)$ are defined and are uniformly exponentially stable for all $\theta \subset D \subset R^p$ where D is some compact domain.

Condition 2.3. The input-noise process e is assumed to be a second order stationary, zero mean, bounded L-mixing process with respect to a pair of families of σ-algebras $(\mathcal{F}_t . \mathcal{F}_t^+)$. We set say:

$$|e(t)| < b.$$

Let us now consider the time-varying system

$$y = H(\theta(t))e \tag{2.2}$$

which we interpret by inserting a time-varying parameter into the time-domain description. We have the following basic assumption:

Condition 2.4. $(\theta(t))$ is a deterministic process with continuous and bounded derivatives; say

$$\delta = \sup_t |\dot{\theta}(t)|$$

for all t and $\theta(t) \in D_0 \subset \text{int} D$ for all t where D_0 is a small compact domain.
It is well-known that the time-varying system is exponentially stable whenever δ is small enough say

$$\delta \leq \delta_0. \tag{2.3}$$

Assume we observe $y = (y(t))$ and we wish to estimate the time-varying parameter $\theta(t)$. An off-line estimator was defined and analyzed in Gerencsér (1988), Section 7 for the case when e is a Gaussian white noise. Now we present a recursive estimation method for the class of models described above. This algorithm we shall call a recursive quasi-maximum likelihood method with forgetting (RMLF). The estimator process will be denoted by $\hat{\hat{\theta}}(t)$, and the method is described by the following set of differential equations.

$$\epsilon(t) = H^{-1}(\hat{\hat{\theta}}(t))y(t) \tag{2.4}$$

$$\dot{V}(t) = \lambda V(t) + \lambda|\epsilon(t)|^2 \tag{2.5}$$

$$\dot{\hat{\hat{\theta}}}(t) = -\mu V_\theta(t), \tag{2.6}$$

with some $\lambda, \mu > 0$. The interpretation of the algorithm is analogous with the scheme proposed by Ljung (1977) and Djereveckii and Fradko (1981). Thus e.g.

$$V_\theta(t) = \epsilon_\theta(t)\epsilon(t)$$

where $\epsilon_\theta(t)$ is defined by the time-varying filter

$$\epsilon_\theta = (H^{-1})_\theta(\hat{\hat{\theta}}(t))y \tag{2.7}$$

and here the subscript $_\theta$ denotes differentiation with respect to θ. All the initial conditions are taken to be zero except in (2.6) where we take $\hat{\hat{\theta}}(0) \in D$.

Before we formulate our basic theorem we need to make a comment on the identifiability of the time-invariant system (2.1). Let $\theta \in D$ and define a process $\overline{\overline{\epsilon}}(t,\theta,\theta^*)$ by

$$\overline{\overline{\epsilon}}(\theta,\theta^*) = H^{-1}(\theta)\overline{\overline{y}}(\theta^*) = H^{-1}(\theta)H(\theta^*)e.$$

Let

$$W(\theta,\theta^*) = \lim_{t\to\infty} \mathrm{E}|\overline{\overline{\epsilon}}(t,\theta,\theta^*)|^2.$$

The asymptotic quasi-likelihood equation then becomes

$$W_\theta(\theta,\theta^*) = 0, \tag{2.8}$$

where the derivative with respect to θ can be shown to exist. An unfortunate feature of the system (2.1) is that $W_\theta(\theta^*,\theta^*) \neq 0$ in general. However if the system is not overparametrized then (2.8) has a unique solution θ^{**} in a small neighborhood of θ^* whenever e is sufficiently white, which can be formalized as follows:

$$M_2(e)\Gamma_2(e) < c$$

with sufficiently small c.

Condition 2.5. For the matrix

$$\Sigma(\theta^*,\theta^*) = \lim_{t\to\infty} \mathrm{E}\overline{\overline{\epsilon}}_\theta(t,\theta^*,\theta^*)\overline{\overline{\epsilon}}_\theta^T(t,\theta^*,\theta^*)$$

we have

$$\Sigma(\theta^*,\theta^*) > cI \qquad c > 0$$

for all $\theta^* \in D$.

Now we can state our basic result as follows:

Theorem 2.1 Under Conditions 2.1 - 2.5 we have with $\lambda = \delta^{1/3}$ and $\mu = \delta^{2/3}$

$$|\hat{\hat{\theta}}(t) - \theta^{**}(t)| < D(t) + o(1)$$

where $D(t)$ is an L-mixing process such that for all $1 \le q < \infty$ we have

$$M_q(D) < C_q\delta^{1/3}$$

wherever δ and D_0 is small enough, e is sufficiently white and sufficiently small and the initial value $\widehat{\theta}(0)$ is sufficiently close to $\theta^{**}(0)$.

Remark Note that the tracking error has the same order of magnitude as the tracking error for the off-line estimator (c.f. Gerencsér, (1988a), Section 7).

Proof: Let us introduce a cost-function associated with an off-line estimator method. (c.f. also Gerencsér (1988a) Section 7). Fix $\theta \in D_0$ and define an estimated noise process $(\bar{\epsilon}(t,(\theta))$ given by

$$\bar{\epsilon}(t,\theta) = H^{-1}(\theta)y$$

with zero initial condition. Then define a cost function $\overline{V}(t,\theta)$ by the equation

$$\dot{\overline{V}}(t,\theta) = -\lambda\overline{V}(t,\theta) + \lambda|\bar{\epsilon}(t,\theta)|^2, \qquad \overline{V}(0,\theta) = 0.$$

(Then the off-line estimator of $\theta^*(t)$ is defined to be that θ which minimizes $\overline{V}(t,\theta)$).

Let us now consider the random field

$$H(t,\theta) = -\mu\overline{V}_\theta(t,\theta).$$

We show that H satisfies all relevant conditions of Section 1 together with

$$G(t,\theta) = -\mu W_\theta(\theta,\theta^*(t)).$$

Condition 1.1 is obvious due to the stability and smoothness of the filters generating $\epsilon\ (t,\theta)$ and $\epsilon_\theta\ (t,\theta)$ and the boundedness of the input noise process e.

We can actually estimate $H - G$ as

$$|H(t,\theta) - G(t,\theta)| < C\mu b^2, \tag{2.9}$$

whenever $\delta < \lambda$. Indeed, the arguments given in Gerencsér (1988a) (7.7) and (7.8) show that

$$|\mathrm{E}\overline{V}_\theta(t,\theta) - W_\theta(\theta,\theta^*(t))| < Cb^2\delta/\lambda \tag{2.10}$$

and

$$|\overline{V}_\theta(t,\theta) - \mathrm{E}\overline{V}_\theta(t,\theta)| < Cb^2.$$

Combination of the last two inequalities give (2.9).

It is also useful to have an estimate for the moments of finite order of $\overline{H} = H - \mathrm{E}H$ which is obtained directly using the moment inequality of Gerencsér (1988a) (Theorem 1.1). We thus get for all $2 \leq q < \infty$

$$M_q(\overline{H}) < Cb^2\mu\lambda^{\frac{1}{2}} \tag{2.11}$$

and similarly for H_θ.

Condition 1.2 is also satisfied since e is L-mixing and stable filtering and multiplication of L-mixing processes give L-mixing processes. (Gerencsér (1988a) Lemma 2.4). To estimate the mixing rate of H we need a better result than the quoted one, and this we shall give in the Appendix. (Lemma 3.1). It is easy to see that the proposition of this lemma extends automatically for parameter dependent processes in the sense that if the input process is parametrized and uniformly L-mixing with respect to this parameter then we still have the estimation given in the lemma. Since the process $\epsilon_\theta\ (t,\theta)\cdot\ \epsilon\ (t,\theta)$ is obviously L-mixing uniformly in θ we get for all $2 \leq q < \infty$

$$\Gamma_q(\overline{H}) < C_q b^2 \mu \lambda^{-1/2} \tag{2.12}$$

when λ is sufficiently small. In the same way we get an analogous inequality for H_θ.

Let us now verify the conditions imposed on $G(t,\theta) = -\mu W_\theta(\theta,\theta^*(t))$. Condition 1.3 is obviously satisfied with

$$|G(t,y)| < C\mu b^2 \quad \text{and} \quad \|\frac{\partial}{\partial y}G(t,y)\| < C\mu b^2. \tag{2.13}$$

To check Condition 1.4 is an easy but tedious analysis. Finally to check Condition 1.5 we refer to Gerencsér (1988d) for a result on the exponential stability of time-varying systems. This gives us that

$$\dot{\theta}(t) = G(t,\theta(t)) \qquad \theta(0) = \varsigma$$

with $\varsigma \in D_y$ is exponentially stable whenever

$$\frac{\partial}{\partial t}G(t,\theta) = -\mu W_{\theta\theta^*}(\theta,\theta^*(t))\dot{\theta}^*(t) \tag{2.14}$$

is small enough. More exactly if $\psi(t)$ denotes the solution of the differential equation

$$\dot{\psi}(t) = \frac{\partial}{\partial \theta}G(t,\hat{\hat{\theta}}(t)) \cdot \psi(t) \qquad \psi(0) = I$$

then we have for $0 \leq s \leq t$

$$\|\psi(t)\psi^{-1}(s)\| < Ce^{-\alpha''(t-s)}$$

where

$$\alpha'' = C\mu - C'(\mu\delta)^{1/2} - C'\delta \tag{2.15}$$

(c.f. Theorem 2) of the quoted paper. The conditions of that theorem are trivially satisfied.

We still have to work out an estimator for the process

$$\delta V_\theta(t) = V_\theta(t) - \overline{V}_\theta(t,\hat{\hat{\theta}}(t))$$

in order to apply Theorem 1.1.

We show that

$$|\delta V_\theta(t)| < C\mu\lambda^{-1}b^2 \tag{2.16}$$

where C depends only on the family of time-invariant systems (2.1). The inequality (2.16) will be proved if we manage to show that

$$\hat{\hat{\theta}}(t) \in D \quad \text{and} \quad |\dot{\hat{\hat{\theta}}}(t)| < \delta_0. \tag{2.17}$$

Indeed if the last inequality holds true then the time-varying system (2.2) is exponentially stable hence $| \epsilon_\theta\ (t)\ \epsilon\ (t)| < Cb^2$ and (2.16) follows.

To show (2.17) we use a continuous-time induction. The proposition is obviously true for $t = 0$ and hence by continuity in some interval $0 \le t \le T \quad T > 0$.

Taking into account the $L_\infty(\Omega, \mathcal{F}, P)$ bound for $H - G$ obtained from (2.9) we get from Section 1 that in $[0, T]$

$$|\hat{\hat{\theta}}(t) - \theta(t)| < Cb^2$$

hence $\hat{\hat{\theta}}(t) \in D_0$ if b is small enough. Thus validity of (2.16) in $[0, T]$ implies

$$\hat{\hat{\theta}}(t) \in D_0 \quad \text{and} \quad |\dot{\hat{\hat{\theta}}}(t)| < C\mu\lambda^{-1}b^2 < \delta_0/2 \tag{2.18}$$

whenever B is small enough. Thus by continuity of $\hat{\hat{\theta}}(t)$ and $\dot{\hat{\hat{\theta}}}(t)$ (2.17) can never be violated.

Now we can apply Theorem 1.1 for the random differential equation below to get an upper bound for $\hat{\hat{\theta}}(t) - \theta(t)$. The differential equation governing $\hat{\hat{\theta}}(t)$ can be written as

$$\dot{\hat{\hat{\theta}}}(t) = (H(t, \hat{\hat{\theta}}(t), \omega) + \delta H(t)) \tag{2.19}$$

with $\delta H(t) = -\mu\delta V_\theta(t)$.

Since

$$|\mathrm{E}H(t, \theta, \omega) - G(t, \theta)| < Cb^2\mu\delta/\lambda$$

by (2.10) and

$$|\delta H(t)| < Cb^2\mu^2\lambda^{-1}$$

we get from Theorem 1.1 after suitable adjustment that

$$|\hat{\hat{\theta}}(t) - \theta(t)| < D(t) + Cb^2(\delta/\lambda + \mu/\lambda)$$

where $D(t)$ is an L-mixing process with respect to $(\mathcal{F}_t, \mathcal{F}_t^+)$ and combining (2.11), (2.12) and (2.14) we get from the quoted theorem

$$D = 0(\mu^{\frac{1}{2}}).$$

Finally we estimate $\theta(t)-\theta^{**}(t)$ by Theorem 1′ of Gerencsér (1988b). Since by (2.14) $\alpha'' > C\mu$ and also

$$\dot{G} \triangleq \sup_{\substack{t \\ y \in D}} |\frac{\partial}{\partial t} G(t,y)| < C\mu\delta$$

by (2.15) we get

$$|\theta(t) - \theta^{**}(t)| < C(\alpha'')^{-2}\dot{G} + 0(1) = C\delta/\mu + o(1).$$

Thus we finally get

$$|\hat{\hat{\theta}}(t) - \theta^{**}(t)| < D(t) + C(\delta/\lambda + \mu/\lambda + \delta/\mu + o(1)).$$

If we now take $\lambda = \delta^{1/3}, \mu = \delta^{2/3}$ then we get the theorem.

3. Appendix

We needed the concept of L-mixing processes which we now shortly describe. Let $\overline{H}(t,x,\omega) = H(t,x,\omega) - G(t,x)$ and let for $1 \le q < \infty$

$$M_q(\overline{H}) = \sup_{\substack{0 \le t < \infty \\ x \in D}} \mathrm{E}^{\frac{1}{q}} |\overline{H}(t,s,\omega)|^q.$$

Let $(\mathcal{F}_t, \mathcal{F}_t^+), t \ge 0$ be a pair of families of σ-algebras, such that $\mathcal{F}_t$ is increasing, $\mathcal{F}_t^+$ is decreasing and continuous from the right and $\mathcal{F}_t, \mathcal{F}_t^+$ are independent for all t. Define

$$\gamma_q(\tau, \overline{H}) = \sup_{\substack{0 \le t < \infty \\ x \in D}} \mathrm{E}^{1/q} |\overline{H}(t+\tau,x,\omega) - \mathrm{E}\overline{H}(t+\tau,x,\omega)|\mathcal{F}_t^+)|^q$$

and

$$\Gamma_q(\overline{H}) = \int_0^\infty \gamma_q(\tau,\overline{H}) d\tau$$

We say that $\overline{H}(t,x,\omega)$ is uniformly L-mixing for $x \in D$ with respect to $\{\mathcal{F}_t, \mathcal{F}_t^+\}$ if $M_q(\overline{H})$ and $\Gamma_q(\overline{H})$ are finite for all $q \ge 1$.

Also we will use the following notation: $\Delta H/\Delta x$ will denote a stochastic process parametrized by $t \ge 0, x.x+h \in D, x \ne x+h$ and defined by $\Delta H/\Delta x(t,x,x+h,\omega) = (H(t,x+h,\omega) - H(t,x,\omega))/|h|$.

We define $M_q(\Delta\overline{H}/\Delta x)$ and $\Gamma_q(\Delta H/\Delta x)$ in an analogous way to $M_q(\overline{H}), \Gamma_q(\overline{H})$, but now supremum is taken over $x, x+h \in D$ $h \ne 0$. Finally we set

$$M_q'(\overline{H}) = M_q(\overline{H}) + M_q(\Delta\overline{H}/\Delta x).$$

$$\Gamma_q'(\overline{H}) = \Gamma_q(\overline{H}) + \Gamma_q(\Delta\overline{H}/\Delta x).$$

Lemma 3.1 Let $(u(t))$ $t \ge 0$ be an L-mixing process with respect to a pair of families of σ-algebras $(\mathcal{F}_t, \mathcal{F}_t^+)$ and let the process $x(t)$ be defined by

$$\dot{x}(t) = -\lambda x(t) + \lambda u(t) \qquad x(0) = 0.$$

Then $x(t)$ is L-mixing with respect to $(\mathcal{F}_t, \mathcal{F}_t^+)$ and we have for all $2 \leq q < \infty$

$$\Gamma_q(x) \leq \lambda^{-1/2} 2(q-1)^{\frac{1}{2}} M_q^{\frac{1}{2}}(u) + \Gamma_q(u).$$

References

Caines, P.E., Lafortune, S., Adaptive Control with Recursive Identification for Stochastic Linear Systems. IEEE Trans on Aut. Cont., Vol. AC-29 (1984), 312-321.

Caines, P.E., Meyn, S.P., A new approach to stochastic adaptive control. Preprint of the Computer Vision and Robotics Laboratory, McGill Research Centre for Intelligent Machines 1986, McGill University Montreal, Quebec.

Djereveckii, D.P., Fradko, A.L., Applied Theory of Discrete Adaptive Control Systems. (In Russian), Nauka, Moscow, 1981.

Gerencsér, L., On a class of mixing processes. Preprint of the Dept. of Math., Chalmers Univ. of Techn. and The Univ. of Göteborg, 1986; 11. To appear in Stochastics, 1988a.

Gerencsér, L., On the exponential stability of the mixture of time-invariant systems. Manuscript, 1988d.

Gerencsér, L., Parameter Tracking of Time-Varying Continuous-Time Linear Stochastic Systems. In modelling, Identification and Robust Control (eds.: Ch. I. Byrnes and A. Lindquist) North Holland, 1986, pp. 581-595. (1986a).

Gerencsér, L., Pathwise stability of random differential equations, Preprint of the Department of Mathematics, Chalmers University of Technology and the University of Göteborg, 1986: 19, b. Revised version submitted to Stochastics, 1988b

Gerencsér, L., Recursive estimation of time-varying parameters. Proc. of the IFAC/IFORS Symposium on Identification and System Parameter Estimation, Beijing, 1988c.

Hartman, Ph., Ordinary Differential Equations. Wiley and Sons, Inc., New York, 1964.

Ljung, L., Analysis of recursive stochastic algorithms. IEEE Trans. Auto. Cont., AC-22 (1977), 551-575.

Pontryagin, L.S., Ordinary Differential Equations. (In Russian), Nauka, Moscow, 1970.

Stochastic analysis of intertemporal economic issues

Guillermo L. Gómez M. [1]
Institute of Mathematics
University of Erlangen-Nürnberg
D-8520 Erlangen, Bismarckstr. 1 1/2

Abstract

In the present paper we shall point out some intertemporal issues arising in the study of economic dynamic systems. The process of social reproduction gives rise to a set of activities of exchange and distribution, whenever ultimate consumption does not occur within the production unit. Suitable choices of allocation of social output between consumption and investment bring about alternative paths of accumulation that together with the prevailing institutions generate alternative distributional paths. We can express preference between alternative accumulation paths and these preferences can be given numerical values. It is possible to disagree on what basis consequences of a choice can be judged as advantageous and on the definition of an adequate objective function and so forth. However, whatever choice we make there are always capital costs involved in running an action and meeting a policy target as well as political and social costs of diverting resources to an alternative choice at the expense of others, i.e. those opportunity costs measuring how many units, let us say, of education or health have to be given up in order to allow for an extra unit of defense and the like. Social costs are also known as accounting or shadow prices which reflect the fact that in general they can not be observed directly as for instance market prices.

The purpose of this study is to illustrate based on a model drawn from our current research how we can use stochastic analysis and control theory to recast these problems in the framework amenable to stochastic optimization techniques. Thereby, we obtain optimality conditions and accounting prices in the form of adjoint variables based on which we get a method of assessing the action at issue and gain interesting insight into the qualitative behaviour of random accumulation from highly neglected perspectives.

Contents

[1]Financial support by the DFG is gratefully acknowledged.

1 Introduction

1.1 General remarks

One major theme in political economy is the analysis of economic growth and development. Considerable efforts are directed to the identification and understanding of the basic processes promoting and hindering development and consequently progress. The ultimate aim of these efforts is however to provide a sound basis for the design of economic policies and actions that enable the steering of the forces carrying the actual processes of development and growth. Today it is unanimously recognized that accumulation and reinvestment of a part of the social product are the chief driving forces behind economic growth and development. These findings drive attention further to those general principles that forward patterns relying on which the distribution of the social product results in and upon those aspects of distribution as well as their associated social behaviour mainly responsible for the allocation of capital and labour income, i.e. profits and wages, to the purposes of consumption and investment.

In a sense, the main themes of political economy are dominated by *the principle of social reproduction* that manifests itself as the social activities around the process of production by means of which goods and services are brought about aimed at satisfying social needs. The latter consisting primarily in the reproduction of the factors of production, first of all of labour power forwarding the wage goods required to sustain the labourers employed in producing the social product and secondly of the means of production used up which include stock of machinery, raw materials and buildings. It is customary to call the fraction of the social product dedicated to meet the requirements of the reproduction of the factors of production, the *costs of production* and *necessary consumption* in contrast to net investments aimed at expanding production. Thus, necessary consumption comprises personal consumption of the direct producers as well as consumption of the social product aimed at keeping production at the prevailing level. One calls *economic surplus* the fraction of the social product that goes beyond the level of necessary consumption. The history of mankind has been a continuous struggle for the increase and improvement of the social production of goods and services beyond the level of necessary consumption but also the struggle for the shares of that surplus of the social product.

Let us point out the role played in this connection by labour, first contributing to the conservation of value of the factors and means of production, and secondly generating economic surplus. The crucial relevance of this observation can only be fully appreciated through an analysis of the potential economic surplus foregone and the opportunity costs deriving from the existence of reserve armies of unemployed. Not to mention social costs in terms of moral deterioration of human capital, social strains, poverty, starvation and other social maladies going even until fachismus.

1.2 Production, distribution and allocation

In order to make this paper self-contained, we shall give a brief introduction to the main issues of the production, distribution and allocation of the social output using a very simple model that we take to some extent from Marglin [15].

Let us characterize production by dint of a simple model in which social output consists only of one commodity produced by means of a technology with fixed coefficients a_0 and a_1, where a_0 represents the labour requirements in man-years per unit of output and a_1 the capital requirements per unit of output. It is intuitively appealing to think of this commodity as corn which helps to visualize the assumption that the only factors of production are capital in the form of seed corn and labour. Assuming that the factors rewards are W for labour and $1+r$ for capital both measured in terms of this year prices and making corn the unit of account with current price $P = 1$, we obtain the following *equation of price formation*

$$1 = Wa_0 + (1+r)a_1, \tag{1}$$

where Wa_0 represents labour costs and $(1+r)a_1$ seed corn costs. W and r are as usually the wage rate and the rate of return. Furthermore, denoting consumption per worker by c and assuming that the requirements of seed corn grow at the rate $(1+g)$, we get for *the allocation of one unit of corn output* the relationship

$$1 = ca_0 + (1+g)a_1, \tag{2}$$

where ca_0 represents the fraction allocated to consumption and $(1+g)a_1$ the fraction allocated to investments. Taking a_0 and a_1 as known, one easily recognizes that to each pair (c, g) and (W, r) corresponds respectively a distribution of corn income between wages and profits and an allocation of corn resources between consumption and investment. Moreover, one finds out easily that the *trade-offs* between *growth and consumption* $\frac{dg}{dc}$ and between *profit and wages* $\frac{dr}{dW}$ are given by the rate $-\frac{a_0}{a_1}$, i.e. these are tantamount to trade-offs between labour and capital requirements as a little reflection shows. This suggests that considering the coefficients a_0 and a_1 as non-purely technological parameters but as dependent on the cultural and institutional structure may lead to a more realistic understanding of the issues associated with the distribution and allocation of resources. A look at eqs. (1) and (2) indicates that the introduction of mechanisms or rules for the determination of W or r on the one hand, and of c or g on the other, brings about alternative perspectives for the distribution of income and the allocation of social output. By the way of different choices in this regard, one obtains various alternatives of economic theorizing, i.e. classic, Recardian, Marxian, neoclassic, Keynesian, etc.. It is beyond the scope of the present study to go deeper into these matters so that we just refer the interested reader to [13,15,19] that expose, evaluate critically and compare the major schools of economic thought.

1.3 Organization of the paper

It is our intention to show that stochastic analysis and control methods can be applied to economic problems related to growth and development planning. Since this paper attempts to reach stochastic control theorists, we focus primarily on the construction of a simple economic policy aimed at eliminating unemployment. Therefore, we start giving a concise orientation in the conceptual framework relevant to our purpose. In section 2 we review and discuss some issues concerning the model construction, the dynamics of accumulation, the socioeconomic background on which the performance criterion rests and the formulation of the optimization problem itself. In section 3 we go over to the actual derivation of the employment policy and analize some of its properties.

2 The process of capital accumulation and its guidance

From the point of view of political economy intertemporal analysis seeks to understand basic characteristics of evolutionary economic processes in society, i.e. the time structure of change, needs, preferences, attitutes, decisions, etc., so that we can construct dynamic models that reflect adequately the known as well as the unknown reality, that enhance our ability to learn, to gather information and to adapt to predictable and nonanticipative events. The ultimate aim of which shall be the design of a sequence of planning actions so as to replace possible but unwanted future events by more probable and desirable ones. Therefore, we shall in the forthcoming sections illustrate the usefulness of some stochastic ideas and techniques constructing a simple employment policy for the labour-surplus economy and obtaining insight into its behaviour over time and in its interaction with uncertainty.

2.1 Stochastic accumulation of capital

For the sake of simplicity, we shall use the fiction that the labour-surplus economy consists of a relatively advanced or productive sector and a reserve army of labour, i.e. underemployed or idle labour. The goal is then to elaborate an investment strategy that increases the growth capacity of the economy so that a given fraction of the reserve army can be effectively absorbed into production. [7,8,10] deal with a variety of issues deeply connected with the labour-surplus economy and development planning to which we refer to for a detailed presentation of the subject matter of this section.

Let us assume that the process of capital accumulation per available labour which we denote by $k = (k_t)_{t\geq 0}$ is governed by the stochastic differential equation

$$dk_t = \left[i_t - (\epsilon_t + \delta_t - \pi_t^2)k_t\right] dt - \pi_t k_t \, dB_t, \tag{3}$$

where the term in brakets known as the *drift coefficient* represents the average economic surplus per available worker at time t and the second term describes fluctuations about it driven by the Brownian motion process $B = (B_t)_{t\geq 0}$ defined as usual. In eq. (3) at the right-hand side the first term of the expression in brackets represents *actual investments per available worker* while the second *required investments per available worker* so that the difference of them indicates the average rate of change of capital stock per available worker. The coefficient of the Brownian motion reflects the deviations from this average due to uncertainty. Eq. (3) shall be referred to as the *stochastic equation of the accumulation of capital per available worker* and has been used under different frameworks to investigate a variety of problems in economic dynamics.

In order to save notation let us rewrite eq. (3) in the following more compact although also less transparent form that we shall use often

$$dk_t = f^\alpha(t, k_t)dt + \sigma^\alpha(t, k_t)dB_t, \tag{4}$$

where the coefficients $f^\alpha(t, k_t) = f(t, k_t, \alpha_t) = i_t - (\epsilon_t + \delta_t - \pi_t^2)k_t$ and $\sigma^\alpha(t, k_t) = \sigma(t, k_t, \alpha_t) = -\pi_t k_t$, and α is the functional $\alpha : t \longrightarrow \alpha_t$ with $\alpha_t = \alpha(s_K(t), \lambda_t)$ a function of class $C_b^{1,1}([0,1]^2 \longrightarrow \mathbf{R}_+^2)$. Here λ_t represents the ratio at time t of actually employed to available labour and $s_K(t)$ the fraction of capital income reinvested at time t. The movements of k_t can be gauged by the level or relative position of the functions i_t and $\eta_t k_t$ at time t, i.e. the actual and required investments per available worker respectively, where the new variable η is defined as $\eta_t = \epsilon_t + \delta_t - \pi_t^2$. The difference $i_t - \eta_t k_t$ at time t portrays roughly the expected rate of change of the level of capital per available worker as a function of k_t, more precisely as a function of the information available at time t on the level of k_t, and provide helpful insights into the prevailing stochastic dynamics of accumulation of capital per available worker. Knowing for instance that at time t the inequality $i_t - \eta_t k_t \geq 0$ holds, means that accumulation indeed takes place and that k_t increases. However, one word of caution is in order since these statements hold only almost surely according to the probability measure prevailing at time t.

That means, the choice of the control α which amounts to the choice of how much to reinvest and what level of employment have to prevail should result in bringing about a drift, see eq. (3), that results in elimination of unemployment and in generation of the required capital endowment per actually employed labour. The interaction of the economic processes set forth by the action of the control policy at stake shall ultimately bring about a drift coefficient $f(t, k_t, \alpha_t)$, as in eq. (4), capable of imposing a pace of accumulation of capital that pushes indeed the economy to higher levels of development within the degree of uncertainty attached to $\sigma^\alpha(t, k_t)$. Since the latter equation integrates essential characteristics of the process of capital accumulation and of its interaction with other crucial processes like unemployment elimination it suggests itself as the *system dynamics*.

As suggested already, we are interested by way of appropriate selections of α to ensure that a desired process of capital accumulation comes about. This poses two further problems, one related to the requirements on α so as to enable it to elicit the desired accumulation, and the second connected with existence and uniqueness questions that [3] deals with fully and to which we refer.

In other words, $f^\alpha(t, k_t,)$ measures locally the average macroeconomic tendency at time t of the rate of change of k_t a fact that we shall stress writing formally for the sake of simplicity $\dot{k}_t = f(t, k_t \alpha_t)$. On the other hand, $\sigma^\alpha(t, k_t)$ represents the associated fluctuations about the just mentioned average. More precisely, we should interprete the drift coefficient as the mean conditional forward derivative below, see [2,16]. In this case denoting the drift coefficient of (3) or equivalently of (4) by $D_+ k_t$ we should set it as in the relationship

$$D_+ k_t = \lim_{h \downarrow 0} E_{s,x} \left[\frac{k_{t+h}(\omega) - k_t(\omega)}{h} | \mathcal{F}_t \right].$$

However, we shall use abusing a little the notation the more suggestive expression $\dot{k}_t$ unless any confusion may arise. Let us denote the output capital ratio by y and recall the national income identity $y_t k_t = i_t + c_t$, then our new interpretation of the drift means that accumulation shall obey in the average the relationship

$$y_t k_t = c_t + (\epsilon_t + \delta_t - \pi_t^2) k_t + \dot{k}_t, \tag{5}$$

which states that in the average the social output per available worker $y_t k_t$ shall be allocated to the purposes of maintaining the level of consumption per available worker c_t, and of the level of capital per available worker, i.e. $(\epsilon_t + \delta_t - \pi_t^2) k_t$, as well as to yield net increases in the level of capital per available worker $\dot{k}_t$.

From an economic point of view the choice made over time of alternative control processes α implies a set of alternative drifts and diffusion coefficients $f^\alpha(t, k_t)$ and $\sigma^\alpha(t, k_t)$, and these in turn entail a set of alternative time paths of accumulation $k^\alpha = (k_t)_{t \geq 0}$. Many such accumulation paths are possible under different properties of the instruments. Before we go over to introduce our criterion of social desirability, we shall specify the requirements we expect the policy instruments to fulfil. Concerning various crucial issues on capital accumulation that we can not consider here, we refer to [5,6].

2.2 Qualitative features of the accumulation dynamics

In this section we shall point out a few interesting features of the qualitative behaviour of the process of accumulation of capital per available worker. With this purpose we shall look into the drift coefficient of eq. (4) in the form given by (5) and draw a phase diagram, see Fig. 1. In Fig. 1 we depict a phase plane $(\dot{k}_t + \underline{c}_t, k_t)$ for any time t arbitrary but fixed, portraying social output per available worker net of required investments per worker against the level of capital per available worker, that we obtain recalling that $\eta_t = \epsilon_t + \delta_t - \pi_t^2$ and rewriting eq. (5) as

$$\dot{k}_t + \underline{c}_t = (y_t - \eta_t)\, k_t,$$

where $\underline{c}_t$ stands for the minimum of personal necessary consumption per available worker resulting from the assumption on the existence of a floor on the admissible wage rate W, i.e. from the assumption that the prevailing wage rate W has to satisfy the inequality $W \geq \overline{W}$ with $\overline{W}$ given. We shall refer to $\underline{c}_t$ as the level of ***necessary personal consumption.*** The latter has not to correspond to a minimum level of subsistence and shall adjust rather to the prevailing niveau of socioeconomic progress of the society. The left-hand side in the foregoing equation hints to the potential allocation of economic surplus per available worker to consumption beyond the level $\underline{c}_t$ and to the expansion of the level of capital accumulation. Any increase of the level of personal consumption above $\underline{c}_t$ slows down directly the pace of capital accumulation. For instance, the level of capital denoted by $k_t^\diamond$ known as the Golden Value of capital per available worker which represents that level of k_t capable of sustaining the maximum level of consumption denoted $c_t^\diamond$ is simultaneously the saturation level of capital, i.e. that level at which $\dot{k}_t = 0$. The Golden Value of capital $k_t^\diamond$ is known to be an unstable equilibrium point.

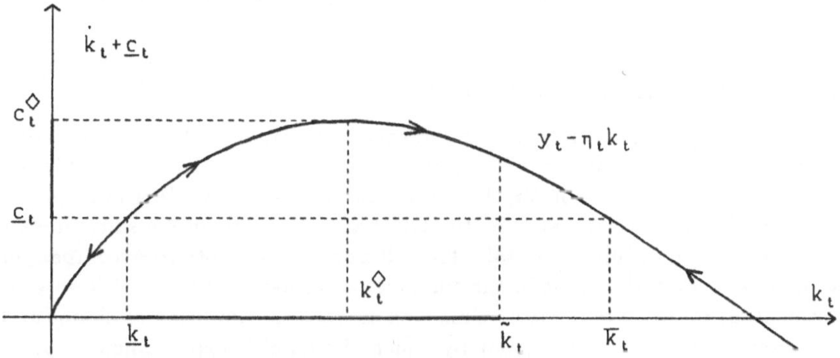

Figure 1: Phase diagram of accumulation of capital per available worker

However, it is the lower level of capital $\underline{k}_t$ attached to $\underline{c}_t$ the one that attracts our attention. Associated with $\underline{c}_t$ there is also the level of capital $\overline{k}_t$. These two levels of capital $\underline{k}_t$ and $\overline{k}_t$ are also equilibrium points, the first unstable while the latter stable as the arrows in Fig. 1 show. From the point of view of the economic policy the qualitative behaviour displayed in the diagram below places some restrictions on the potential admissible initial levels of capital per available worker. For instance, accumulation paths starting from levels of capital at the left of $\underline{k}_t$ are driven by the system dynamics to the zero level of capital. On the other hand, not any path starting from initial levels of capital at the right of $\underline{k}_t$ moves towards the upper level $\overline{k}_t$, since the Brownian motion acting on the accumulation path may still drive it out of the stable region. Finally, let us point out that the level of capital $\tilde{k}_t$ corresponds to that level associated with the average rate of return on capital beyond which further investments are not longer worth. This leaves us with a relevant range of capital k_t given by $[\underline{k}_t, \tilde{k}_t]$. All these features have to be taken into account designing a control strategy and point to the various difficulties of economic and mathematical nature inherent to the problem at stake.

2.3 Social supply price of investment

The search for a convenient level of investment leads to the question of the cost in terms of consumption that this policy entails as well as its effects on the level of employment. Therefore, we associate with the given technology a path $t \longrightarrow {}_{*}P_K(t)$ with ${}_{*}P_K(t)$ representing the so-called ***accounting price of investment*** which is given by the relationship

$$\begin{aligned} {}_{*}P_K(t) &= -\left(\frac{dc}{di}\right)_{\mathcal{R}}, \\ &= -\frac{\partial_\lambda c_t}{\partial_\lambda i_t}, \end{aligned} \tag{6}$$

which indicates how much consumption $c_t = c(k_t, \lambda_t)$ at any time t, under the given technology and institutions, has the economy to sacrifice in order to forward an additional unit of investment $i_t = i(k_t, \lambda_t)$ along the employment path $t \longrightarrow \lambda_t$. Therefore, ${}_{*}P_K$ represents a ***social supply price of investment.*** Institutions are represented here by the assumption $\mathcal{R}$ that says that, if unemployment prevails the level of aggregate output cannot be determined independently of the level of employment. Hence, we assume that the control board is able to require capitalists to reinvest any fraction s_K of capitalist income as long as it remains below $\bar{s}_K$. The latter being a result of negotiations between capitalists and the control board. See [8,10,14].

2.4 On the performance criterion

It is natural to ask about the willingness of the society to give up the required amount of consumption in order to implement the investment policy at issue. To answer this, we shall resort to a direct assessment based on the flow of instantaneous social utility of aggregate consumption per available labour generated by the alternative investment-consumption mix and on the social value of the terminal capital. This alternative amounts to accompanying any expansion of employment with a shift in the composition of national product from consumption to investment, or in other words from claims on current consumption to claims on future consumption, equal at least to the amount by which the employment-induced expansion of consumption exceeds the expansion of the social output. Any evaluation of this shift shall rely on the assumption of the existence of a ***system of social preferences*** and ***value judgements*** reflecting the prevailing views and attitudes towards poverty, unemployment, redistribution of consumption, social justice, etc. on the moulding of which the control board or government should participate to the limits of political strength and courage, see [14]. At any rate, a basic hypothesis of sociological theorizing says that the value-judgements individuals depend upon are in a large part the internalization of those inculcated in them by society.

For that reason we introduce a social welfare criterion J by means of a map $\alpha. \longrightarrow J(\alpha.)$ with J given by the functional

$$J(\alpha.) = E_{s,x}\left\{\int_0^T [U(c_t,\alpha_t) - U^\diamond]\,dt + g(k_T)\right\} \tag{7}$$

where $U(\cdot,\alpha_t)$ and $g(\cdot)$ are strict concave functionals of class $\mathbf{C}^1(\mathbf{R}^d \longrightarrow \mathbf{R}_-)$, and $T \in \mathbf{R}_+$ is a given terminal time. In (7) U describes the social utility and g measures the social value attached to the terminal capital, more explicitly, g penalizes terminal deviations from $k_t^\diamond$, a level of capital per available worker that once the economy attains it makes it able to sustain. Further, $U^\diamond$ stands for the social utiliy at the maximum level of sustainable consumption per available labour, i.e. $U - U^\diamond$ measures deviations from the bliss value $U^\diamond$, see [12]. The functional $J(\alpha.)$ shall reflect or convey a measure of social desirability of the alternative forwarded by the control decision $\alpha.$. Hence, our control problem can be formulated as follows. Consider the *completely observable control problem*

$$\sup_{\alpha\in\mathcal{A}} J(\alpha) \tag{8}$$

subject to eqs. (3) and (5) and appropriate constraints. In (8) $J(\alpha.)$ is given by (7) and $\mathcal{A}$ stands for the set of admissible controls α a concept related to the already mentioned constraints and further some mathematical requirements, see [10]. We shall use also the notation $U^\alpha(\cdot) = U(\cdot,\alpha_t)$.

3 Designing an optimal policy of labour allocation

Stochastic control theory delivers the necessary machinery by means of which to approach the problem stated in (8). At the risk of oversimplification, we shall recall the Maximum Principle as follows.

Stochastic Maximum Principle. A necessary and sufficient condition for the optimality of problem (8) is that there exist the adjoint processes $p = (p_t)_{t\geq 0}$ and $\rho = (\rho_t)_{t\geq 0}$ so that they satisfy

$$\begin{aligned} -dp &= \nabla_y H_t^{\alpha^*}(k_t,p_t,\rho_t)dt - \rho_t dB_t, \\ p_T &= \nabla_k g(k_T), \end{aligned} \tag{9}$$

where $H_t(k_t,p_t,\rho_t)$ is given by (12) and (18) below.

Before we begin constructing the optimal policy of labour allocation, let us recall that for any arbitrary but fixed time t, the drift and diffusion coefficients of eq. (3) or equivalently of eq. (4) in section 2.1 are given by

$$f^\alpha(t,k_t) = i_t - (\epsilon_t + \delta_t - \pi_t^2)k_t \tag{10}$$

$$\sigma^\alpha(t,k_t) = -\pi_t k_t. \tag{11}$$

Let us assume that the time horizon is given by $[0,T]$, with $T \in \mathbf{R}_+$ given. Then, applying the stochastic maximum principle due to [1] and criterion (7) we obtain the Hamiltonian

$$H_t^\alpha(k_t,p_t,\rho_t) = (U(c_t,\alpha_t) - U^\diamond) + p_t f^\alpha(t,k_t) + \rho_t\sigma^\alpha(t,k_t), \tag{12}$$

or equivalently in its more explicit form

$$H_t^\alpha(k_t,p_t,\rho_t) = (U^\alpha(c_t)-U^\circ)+p_t[i_t-(\epsilon_t+\delta_t-\pi_t^2)k_t] -\rho_t\pi_t k_t. \qquad (13)$$

The Hamiltonian in its form (12) or (13) articulates, in the form of isoquants, the existing system of social preferences and value-judgements and orders completely alternative combinations of investment i and consumption c. Therefore, abusing a little the notation we may write (13) simply as $H(i_t,c_t)$. Regarding questions related to the determination of such social orderings [17,18] are excellent references. On the other hand, the Hamiltonian as given by (12) represents total social utility at any time $t, t \in [0,T]$. We shall call the first and second term of (12) or (13), i.e. $U(c_t,\alpha_t)-U^\circ$ and $p_t f^\alpha(t,k_t)$, the potential and the kinetic social utility respectively, since they resemble the concepts of potential and kinetic energy in classical physics. The last term $-\rho_t\pi_t k_t$ accounts for the social disutility or social cost due to uncertainty or risk associated with changes generated by the control policy at issue.

The logic underlying the Hamiltonian shall be understood as follows. As we have seen, the social utility functional $U^\alpha - U^\circ$ is responsible for social preferences, attitudes, value-judgements and so forth. Thus, if the economy control policy is one under which the representative man is expected to postpone certain amount of current consumption, which amounts to giving up certain quantity of current utility, then the second term $p_t f^\alpha(t,k_t)$ transforms this amount of current consumption, in the form of potential future consumption or better in the form of investment, into future utility. The third term $\rho_t\pi_t k_t$ makes up for the utility changes due to adjustments to the new configuration.

3.1 Static characterization of the optimal control policy

Let us assume that an optimal control exists and denote it by α^*. Further, let $k^* = (k_t^*)_{t\geq 0}$, $c^* = (c_t^*)_{t\geq 0}$, $i^* = (i_t^*)_{t\geq 0}$ etc., denote the associated optimal trajectories. Then, using the stochastic maximun principle as in [10] we obtain for (12) the relationship

$$H_t(k_t^*,\alpha_t^*,p_t,\rho_t) = (U(c_t^*,\alpha_t^*)-U^\circ)+p_t f^{\alpha^*}(t,k_t^*)-\rho_t\pi_t k_t^*. \qquad (14)$$

Next,recalling that c_t and i_t are functionals of k_t, see (5,) and that the Hamiltonian has a maximum at $i^* = i_t^*$, then the control set R^1A is all of $\mathbf{R}_+$, and since H is differentiable in i^*, we must have

$$0 = \frac{dH_t}{di_t^*} = -\nabla_c U \cdot \left[(-\frac{dc^*}{di_t^*})\right]_{\mathcal{R}} + p_t.$$

Here, we have used the fact that the Hamiltonian associated with an optimal policy is constant. Taking into account eq. (6), we get

$$p_t = \nabla_c U(c_t^*,\alpha_t^*)._* P_K(t) \quad a.e.\ t, \quad P\otimes dt - a.s. \qquad (15)$$

Hence, eq. (15) holds for all $t\in[0,T]$ with possible exceptions on $dP\otimes dt$–null sets. For that reason it is a moment-to-moment relation known in dynamic economics as the *dynamic efficiency condition*. They says that $P-a.s.$ at any t the social utility of the representative man, derived from the decision of the economy to invest according to i^*, should equate the consumption utility loss of the representative man associated with the consumption she or he has to sacrifice in order to further investments as the control α^* requires. Even more, taking into account that $_*P_K(t) \geq 1$ for any $t\in[0,T]$ as a simple computation shows, the relationship (15) tells us that to forward the economic policy represented by α^* future utility claims, as described by p_t, shall be higher than the current utility associated with the fraction of current consumption that has to be postponed, i.e.

$$p_t \geq \nabla_c U(c_t^*, \alpha_t^*) \quad a.e.\ t, \quad P \otimes dt - a.s. \tag{16}$$

The dynamic efficiency condition given by eq. (15) amounts to the well-known tangency condition between the investment-consumption transformation functional and the associated utility substitution functional articulated by means of the family of Hamiltonians $H(i_t, c_t)$, or utility isoquants in the terminology of economic theory, which in turn defines a ***social demand price of investments*** measured in terms of current consumption. The tangency condition, which follows easily from eq. (15), can be written as

$$_{*}P_{\mathrm{K}}(t) = \frac{p_t}{\nabla_c U(c_t^*)} = \frac{\nabla_i H_t}{\nabla_c H_t} \quad a.e.\ t, \quad P \otimes dt - a.s. \tag{17}$$

Eq. (17) becomes evident recalling that $_{*}P_{\mathrm{K}}(t)$ is given by eq. (6) and taking into account eqs. (12) and (10) on the one hand, and eq. (13) on the other.

3.2 The Marglin-Pontryagin path of labour allocation

Further, we obtain in [10] that at the optimal control α^* the relationship

$$H_t(k_t^*, p_t, \rho_t) = \max_{\alpha \in \mathcal{A}} H_t^{\alpha}(k_t, p_t, \rho_t). \tag{18}$$

holds $a.e.\ t, \quad P \otimes dt - a.s.$. Hence, for fix t and taking into account the differentiability of $H_t^{\alpha}(k_t, p_t, \rho_t)$, see [10], one gets from (18) the following ***static first-order conditions of optimality*** that describe fully the following three phases the economy undergoes before entering a state of laissez-faire at which full employment prevails.

- **Phase I**

$$\text{If} \quad \lambda_t^* < 1, \quad \text{and} \quad s_{\mathrm{K}}^*(t) = \bar{s}_{\mathrm{K}}, \quad \text{then} \quad \frac{p_t}{\nabla_c U(c_t^*)} = \frac{(s_{\mathrm{K}} - s_{\mathrm{L}})W + (1 - s_{\mathrm{K}})y_t'}{(s_{\mathrm{K}} - s_{\mathrm{L}})W - s_{\mathrm{K}} y_t'} \quad \text{a.e. } t,\ P \otimes dt - \text{a.s.} \tag{19}$$

- **Phase II**

$$\text{If} \quad \lambda_t^* = 1, \quad \text{and} \quad s_{\mathrm{K}}^*(t) = \bar{s}_{\mathrm{K}}, \quad \text{then} \quad 1 \leq \frac{p_t}{\nabla_c U(c_t^*)} \leq \frac{(s_{\mathrm{K}} - s_{\mathrm{L}})W + (1 - s_{\mathrm{K}})y_t'}{(s_{\mathrm{K}} - s_{\mathrm{L}})W - s_{\mathrm{K}} y_t'} \quad \text{a.e. } t,\ P \otimes dt - \text{a.s.} \tag{20}$$

- **Phase III**

$$\text{If} \quad \lambda_t^* < 1, \quad \text{and} \quad s_{\mathrm{K}}^*(t) < \bar{s}_{\mathrm{K}}, \quad \text{then} \quad \frac{p_t}{\nabla_c U(c_t^*)} = 1 \quad \text{a.e. } t,\ P \otimes dt - \text{a.s.} \tag{21}$$

The Hamiltonian, together with the initial conditions and constraints, determines whether the economy finds itself in Phase I, II or III. As one can show, the economy optimally develops by moving from Phase I to Phase II and from Phase II to Phase III, when it starts from a capital intensity k_0 which is low enough but appropriately larger than $\underline{k}_t$. However, the economy not always has to begin with Phase I. A sufficiently large initial endowment of capital k_0 may put the economy also in Phase II or even in Phase III.

The phases obtained as the necessary conditions of (18) shall be interpreted as follows. First of all, the combination of events like employment, i.e. $\lambda_t^* = 1$, or unemployment, i.e. $\lambda_t^* < 1$, and a binding investment policy, i.e. $s_K^*(t) = \bar{s}_K$, or not i.e. $s_K^*(t) < \bar{s}_K$, characterizes the case to hold and this is indicated at the left-hand side under the corresponding phase. Then, according to the phase, if optimality dictates full employment or unemployment and a binding or a nonbinding investment policy, the relative social desirability of the couple $(s_K^*(t), \lambda_t^*)$ or alternatively of the resulting investment-consumption mix (i_t^*, c_t^*) has to be measured by means of the corresponding weight $_*P_K(t)$ resulting from the first-order conditions, associated with eqs. (18) and (17). A Marglin-Pontryagin path of labour allocation is one satisfying the eqs. (19) to (21).

3.3 On the dynamic characterization of the optimal control policy

At this stage we like to point out that the Maximum Principle as presented in [10] enables us splitting the intertemporal optimization problem (8) into a static, i.e. eq. (18), and a dynamic one. In other words, it allows a time decentralization of the decision process and this is extremely convenient for applications. We have considered the static features in the foregoing section. Concerning the dynamic aspects of optimality and the tranversality condition we shall refer to [9,10].

References

[1] Bensoussan, A. Stochastic Maximum Principle for Distributed Parameter Systems. *Journal of the Franklin Institute.* Vol. 315, No. 5/6, pp. 387-406. 1983

[2] Blanchard, Ph., Ph. Combe and W. Zheng. *Mathematical and Physical Aspects of Stochastic Mechanics.* Springer Verlag, Berlin. 1987

[3] Davis, M. H. A. and G. L. Gómez M. The semi-martingale approach to the optimal resource allocation in the controlled labour-surplus economy. In *Lecture Notes in Mathematics* Nr. 1250, pp. 36-74. Ed. by S. Albeverio et al. Springer Verlag, Berlin. 1987

[4] Foley, D. K. *Understanding Capital: Marx's Economic Theory.* Harvard University Press. Cambridge, Mass., 1986

[5] Frank, A. G. *World Accumulation 1492-1780.* Monthly Review Press, New York. 1978

[6] Galeano, E. *Open Veins of Latin America: Five Centuries of the Pillage of a Continent.* Monthly Review Press, New York. 1973

[7] Gómez M., G. L. Modelling the economic development by means of impulsive control techniques. *Mathematical Modelling in Sciences and Technology.* pp. 802-806. Eds. X. J. Avula and R. E. Kalman. Pergamon-Press, New York. 1984

[8] Gómez M., G. L. The intertemporal labour allocation inherent in the optimal stopping of the dual economy: the dynamic case. *Methods of Operations Research.* Vol. 49, pp. 523-543. 1985

[9] Gómez M., G. L. Discounted values and stochastic rates arising in control theory. *Methods of Operations Reasearch,* 57, pp. 379-392. 1987

[10] Gómez M., G. L. Attainability and Reversibility of a Golden Age for the Labour Surplus Economy: A Stochastic Variational Approach. In *Stochastic Processes in Physics and Engineering,* pp. 107-48. Ed. by Albeverio et al. D. Reidel Publishing Co. Dordrecht, 1988

[11] Gómez M., G. L. *Lectures on Economic Development.* Work in progress. 1988

[12] Koopmans, T. On the concept of optimal economic growth. *The Econometric Approach to Planning.* Rand McNally, Chicago. 1966

[13] Lichtenstein, P. M. *An Introduction to Post-Keynesian and Marxian Theories of Value and Price.* The Macmillan Press, London. 1983

[14] Marglin, S. A. *Value and Price in the Labour-Surplus Economy.* Oxford University Press, London. 1976

[15] Marglin, S. A. *Growth, Distribution and Prices.* Harvard University Press. Mass. 1984

[16] Nelson, E. *Dynamical Theories of Brownian Motion.* Princeton University Press. Princeton. 1967

[17] Sen, A. *Choice, Welfare and Measurement.* Basil Blackwell, Oxford. 1982

[18] Sen, A. *Resources, Values and Development.* Harvard University Press, Cambridge, Mass. 1984

[19] Wolff, R. D. and S. A. Resnick. *Economics: Marxian Versus Neoclassical.* The Johns Hopkins University Press. Baltimore. 1987

OLS–ESTIMATION AND RATIONALITY IN LINEAR MODELS WITH FORECAST FEEDBACK

Th. Kottmann
Institut für Ökonometrie und Operations Research
Universität Bonn
Adenauerallee 24 – 42
D–5300 Bonn 1

In macroeconomics one is often concerned with the dynamic behaviour of a quantity y whose evolution over time does not only depend (linearly) on exogenous and disturbance variables, but also on forecasts of present and/or future values of y. For instance in many models supply of and demand for and thus the price of a specific commodity at time t depends (among other things) on its expected price at time t+1.

Models of this kind are called linear models with forecast feedback (LMFF). The formal specification is given by

(1) $$y_t = x_t'm + \sum_{i=1}^{r} a_i F_i(y_{t+k_i}|I_t^i) + u_t \quad (t\geq 1)$$

with

$(x_t)_t$ observable n–dimensional stochastic process;

$(u_t)_t$ unobservable scalar disturbance process;

$m \in \mathbb{R}^n$; $a_i \in \mathbb{R}$ $(i = 1,\dots,r)$; $k_i \in \mathbb{N}_0$ $(i = 1,\dots,r)$;

I_t^i information at time t, the set of all variables that the economic agents actually use in calculating $F_i(y_{t+k_i}|I_t^i)$;

$F_i(y_{t+k_i}|I_t^i)$ (subjective) forecast of y_{t+k_i} based on information I_t^i.

This specification formalizes the situation where forecasts of different groups of economic agents are relevant who may differ in the corresponding sets of information available to them and in the corresponding forecast or planning horizons. The weighting coefficient a_i may then be interpreted as a measure of the (relative) size or importance of group i.

In order to complete model (1) one has to specify the employed information processes $(I_t^i)_t$ and the forecast terms $F_i(y_{t+k_i}|I_t^i)$. Since 1961 ([1]) LMFFs are analysed under the

Assumption of rational expectations:

$$I_t^i = \{x_t, x_{t-1},\dots,x_1; y_{t-1}, y_{t-2},\dots,y_1\}, \quad F_i(y_{t+k_i}|I_t^i) = EL(y_{t+k_i}|I_t) \quad \forall i \; \forall t$$

(EL denoting wide sense conditional expectations, cf. [2]), i.e. forecasts are assumed to be rational in the sense that exactly all relevant variables specified by (1) are used to obtain the statistically most efficient prediction.

In macroeconomic LMFFs the forecasting individuals are typical market participants without statistical or econometric abilities above average. Since on the other hand the calculation of rational expectations requires almost full knowledge of structure and parameters of (1), the assumption of rational expectations has to be (and actually is) justified by the argument that the forecasting agents will somehow learn to become rational.

Up to now, possible learning procedures based on ordinary least squares (OLS) estimation have been investigated under very restrictive stochastic assumptions only (e.g. $(x_t,u_t)_t$ iid., $r = 1$, $k_1 = 0$; cf. [3], [4]).

The aim of this paper is to specify a conceptionally simple learning (i.e. forecast) procedure which does not require any knowledge of the parameters of the LMFF and which under acceptable stochastic assumptions gives rational expectations in the limit.

This learning procedure is given by the

Assumption of OLS expectations:

The individuals forecasting by $F_i(y_{t+k_i}|I_t^i)$ start from the auxiliary model

(2) $$y_{t+k_i} = z_t^{i\prime}\gamma^i + v_t^i \quad ,$$

where

$(z_t^i)_t$ observable n_i–dim. stoch. process;

$(v_t^i)_t$ scalar iid. stoch. process with $Ev_t^i = 0$;

$\gamma^i \in \mathbb{R}^{n_i}$.

γ^i is estimated by γ_t^i using OLS:

(3) $$\gamma_t^i = \begin{cases} \left[\sum_{j=1}^{t-k_i-1} z_j^i z_j^{i\prime}\right]^{-1} \sum_{j=1}^{t-k_i-1} z_j^i y_{j+k_i} & \text{if } t \geq k_i+2,\ (.)^{-1} \text{ exists} \\ \in \mathbb{R}^{n_i} & \text{otherwise} \end{cases}$$

Then

(4) $$F_i(y_{t+k_i}|I_t^i) := z_t^{i\prime}\gamma_t^i$$

with

$$I_t^i = \{z_t^i, z_{t-1}^i, \ldots, z_1^i; y_{t-1}, \ldots, y_1\}.$$

Thus the forecasting agents do not realize or take into account that (1) is the correct model specification. Instead they adopt an ordinary linear model with explanatory variables $(z_t^i)_t$, and they estimate the unknown (hypothetical) parameter γ^i by the standard OLS method (which

would be adequate if the postulated specification was correct).
We shall use the following

Stochastic assumptions:

There exist matrices V^i $\forall i \leq r$, $V_{k_1,\ldots,k_r}$ and $W_{k_1,\ldots,k_r}$ and a vector $s_{k_1,\ldots,k_r}$ such that

(A.1) $$\frac{1}{t}\sum_{j=1}^{t} z_j^i z_j^{i\prime} \longrightarrow V^i \text{ a.s. } \forall i \leq r,\ V^i \text{ regular } \forall i \leq r;$$

(A.2) $$\frac{1}{t}\sum_{j=1}^{t} \left[\begin{bmatrix} z_j^1 \\ \vdots \\ z_j^r \end{bmatrix} (z_{j+k_1}^{1\prime},\ldots,z_{j+k_r}^{r\prime})\right] \longrightarrow V_{k_1,\ldots,k_r} \text{ a.s.};$$

(A.3) $$\frac{1}{t}\sum_{j=1}^{t} \begin{bmatrix} z_j^1 x'_{j+k_1} \\ \vdots \\ z_j^r x'_{j+k_r} \end{bmatrix} \longrightarrow W_{k_1,\ldots,k_r} \text{ a.s.};$$

(A.4) $$\frac{1}{t}\sum_{j=1}^{t} \begin{bmatrix} z_j^1 u_{j+k_1} \\ \vdots \\ z_j^r u_{j+k_r} \end{bmatrix} \longrightarrow s_{k_1,\ldots,k_r} \text{ a.s.}$$

If we assume

(A.5) $((z_t^1,\ldots,z_t^r,x_t,\ u_t)_t$ is a stationary, ergodic, square integrable stochastic process with $Ez_t^i z_t^{i\prime}$ regular for all $i \leq r$,

(A.1) – (A.4) hold with the limit quantities replaced by the corresponding theoretical moments.

These (ergodicity) assumptions reflect the kind of long–run stationarity of exogenous and auxiliary processes that is needed for our results. All we need are convergence properties of various empirical moments. Thus we use the same kind of assumptions as those used in standard linear regression theory when proving consistency of the OLS estimators.

For model (1) – (4) with the above assumptions we want to answer the following

Questions:

What are the possible limit parameters $\gamma \in \mathbb{R}^{n_1+\ldots+n_r}$ to which $(\gamma_t)_t = (\gamma_t^1,\ldots,\gamma_t^r)_t$ may converge pathwise with positive probability?

How may the possible limit parameters be interpreted in terms of rational expectations?

For what values of the weighting coefficients a_i can we actually prove almost sure convergence of the estimation processes?

We obtain the following

Results:

Generically, possible limit points of $(\gamma_t^1,...,\gamma_t^r)$ in the sense of pathwise convergence are uniquely determined. If the forecast feedback is not dominant, $(\gamma_t^1,...,\gamma_t^r)$ does converge a.s. to this unique limit parameter $(\gamma^1,...,\gamma^r)$.

This limit parameter is constrained rational in the following sense: If the forecasts in (1) are calculated using the limit parameters. i.e.

$$F_i(y_{t+k_i}|I_t^i) = z_t^{i\prime}\gamma^i \quad \forall i \leq r \text{ in } (1),$$

then $(y_t)_t$ satisfies

$$F_i(y_{t+k_i}|I_t^i) = EL(y_{t+k_i}|z_t^i) \quad \forall i \leq r.$$

Formally, we have the following

THEOREM:

(i) *Under (A.1) – (A.4) each* $\gamma = (\gamma^1,...,\gamma^r) \in \mathbb{R}^{n_1+...+n_r}$ *with* $P((\gamma_t^1,...,\gamma_t^r) \to (\gamma^1,...,\gamma^r)) > 0$ *satisfies*

$$\left[I - V_{k_1,...,k_r} \cdot \text{diag}(a_i(V^i)^{-1})\right] (\text{diag}(V^i)) \cdot \gamma = W_{k_1,...,k_r} m + s_{k_1,...,k_r}.$$

If $I - V_{k_1,...,k_r} \cdot \text{diag}(a_i(V^i)^{-1})$ *is regular, then*

$$\gamma = (\text{diag}(V^i))^{-1}\left[I - V_{k_1,...,k_r} \cdot \text{diag}(a_i(V^i)^{-1})\right]^{-1}(W_{k_1,...,k_r} m + s_{k_1,...,k_r})$$

is the unique possible limit point.

(ii) *Suppose (A.1) – (A.4) hold. Let* γ *be as above. If* $\text{Re}(\lambda) > 0$ *for all eigenvalues* λ *of* $\left[I - V_{k_1,...,k_r} \cdot \text{diag}(a_i(V^i)^{-1})\right]$, *then*

$$(\gamma_t^{1\prime},...,\gamma_t^{r\prime})' \to \gamma \text{ a.s.}$$

Convergence holds in particular if $\sum_{i=1}^{r} |a_i| < 1$.

(iii) *Suppose (A.5) and the stability assumption in (ii) hold. Let* $\gamma = (\gamma^{1\prime},...,\gamma^{r\prime})'$ *be as in (i).*

Define $(\bar{y}_t)_t$, *the process with limit expectations, by*

$$\bar{y}_t := x_t' m + \sum_{i=1}^{r} a_i z_t^{i\prime}\gamma^i + u_t.$$

Then

$$z_t^{i\prime}\gamma^i = EL(\bar{y}_{t+k_i}|z_t^i) \text{ for all } i,$$

and $(\bar{y}_t)_t$ *satisfies*

$$\bar{y}_t = x_t'm + \sum_{i=1}^{r} a_i EL(\bar{y}_{t+k_i} | z_t^i) + u_t.$$

The complete proof of the Theorem can be found in [5] together with a discussion of (economically) interesting special cases. Here we will only sketch the proof of the convergence result (ii).

We will make use of the following result of H. Walk ([6]) on the a.s. convergence of Robbins–Monro type stochastic approximation algorithms with ergodic inputs (given here in a deterministic formulation):

THEOREM:

Let $S, S_1, S_2, \ldots$ *be real* $n \times n$*–matrices*, $q, q_1, q_2, \ldots \in \mathbb{R}^n$. *Define the sequence* $(x_t)_t$ *in* $\mathbb{R}^n$ *recursively by*

$$x_1 \in \mathbb{R}^n \textit{ arbitrary}, \quad x_{t+1} := x_t - \tfrac{1}{t}(S_t x_t - q_t).$$

Suppose

(a) $\frac{1}{t}(S_1 + \ldots + S_t) \to S$ *for* $t \to \infty$;

(b) $\frac{1}{t}(q_1 + \ldots + q_t) \to q$ *for* $t \to \infty$;

(c) *all eigenvalues of* S *have strictly positive real parts.*

Then $x_t \to S^{-1}q$ *for* $t \to \infty$.

Actually, Walk proved a considerably more general result, but this version will suffice for our purposes.

Sketch of proof of (ii):

Define the following variables:

(5) $c_t^i := \frac{1}{t} \sum_{j=1}^{t} z_j^i y_{j+k_i} \qquad j=1,\ldots,r;$

(6) $B_t^i := \frac{1}{t} \sum_{j=1}^{t} z_j^i z_j^{i\prime} \qquad j=1,\ldots,r.$

Then $\gamma_t^i = (B_{t-k_i-1}^i)^{-1} c_{t-k_i-1}^i$.

We show $(\gamma_t^{1\prime}, \ldots, \gamma_t^{r\prime})' \to \gamma$ on Ω_1, the set of paths satisfying (A.1) – (A.4). Since convergence of B_t^i for $t \to \infty$ is assured on Ω_1 by (A.1) for all i, it will suffice to prove convergence of $(c_t^i)_t$ for all i. This will be achieved by representing (c_t^i) in the recursive form of Walk's theorem and by verifying its assumptions.

(5) gives the recursive representation

$$\begin{aligned} c_t^i &= (1-\tfrac{1}{t})\, c_{t-1}^i + \tfrac{1}{t} z_t^i y_{t+k_i} \\ &= (1-\tfrac{1}{t}) c_{t-1}^i + \tfrac{1}{t} z_t^i \Big(x'_{t+k_i} m + \sum_{l=1}^{r} a_l z^{l\prime}_{t+k_i} \gamma^l_{t+k_i} + u_{t+k_i}\Big) \\ &= (1-\tfrac{1}{t}) c_{t-1}^i + \tfrac{1}{t} \Big(z_t^i x'_{t+k_i} m + \sum_{l=1}^{r} a_l z_t^i z^{l\prime}_{t+k_i} (B^l_{t-1})^{-1} c^l_{t-1} + z_t^i u_{t+k_i}\Big). \end{aligned}$$

Thus

$$\begin{bmatrix} c_t^1 \\ c_t^2 \\ \vdots \\ c_t^r \end{bmatrix} = \begin{bmatrix} c_{t-1}^1 \\ c_{t-1}^2 \\ \vdots \\ c_{t-1}^r \end{bmatrix} - \frac{1}{t} \left[S_t \begin{bmatrix} c_{t-1}^1 \\ c_{t-1}^2 \\ \vdots \\ c_{t-1}^r \end{bmatrix} - q_t \right]$$

with

$$S_t = I - \left[\begin{bmatrix} z_t^1 \\ z_t^2 \\ \vdots \\ z_t^r \end{bmatrix} (z^{1\prime}_{t+k_1}, z^{2\prime}_{t+k_2}, \ldots, z^{r\prime}_{t+k_r}) \begin{bmatrix} a_1 (B^1_{t-1})^{-1} & & & \\ & a_2 (B^2_{t-1})^{-1} & & \\ & & \ddots & \\ & & & a_r (B^r_{t-1})^{-1} \end{bmatrix} \right]$$

and

$$q_t = \begin{bmatrix} z_t^1 x'_{t+k_1} \\ z_t^2 x'_{t+k_2} \\ \vdots \\ z_t^r x'_{t+k_r} \end{bmatrix} m + \begin{bmatrix} z_t^1 u_{t+k_1} \\ z_t^2 u_{t+k_2} \\ \vdots \\ z_t^r u_{t+k_r} \end{bmatrix}.$$

The assumptions of Walk's Theorem are satisfied on Ω_1:

(a): Define

$$\tilde{S}_j = I - \left[\begin{bmatrix} z_j^1 \\ z_j^2 \\ \vdots \\ z_j^r \end{bmatrix} (z^{1\prime}_{j+k_1}, z^{2\prime}_{j+k_2}, \ldots, z^{r\prime}_{j+k_r}) \begin{bmatrix} a_1 (V^1)^{-1} & & & \\ & a_2 (V^2)^{-1} & & \\ & & \ddots & \\ & & & a_r (V^r)^{-1} \end{bmatrix} \right]$$

Then $\frac{1}{t}(\tilde{S}_1 + \ldots + \tilde{S}_t) \to I - V_{k_1,\ldots,k_r} \cdot \mathrm{diag}(a_i (V^i)^{-1})$.

Now $\|\frac{1}{t}(\tilde{S}_1 + \ldots + \tilde{S}_t) - \frac{1}{t}(S_1 + \ldots + S_t)\| \leq$

$$\leq \frac{1}{t} \sum_{j=1}^{t} \left\| \begin{bmatrix} z_j^1 \\ z_j^2 \\ \vdots \\ z_j^r \end{bmatrix} \right\| \left\| \begin{bmatrix} z^1_{j+k_1} \\ z^2_{j+k_2} \\ \vdots \\ z^r_{j+k_r} \end{bmatrix} \right\| \|\mathrm{diag}\,(a_i((B^i_{j-1})^{-1} - (V^i)^{-1}))\|$$

By (A.1), $\|\mathrm{diag}\,(a_i((B^i_{j-1})^{-1} - (V^i)^{-1}))\| \to 0$ for $j \to \infty$, and $\|\frac{1}{t}(\tilde{S}_1 + \ldots + \tilde{S}_t) - \frac{1}{t}(S_1 + \ldots + S_t)\|$ $\to 0$ for $t \to \infty$, thus (a);

(b): by (A.3) and (A.4);

(c): by assumption in (ii).

This proves the first part of (ii). Now let $M := V_{k_1,\dots,k_r} \cdot \mathrm{diag}(a_i(V^i)^{-1})$. By employing an appropriate equivalence transformation and repeatedly applying the Cauchy–Schwarz inequality we prove:

(*) If $\Sigma|a_i| < 1$, then for all eigenvalues λ of M we have $|\lambda| < 1$.

The eigenvalues of $I - M$ are $1 - \lambda$, λ eigenvalue of M; thus with (*), $\mathrm{Re}(1-\lambda) = 1 - \mathrm{Re}(\lambda) \geq 1 - |\lambda| > 0$.

Let $\alpha_1,\dots,\alpha_r$ be arbitrary positive real numbers (to be chosen appropriately below).Since V^i is positive definite symmetric for $i = 1,\dots,r$, there are regular matrices Γ^i with $\Gamma^{i\prime}\Gamma^i = (V^i)^{-1}$. Then M is equivalent to $\tilde{M}$ with

$$\tilde{M} = \mathrm{diag}(\alpha_i\Gamma^i)\cdot M\cdot \mathrm{diag}((\alpha_i\Gamma^i)^{-1}) = \mathrm{diag}(\alpha_i\Gamma^i)\cdot V_{k_1,\dots,k_r}\cdot \mathrm{diag}(\tfrac{a_i}{\alpha_i}\Gamma^{i\prime})$$

$$= \lim \frac{1}{t}\sum_{j=1}^{t}\left[\begin{bmatrix}\alpha_1\Gamma^1 z_j^1\\ \vdots \\ \alpha_r\Gamma^r z_j^r\end{bmatrix}\left(\tfrac{a_1}{\alpha_1}(\Gamma^1 z^1_{j+k_1})',\dots,\tfrac{a_r}{\alpha_r}(\Gamma^r z^r_{j+k_r})'\right)\right]$$

$$= \lim \frac{1}{t}\sum_{j=1}^{t}\left[\begin{bmatrix}\alpha_1\tilde{z}_j^1\\ \vdots \\ \alpha_r\tilde{z}_j^r\end{bmatrix}\left(\tfrac{a_1}{\alpha_1}\tilde{z}^{1\prime}_{j+k_1},\dots,\tfrac{a_r}{\alpha_r}\tilde{z}^{r\prime}_{j+k_r}\right)\right]$$

with $\tilde{z}^i_j = \Gamma^i z^i_j$.

We have $\lim \frac{1}{t}\sum_{j=1}^{t} \tilde{z}^i_j\tilde{z}^{i\prime}_j = \Gamma^i(\lim \frac{1}{t}\sum_{j=1}^{t} z^i_j z^{i\prime}_j)\Gamma^{i\prime} = \Gamma^i V^i \Gamma^{i\prime} = I_{n_j\times n_j}$ by (A.1) and definition of Γ^i $\forall i$.

Let $\lambda \in \mathbb{C}$ be any eigenvalue, $x = (x^{1\prime},\dots,x^{r\prime})' \in \mathbb{C}^{n_1+\dots+n_r}$ an eigenvector for λ of length 1. Then $\bar{x}'\tilde{M}x = \lambda\bar{x}'x = \lambda$, i.e.

$$\lambda = \lim \frac{1}{t}\sum_{j=1}^{t}\Big(\sum_{i=1}^{r}\alpha_i\bar{x}^{i\prime}\tilde{z}^i_j\Big)\cdot\Big(\sum_{p=1}^{r}\frac{a_p}{\alpha_p}x^{p\prime}\tilde{z}^p_{j+k_p}\Big) =$$

$$= \sum_{i,p=1}^{r} a_p\frac{\alpha_i}{\alpha_p}\cdot\lim\frac{1}{t}\sum_{j=1}^{t}(\bar{x}^{i\prime}\tilde{z}^i_j)(x^{p\prime}\tilde{z}^p_{j+k_p}).$$

By Cauchy–Schwarz inequality

$$\frac{1}{t}\sum_{j=1}^{t}(\bar{x}^{i\prime}\tilde{z}^i_j)(x^{p\prime}\tilde{z}^p_{j+k_p}) \leq \sqrt{\frac{1}{t}\sum_{j=1}^{t}\bar{x}^{i\prime}\tilde{z}^i_j\tilde{z}^{i\prime}_j x^i}\cdot\sqrt{\frac{1}{t}\sum_{j=1}^{t}\bar{x}^{p\prime}\tilde{z}^p_{j+k_p}\tilde{z}^{p\prime}_{j+k_p}x^p} =$$

$$= \sqrt{\bar{x}^{i\prime}\Big(\frac{1}{t}\sum_{j=1}^{t}\tilde{z}^i_j\tilde{z}^{i\prime}_j\Big)x^i}\cdot\sqrt{\bar{x}^{p\prime}\Big(\frac{1}{t}\sum_{j=1}^{t}\tilde{z}^p_{j+k_p}\tilde{z}^{p\prime}_{j+k_p}\Big)x^p}.$$

Thus $\left|\lim\frac{1}{t}\sum_{j=1}^{t}(\bar{x}^{i\prime}\tilde{z}^i_j)(x^{p\prime}\tilde{z}^p_{j+k_p})\right| \leq \sqrt{\bar{x}^{j\prime}x^j}\cdot\sqrt{\bar{x}^{l\prime}x^l}$.

For $\epsilon > 0$ arbitrary define $a_i^+ := |a_i| + \epsilon$ and choose $\alpha_i = \sqrt{a_i^+}$. Then we have

$$|\lambda| \leq \sum_{i,p=1}^{r} |a_p| \cdot \frac{\alpha_i}{\alpha_p} \cdot \sqrt{\bar{x}^{i\prime} x^i} \cdot \sqrt{\bar{x}^{p\prime} x^p} \leq \sum_{i,p=1}^{r} a_p^+ \cdot \frac{\sqrt{a_i^+}}{\sqrt{a_p^+}} \cdot \sqrt{\bar{x}^{i\prime} x^i} \cdot \sqrt{\bar{x}^{p\prime} x^p} =$$

$$= \left[\sum_{i=1}^{r} \sqrt{a_i^+} \sqrt{\bar{x}^{i\prime} x^i}\right]^2 \leq \left(\sum_{i=1}^{r} a_i^+\right)\left(\sum_{i=1}^{r} \bar{x}^{i\prime} x^i\right) = \sum_{i=1}^{r} a_i^+ = \sum_{i=1}^{r} |a_i| + r\epsilon,$$

again by Cauchy–Schwarz inequality.

Thus if $\sum_{i=1}^{r} |a_i| = \delta < 1$, for $\epsilon < \frac{1-\delta}{r}$ we have $|\lambda| < 1$, qed.

By (iii) of the Theorem, the limit parameter is only constrained rational since the assumption of rational expectations is only satisfied with a modified information process. But for some configurations (full) rationality in the limit is possible. Suppose for instance that $(x_t)_t$ is a stable stationary AR(p)–process and that each z_t^i consists of sufficiently many lagged x–values, i.e.

$$z_t^i = (x_t, x_{t-1}, \ldots, x_{t-p_i-1})' \text{ with } p_i \geq \max\{k_1, \ldots, k_r, p\} \quad \forall\, i.$$

Then the limit parameter is rational.

It should be noted that in contrast to all existing approaches to be found in the literature our methods and results can be generalized to simultaneous equations LMFFs, i.e. to the case of a multivariate y in (1) where instead of the weighting parameters a_i weighting matrices A_i appear in the feedback part of (1) (cf. [7]). From the view of our starting point this is quite satisfactory since many macroeconomic models with forecast feedback are simultaneous equations models (cf. [8]).

However, unfortunately our approach does not apply to LMFFs with lagged endogenous variables either in model equation (1) or in the auxiliary models. This is because in such models the ergodicity asssumptions (A.1) – (A.4) must be proved for the endogenous process which is just as difficult as to prove convergence of the parameter estimates. The problem whether for these models OLS estimation forecasts show any stable asymptotic behaviour is up to now completely open.

<u>References:</u>

[1] J.F. Muth (1961): "Rational Expectations and the Theory of Price Movements", *Econometrica* 29, 315–335.

[2] J.L. Doob (1953): *Stochastic Processes.* New York: Wiley & Sons.

[3] M.M. Bray, N.E. Savin (1986): "Rational Expectations Equilibria, Learning and Model Specification", *Econometrica* 54, 1129–1160.

[4] C. Fourgeaud, C. Gourieroux, J. Pradel (1986): "Learning Procedures and Convergence to Rationality", *Econometrica* 54, 845–868.

[5] Th. Kottmann (1988): "OLS–Estimation and Rationality in Linear Models with Forecast Feedback", *Discussion Paper B–96*, Universität Bonn.

[6] H. Walk (1985): "Almost Sure Convergence of Stochastic Approximation Processes", *Statistics and Decisions, Suppl. Issue* No. 2, 137–141.

[7] Th. Kottmann (1989): "Simultaneous Equations Linear Models with Forecast Feedback", *Discussion Paper B–106*, Universität Bonn.

[8] M.H. Pesaran (1988): *The Limits to Rational Expectations.* New York: Basil Blackwell.

INVARIANCE OF CONES AND COMPARISON RESULTS FOR SOME CLASSES OF DIFFUSION PROCESSES

Pawel Kröger
Mathematisches Institut der Universität
Bismarckstr. 1 1/2, D-8520 Erlangen

0. Introduction

Let $L = L_t + \frac{\partial}{\partial t} = \sum_{i,j=1}^{d} a_{ij}(t,x)\frac{\partial^2}{\partial x_i \partial x_j} + \sum_{i=1}^{d} b_i(t,x)\frac{\partial}{\partial x_i} + \frac{\partial}{\partial t}$

be a second-order parabolic differential operator in nondivergence form. We will generally assume that the coefficients a_{ij} and b_i of L are continuous functions from $[0,\infty) \times \mathbb{R}^d$ to $\mathbb{R}$ and that there exists a positive constant δ such that

$$\sum_{i,j=1}^{d} a_{ij}(t,x)\xi_i\xi_j \geq \delta \cdot |\xi|^2 \quad \text{for all} \quad (t,x) \in [0,\infty) \times \mathbb{R}^d, \quad \xi \in \mathbb{R}^d .$$

By [S/V], Theorem 10.1.3, for each starting point $(t,x) \in [0,\infty) \times \mathbb{R}^d$ there exists at most one solution of the martingale problem (see [S/V], Section 6.0). To ensure the existence of a solution of the martingale problem we impose on the coefficients of L the following (rather crude) condition which prevents explosion (see [S/V], Theorem 10.2.2):

There exists a positive constant C with

$$|a_{ij}(t,x)| \leq C \cdot (1 + |x|^2) ,$$
$$\langle x, (b_i(t,x))_i \rangle \leq C \cdot (1 + |x|^2)$$

for all (t,x) and i,j.

Under these assumptions the family of all solutions of the martingale problems with different starting points defines a strong Markov process.

Notation: Given $t_1,t_2 \in [0,\infty)$ with $t_1 \leq t_2$ and a continous real valued function f on $\mathbb{R}$ which is bounded from below by an appropriate affine linear function, we define the function $P^{t_1,t_2}f$ from $\mathbb{R}^d$ to $\mathbb{R} \cup \{+\infty\}$ by

$$P^{t_1,t_2}f(x) := E_{t_1,x}[f(t_2,x(t_2))] \quad \text{for each} \quad x \in \mathbb{R}^d ;$$

here $E_{t_1,x}$ denotes the expectation operator with respect to the unique solution $P_{t_1,x}$ of the martingale problem starting at (t_1,x). Given another parabolic differential operator $\tilde{L}$ with similar conditions on the coefficients, we define $\tilde{P}^{t_1,t_2}f$ in an analogous way.

The aim of this paper is to find conditions on $L,\tilde{L}$ and an appropriate set $\tilde{K}$ of functions on $\mathbb{R}^d$ such that

$$(L_t - \tilde{L}_t)f \underset{(\text{resp.},\leq)}{\geq} 0 \quad \text{for all } f \in \tilde{K} \text{ and all } t \in [0,\infty)$$

implies

$$P^{t_1,t_2}f \underset{(\text{resp.},\leq)}{\geq} \tilde{P}^{t_1,t_2}f \quad \text{for all } f \in \tilde{K} \text{ and all } t_1,t_2 \in [0,\infty)$$

with $t_1 \leq t_2$. Actually, we will spend the main efforts to the task of showing that we can choose a convex cone $\tilde{K}$ in such a way that the transition maps $\tilde{P}^{t_1,t_2}$ leave the set $\tilde{K}$ invariant if $\tilde{L}$ satisfies appropriate conditions which are not too restrictive. The proofs of the comparison results of the above type will be accomplished by an application of a comparison lemma for diffusion processes (see [K] and [K1], cf. also [P]).

Applying the above results for appropriate functions f, it is possible to give estimates for transition probabilities $P(t_1,x;t_2,\Gamma)$ with $\Gamma \subset \mathbb{R}^d$. Combining these estimates with results about the local behaviour of solutions of parabolic differential equations (we could apply for instance the Harnack inequality proven by Krylov and Safonov in [K/S]), we obtain pointwise estimates for solutions of parabolic differential equations.
We emphasize that we do not need any assumptions on the moduli of continuity of the coefficients a_{ij} and b_i of L.

Previous papers on comparison of diffusion processes only treat the case that both processes have the same diffusion term (cf. Skorochod's result in [Sk] for $d=1$) or the case that one of the diffusion processes is closely related to the Brownian motion (cf. Aronson's estimates for the fundamental solution of a parabolic equation in [A]; cf. also Hajek's generalization [H] of Skorochod's result [Sk] to pairs of one-dimensional diffusion processes under the condition that one of them is related to the Brownian motion via a random time change). Our intention is to make an attempt to remove some of the above restrictions.

1. Convex combinations of differential generators

The following theorem enables us under some circumstances to treat the diffusion and the drift term of the process separately.

Theorem 1: Assume that the coefficients of L_t and $\tilde{L}_t$ are uniformly bounded. Let K be a closed subcone of the space of all continous functions from $\mathbb{R}^d$ to $\mathbb{R}$ equipped with the topology of pointwise convergence. Suppose that K is invariant with respect to the transition maps P^{t_1,t_2} and $\tilde{P}^{t_1,t_2}$ for all t_1,t_2. Let λ and $\tilde{\lambda}$ be nonnegative constants. Then K is invariant with respect to the transition maps of the (unique) diffusion process with differential generator

$$\lambda L_t + \tilde{\lambda}\tilde{L}_t + \frac{\partial}{\partial t} \quad .$$

Remark: We do not need the assumption $\lambda + \tilde{\lambda} = 1$, thus scalar multiplication of the differential generator by a constant factor is included.

Sketch of the proof: Given t_1,t_2 with $t_1 \leq t_2$, set $T := t_2 - t_1$. Obviously K is invariant with respect to the composite map

$$P^{t_1,t_1+\lambda\cdot\frac{T}{n}}\tilde{P}^{t_1,t_1+\tilde{\lambda}\cdot\frac{T}{n}}P^{t_1+\frac{T}{n},t_1+(1+\lambda)\cdot\frac{T}{n}}\tilde{P}^{t_1+\frac{T}{n},t_1+(1+\tilde{\lambda})\cdot\frac{T}{n}}$$

$$\cdots P^{t_2-\frac{T}{n},t_2-(1-\lambda)\cdot\frac{T}{n}}\tilde{P}^{t_2-\frac{T}{n},t_2-(1-\tilde{\lambda})\cdot\frac{T}{n}}$$

for each $n \in \mathbb{N}$. The assertion follows from [S/V], Theorem 11.3.3 if we let n tend to infinity.

2. The one-dimensional case

In this section we will always assume that the dimension of the underlying space $\mathbb{R}^d$ is equal to 1.

Theorem 2: Suppose that the function $x \to \tilde{b}(t,x)$ is convex for each $t \in [0,\infty)$. Then the cone of all increasing convex functions on $\mathbb{R}$ with at most polynomial growth at infinity is invariant with respect to the transition maps $\tilde{P}^{t_1,t_2}$ for all t_1,t_2.

In order to prove the theorem we need the following auxiliary result about Markov chains on appropriate subsets of $\mathbb{R}$. Once Lemma 3 is established, the proof of the theorem will be accomplished using Stroock and Varadhan's limit theorems (see [S/V], Section 11.2).

For each constant α with $1 < \alpha \le 2$ we define the set S_α by

$$S_\alpha := \{\pm\alpha^k \mid k \in \mathbb{N}_0\} \cup \{k/[\tfrac{1}{\alpha-1}] \mid k \in \mathbb{Z} \text{ with } |k| < [\tfrac{1}{\alpha-1}]\} ;$$

here $[\lambda]$ stands for the integer part of the real number λ.
Given $n \in S_\alpha$, we set $n^+ := \min\{m \in S_\alpha \mid m > n\}$ and $n^- := \max\{m \in S_\alpha \mid m < n\}$.

<u>Lemma 3</u>: Let $X = \{X_0, X_1, X_2, \ldots\}$ be a Markov chain on S_α such that $P(X_{i+1} = n \mid X_i = n) \ge \frac{3}{4}$ and $P(X_{i+1} = m \mid X_i = n) = 0$ for all $m, n \in S_\alpha$ with $m \notin \{n^-, n, n^+\}$. Suppose that $P^{i,i+1}f$ is convex for every increasing affine linear function f on S_α and all $i \in \mathbb{N}_0$.
Then $P^{i,j}f$ is convex for every increasing convex function f on $\mathbb{R}$ and all $i, j \in \mathbb{N}_0$ with $i \le j$.

<u>Sketch of the proof</u>: By induction we can reduce the assertion to the case $j-i = 1$. Since each inreasing convex function on S_α can be represented as the pointwise limit of a sequence of convex combinations of functions of the type $x \to (x-n)_+$ for $n \in S_\alpha$ ($(x-n)_+$ is by definition $\max\{0, (x-n)\}$) and some constant functions, we may and will assume that the function f under consideration is the function $x \to (x-n)_+$ for some $n \in S_\alpha$. The assumptions on X yield

$P^{i,i+1}f(m) = 0$ for $m \in S_\alpha$ with $m < n$,

$P^{i,i+1}f(m) = P^{i,i+1}\mathrm{id}(m)$ for $m \in S_\alpha$ with $m > n$ (id denotes the function $x \to x$)

and

$$P^{i,i+1}f(n) \le \frac{1}{4}\cdot n^+ \le \frac{n-n^-}{n^+-n^-}\cdot\frac{3}{4}\cdot n^+ \le \frac{n-n^-}{n^+-n^-}\cdot P^{i,i+1}f(n^+) + \frac{n^+-n}{n^+-n^-}\cdot P^{i,i+1}f(n^-) .$$

Hence $P^{i,i+1}f$ is convex.

Now we turn to the task of proving Theorem 2. Since $\tilde{b}$ is increasing, for each $x > 0$ holds $\langle\tilde{b}(t,x),x\rangle \ge -\langle\tilde{b}(t,-x),-x\rangle$ and hence $|\langle\tilde{b}(t,x),x\rangle| \le C\cdot(1+|x|^2)$ for all $(t,x) \in [0,\infty)\times\mathbb{R}$. We restrict ourselves to a bounded time interval $[0,T]$. Without loss of generality we may assume that $|\tilde{b}(t,x)| \le C$ for all $t \in [0,T]$ and $x \in [-1,1]$ and that $C \ge 1$.
For each α with $1 < \alpha < 2$ we choose a positive number h with $h < \frac{(\alpha-1)^2}{100\cdot C}$.
Given $(k\cdot h, n) \in [0,T]\times\mathbb{R}$ with $k \in \mathbb{N}_0$ and $n \in S_\alpha$, let $\Pi_h((k\cdot h,n),.)$ be the probability measure which is concentrated on the set $\{((k+1)\cdot h, m) \mid m = n^-, n, n^+\}$ and satisfies

$$\frac{1}{h}\int_{\mathbb{R}} (y-n)^2\ \Pi_h((k\cdot h,n),((k+1)\cdot h,dy)) = a(k\cdot h,n) + 2\cdot C\cdot(\alpha-1)\cdot(1+n^2)\ ,$$

$$\frac{1}{h}\int_{\mathbb{R}} (y-n)\ \Pi_h((k\cdot h,n),((k+1)\cdot h,dy)) = b(k\cdot h,n)$$

(cf. [S/V], p. 267 below).

Since $h < \frac{(\alpha-1)^2}{100\cdot C}$, we get

$$\Pi_h((k\cdot h,n),((k+1\cdot h,n)) > \frac{3}{4}.$$

Thus we can apply Lemma 3 to the Markov chain which is defined by the family of transition probabilities $\Pi_h((k\cdot h,n),.)$ for all $k \in \mathbb{N}_0$ with $k\cdot h \leq T$ and all $n \in S_\alpha$. The assertion of the theorem follows from [S/V], Theorem 11.2.3, if we let α tend to 1 and h tend to 0.

Theorem 4: Let $d=1$ and let L and $\tilde{L}$ be parabolic differential operators such that

$$a(t,x) \leq \tilde{a}(t,x) \quad\text{and}\quad b(t,x) \leq \tilde{b}(t,x) \quad \text{for all} \quad t\in[0,\infty),\ x\in\mathbb{R}.$$

Suppose that the function $x \to b(t,x)$ is convex for each t or that the function $x \to \tilde{b}(t,x)$ is convex for each t.
Then $P^{t_1,t_2}f \leq \tilde{P}^{t_1,t_2}f$ for every increasing convex function f on $\mathbb{R}$ with at most polynomial growth at infinity and all t_1,t_2.

Theorem 4 can be proved in a similar way as Theorem 2 using the following auxiliary result.

Lemma 5: Let X be a Markov chain which satisfies the conditions of Lemma 3. Let $\tilde{X}$ be another Markov chain on the same subset S_α of $\mathbb{R}$ such that $P^{i,i+1}f \underset{(\text{resp.},\geq)}{\leq} \tilde{P}^{i,i+1}f$ for each increasing convex function f on S_α and all $i \in \mathbb{N}_0$. Then $P^{i,j}f \underset{(\text{resp.},\geq)}{\leq} \tilde{P}^{i,j}f$ for each increasing convex function f on S_α and all i,j.

Lemma 5 immediately follows from Lemma 3 by induction.

With the same method as above we can also prove that the cone of all convex functions (resp., increasing functions) with at most polynomial growth at infinity is invariant under the transition maps P^{t_1,t_2} presupposed that $\tilde{b} \equiv 0$ (resp., without special assumptions on $\tilde{a}$ or $\tilde{b}$), (cf. also [N] and [W], Section 27 for invariance results for parabolic differential equations). Again, we can derive from these invariance results related comparison results.

3. The multidimensional case

The main purpose of the present section is to characterize those diffusion processes on $\mathbb{R}^d$ such that all transition maps $\tilde{P}^{t_1,t_2}$ leave the cone of all convex functions from $\mathbb{R}^d$ to $\mathbb{R}$ with at most polynomial growth at infinity invariant. First we will derive necessary conditions on the coefficients $\tilde{a}_{ij}$, $\tilde{b}_i$. For technical reasons we will restrict ourselves to the case $\tilde{a}_{ij}, \tilde{b}_i \in C^2(\mathbb{R}^d)$.

The fact that $\tilde{P}^{t_1,t_2}f$ is convex for each affine linear function f on $\mathbb{R}^d$ immediately yields a condition on the drift term. Actually, since $-f$ is affine linear as well, we obtain that $\tilde{P}^{t_1,t_2}f$ is affine linear. Hence, $\lim_{\varepsilon \downarrow 0} \frac{1}{\varepsilon} (\tilde{P}^{t,t+\varepsilon}f - f)$ is affine linear for every t and every affine linear f. Hence, according to the definition of the differential generator, $\langle \tilde{b}(t,.), \nabla f \rangle$ is affine linear with respect to the space variable for every t and every affine linear f. In other words, $x \to \tilde{b}(t,x)$ is affine linear for any t. It is easy to see that a drift term of this type does not influence the invariance or non-invariance of the cone of all convex functions (with the usual growth condition) under the transition maps. In order to simplify the notation, in the sequel we will restrict ourselves to the case $\tilde{b} \equiv 0$.

Now we are going to find necessary conditions on the behaviour of the diffusion coefficients $\tilde{a}_{ij}$ in a neighbourhood of a fixed point, say $t=1$ and x equal to the origin of the Cartesian coordinate system in $\mathbb{R}^d$. Let a unit vector $z \in \mathbb{R}^d$ and a convex function f be given such that the directional derivative of the second order f_{zz} vanishes at the origin. From the invariance of the cone of all convex functions follows that

$$\left. \frac{\partial (\tilde{P}^{t,1}f)_{zz}}{\partial t} \right|_{(1,0)} \leq 0 .$$

Differentiating Kolmogorov's backward equation twice with respect to z, we are led to

$$\left(\mathrm{trace}((\tilde{a}_{ij})Hf)\right)_{zz}\Big|_{(1,0)} \geq 0 \quad ;$$

here Hf stands for the Hessian matrix of the function f. Thus

$$\mathrm{trace}\left((\tilde{a}_{ij})_{zz}Hf + 2\cdot(\tilde{a}_{ij})_z H(f_z) + (\tilde{a}_{ij})H(f_{zz})\right)\Big|_{(1,0)} \geq 0 .$$

Now we are going to specify f. Let F and G be symmetric linear endomorphisms of the orthogonal complement $z^{\perp}$ of z in $\mathbb{R}^d$. Furthermore, we will assume that F is positive definite (in particular, F must be non-degenerate). Given an arbitrary positive constant ε, we can define a convex smooth function f on $\mathbb{R}^d$ such that

$$Hf(y+\lambda z) = \begin{pmatrix} F & Gy \\ Gy & \langle GF^{-1}Gy,y\rangle \end{pmatrix} + \begin{pmatrix} \varepsilon F + \lambda G + \lambda^2 GF^{-1}G & 2\lambda G^{-1}FGy \\ 2\lambda G^{-1}FGy & \varepsilon\lambda^2 \end{pmatrix}$$

for all $y,\lambda z$ in an appropriate neighbourhood of 0 with $y \perp z$ (verify that Hf is positive semidefinite in an appropriate neighbourhood of 0). If we insert the so-defined matrices Hf, $H(f_z)$ and $H(f_{zz})$ in the above inequality, we obtain the necessity part of the following theorem.

Theorem 6: Suppose $\tilde{a}_{ij}$, $\tilde{b}_i \in C^2(\mathbb{R}^d)$ for all i,j. The transition maps $\tilde{P}^{t_1,t_2}$ leave the cone of all convex functions from $\mathbb{R}^d$ to $\mathbb{R}$ with at most polynomial growth at infinity invariant for any t_1,t_2, if and only if the following conditions are satisfied:

1) For every $z \in \mathbb{R}^d$ and every pair F,G of symmetric linear endomorphisms of $z^{\perp} \subset \mathbb{R}^d$ with F positive definite the following holds:

$$\mathrm{trace}\left(P_z^{\perp}((\tilde{a}_{ij})_{zz}F + 2\cdot(\tilde{a}_{ij})_z G + 2\cdot(\tilde{a}_{ij})GF^{-1}G)P_z^{\perp}\right)\Big|_{(t,x)} \geq 0$$

for all $(t,x) \in (0,\infty) \times \mathbb{R}^d$; $P_z^{\perp}$ is the orthogonal projection on $z^{\perp}$.

2) The map $x \to \tilde{b}(t,x)$ is affine linear for any t.

Remarks: 1.) If d=1, the first condition of the theorem becomes meaningless. In fact, according to the result mentioned at the end of Section 2, in the case d=1 we simply have to drop Condition 1.

2.) If $d > 1$, the first condition of the theorem in particular yields $\langle(\tilde{a}_{ij})_{zz}y,y\rangle \geq 0$ for every y with $y \perp z$. Therefore the function $\lambda \to \langle(\tilde{a}_{ij})y,y\rangle\,(t,x_0 + \lambda z)$ is convex for every y with $y \perp z$ and every $(t,x_o) \in (0,\infty) \times \mathbb{R}^d$. The necessity of this condition becomes obvious, if we repeat the arguments of the proof that Condition 1 is necessary with $x \to \langle x-x_o,y\rangle^2$ in place of the function f. Actually, Condition 1 of the theorem is much stronger. This reflects the fact that the "part of the curvature" of a convex function "which is perpendicular" to a fixed straight line is in general not convex but only "convex with respect to the harmonic mean" "along the line". A simple example for this fact is given by the function $x \to |x|$.

3.) The construction of matrices $(\tilde{a}_{ij})$ which satisfy not only the condition for non-explosion from the introduction but also the Condition 1 of the theorem will be postponed to the end of the paper (see Proposition 8).

Sketch of the proof of the sufficiency part of the theorem:

1. Step: Let $f \in C^4(\mathbb{R}^d)$ be a convex function such that the directional derivative $f_{zz}(0)$ vanishes.

Then $\frac{\partial}{\partial t}(\tilde{P}^{t,t_2} f)_{zz}(t_2,0) \leq 0$ for every $t_2 > 0$.

Proof: First we will assume that $f_{xx}(0) > 0$ for every x with $x \perp z$ and $x \neq 0$. Since the function f_{zz} has a local minimum at 0, the Taylor expansion of f_{zz} at 0 has the form

$$f_{zz}(x) = \langle Dx,x\rangle + o(|x|^2)$$

for a positive semidefinite linear $D : \mathbb{R}^d \to \mathbb{R}^d$. Furthermore, we can conclude from the convexity of f that there exist symmetric linear endomorphisms F and G of $z^\perp$ such that (assume that z is the last coordinate vector in $\mathbb{R}^d$)

$$H f(x) = \begin{pmatrix} F + o(|x|) & G P_z^\perp x + o(|x|^2) \\ G P_z^\perp x + o(|x|^2) & \langle Dx,x\rangle + o(|x|^3) \end{pmatrix}$$

for every x. The assumptions that $Hf(0)$ is positive semidefinite and that $f_{xx}(0) > 0$ for every x with $x \perp z$ and $x \neq 0$ imply that F is positive definite. Moreover, from $Hf(x)$ positive semidefinite for any x follows

$$\langle G F^{-1} G P_z^\perp, P_z^\perp x\rangle \leq \langle Dx,x\rangle \quad \text{for any } x.$$

On the other hand, we have by definition

$$H f(0) = F P_z^\perp, \quad H(f_z)(0) = G P_z^\perp, \quad H(f_{zz})(0) = 2\cdot D.$$

Taking into account Condition 1 of the theorem, we obtain

$$\operatorname{trace}\left((\tilde a_{ij})Hf\right)_{zz}\Big|_{(t_2,0)}$$

$$= \operatorname{trace}\left(P_z^{\perp}((\tilde a_{ij})_{zz}F + 2\cdot(\tilde a_{ij})_z G)P_z^{\perp} + 2\cdot(\tilde a_{ij})D\right)\Big|_{(t_2,0)}$$

$$= \operatorname{trace}\left(P_z^{\perp}((\tilde a_{ij})_{zz}F + 2\cdot(\tilde a_{ij})_z G + 2\cdot(\tilde a_{ij})GF^{-1}G)P_z^{\perp}\right)\Big|_{(t_2,0)}$$

$$+ \operatorname{trace}\left((\tilde a_{ij})\cdot 2\cdot\left(D - P_z^{\perp}GF^{-1}G\,P_z^{\perp}\right)\right)\Big|_{(t_2,0)}$$

$$\geq 0.$$

By Kolmogorov's backward equation, the assertion is herewith established under the additional assumption $f_{xx}(0) > 0$ for all x with $x \perp z$ and $x \neq 0$. To get rid of this assumption, we replace in the general situation the function f by the function $x \to f(x) + \varepsilon\cdot|P_z^{\perp}x|^2$ for an arbitrary $\varepsilon > 0$. The assertion follows if we let ε tend to zero.

<u>2. Step</u>: Suppose that $(\tilde a_{ij})$ is smooth and that there exist constants $\tilde C, T$ and K with $(\tilde a_{ij}(t,x)) = \tilde C\cdot|x|^2\cdot \mathbb{1}$ for all $t \leq T$ and all x with $|x| \geq K$. Let f be a convex function with $f(x) = |x|^2$ for all x with $|x| \geq K$.

Then there exists a constant R with

$$\left(\tilde P^{t_1,t_2}f\right)_{zz}(x) > 0 \quad \text{for all } t_1 \leq t_2 \leq T,\ 0 \neq z \in \mathbb{R}^d \text{ and } x \text{ with } |x| \geq R.$$

<u>Sketch of the proof</u>: Set $(\hat a_{ij}(t,x)) := \tilde C\cdot|x|^2\cdot\mathbb{1}$ for all $t \leq T$ and $x \in \mathbb{R}^d$. We can define transition maps $\hat P^{t_1,t_2}$ in a similar way as P^{t_1,t_2} (although $(\hat a_{ij}(t,.))$ is degenerate at zero). Set $\hat f(x) := |x|^2$ for every $x \in \mathbb{R}^d$. It is easy to see that $\hat P^{t_1,t_2}\hat f = \exp((t_2-t_1)\cdot\tilde C)\cdot\hat f$. A perturbation result yields (cf. [K], [K1], [P]):

$$\left(\tilde P^{t_1,t_2}f\right)_{zz}(x) - \left(\hat P^{t_1,t_2}\hat f\right)_{zz}(x) = -\int_{t_1}^{t_2}\left(\hat P^{t_1,t}(\hat L_t-\tilde L_t)\tilde P^{t,t_2}f\right)_{zz}(x)\,dt$$
$$+\left(\hat P^{t_1,t_2}(f-\hat f)\right)_{zz}(x).$$

The proof of the assertion will be complete once we establish a convenient bound for the modulus of the right-hand side of this equation. Such a bound exists for all x with $|x| \geq R$ and an appropriate R because the supports of the functions $(\hat L_t-\tilde L_t)\tilde P^{t,t_2}f$ and $f-\hat f$ are contained in $\{x \mid\ |x| < K\}$.

<u>3. Step</u>: Suppose that $(\tilde{a}_{ij})$ satisfies the same assumptions as in the second step. Let $\varepsilon > 0$ be an arbitrary constant. Let $f \in C^4(\mathbb{R}^d)$ be a strict convex function and let u be a solution of the terminal value problem

$$\frac{\partial u}{\partial t} + \tilde{L}_t u + \varepsilon \cdot |.|^2 = 0 \qquad \text{for } t < t_2 \text{ with } t_2 \leq T \text{ fixed}$$

and $u(t_2,x) = f(x)$ for all $x \in \mathbb{R}^d$.
Then the function $x \to u(t,x)$ is convex for every $t \leq t_2$.

<u>Sketch of the proof</u> : Suppose that the set

$M = \{(t,x) \mid t \leq t_2,\ x \in \mathbb{R}^d$ such that there exists a $z \in \mathbb{R}^d$ with $u(t,.)_{zz}|_x = 0\}$ is not empty. Since

$$u(t,x) = \tilde{P}^{t,t_2} f(x) + \varepsilon \cdot \int_t^{t_2} \tilde{P}^{t,s}\hat{f}(x)ds$$

(where $\hat{f}(x) = |x|^2$ for $x \in \mathbb{R}^d$), it follows from the second step that $u_{zz}(t,x) > 0$ for all x from the complement of an appropriate compact set. Hence there exists a point $(t_o,x_o) \in M$ with the property that $(t,x) \notin M$ for every t with $t > t_o$ and any $x \in \mathbb{R}^d$. In particular, the function $x \to u(t_o,x)$ is convex. If we slightly modify the argument of the first step, we obtain $\frac{\partial}{\partial t} u(t,.)_{zz}\ (t_o,x_o) < 0$. This leads to a contradiction.

<u>Final step</u>: First we let ε tend to 0. Since each $(\tilde{a}_{ij})$ which satisfies the conditions of the theorem, can be approximated uniformly on compact sets by matrices of functions, which satisfy the conditions of the second step (and of the theorem as well), the proof of the theorem can be completed by an application of a limit theorem (see [S/V], Theorem 11.1.4).

<u>Theorem 7</u>: Let L and $\tilde{L}$ be parabolic differential operators such that

$$(a_{ij}(t,x))_{i,j} \underset{(\text{resp.},\geq)}{\leq} (\tilde{a}_{ij}(t,x))_{i,j} \quad \text{and} \quad (b_i(t,x))_i = (\tilde{b}_i(t,x))_i$$

for all $t \in [0,\infty)$ and $x \in \mathbb{R}^d$. Suppose that the transition maps $\tilde{P}^{t_1,t_2}$ leave the cone of all convex functions from $\mathbb{R}^d$ to $\mathbb{R}$ with at most polynomial growth at infinity invariant for any t_1,t_2.

Then $P^{t_1,t_2} f \underset{(\text{resp.},\geq)}{\leq} \tilde{P}^{t_1,t_2} f$ for every convex function f from $\mathbb{R}^d$ to $\mathbb{R}$ and any t_1,t_2.

Theorem 7 is an immediate consequence of a comparison result similar to Corollaire 1 in [K1] (see also Folgerung 2.2 in [K]).

The following proposition, which we state here without proof, provides a rich choice of examples of differential generators which satisfy the conditions of Theorem 6 and Theorem 7.

Proposition 8: Given a $d\times d$-matrix (c_{ij}) and a vector $z\in\mathbb{R}^d$ with $z\neq 0$, we define $(c_{ij})|_{z^\perp} := P_z^\perp (c_{ij}) P_z^\perp$.

1) Let $((\tilde{a}_{ij})|_{z^\perp})^{-1}$ be the unique endomorphism of $z^\perp$ such that $((\tilde{a}_{ij})|_{z^\perp})^{-1}(\tilde{a}_{ij})|_{z^\perp} y=y$ for every $y\in z^\perp$. Assume that for every $z\in\mathbb{R}^d$ the following holds:

$$(\tilde{a}_{ij})_z|_{z^\perp}((\tilde{a}_{ij})|_{z^\perp})^{-1}(\tilde{a}_{ij})_z|_{z^\perp} \leq 2\cdot(\tilde{a}_{ij})_{zz}|_{z^\perp}.$$

Then Condition 1 of Theorem 6 is satisfied.

2) Suppose that $(\tilde{a}_{ij})|_{z^\perp}$, $(\tilde{a}_{ij})_z|_{z^\perp}$ and $(\tilde{a}_{ij})_{zz}|_{z^\perp}$ pairwise commute for every $z\in\mathbb{R}^d$.

Then Condition 1 of Theorem 6 is equivalent to the condition that the function

$$\lambda \to \langle(\tilde{a}_{ij})^{1/2}y,y\rangle\ (t,x_o+\lambda z)$$

is convex for all $x_o, y, z\in\mathbb{R}^d$ with $y\perp z$.

Remark: The presuppositions of the second part of the proposition are in particular satisfied if $d=2$.

References

[A] Aronson, D.G. Bounds for the fundamental solution of a parabolic equation, Bull. Amer. Math. Soc. 73 (1967), 89o-896.

[B] Borell, C. Convex measures on locally convex spaces, Ark. Mat. 12 (1974), 239-252.

[B1] Borell, C. Greenian potentials and concavity, Math. Ann. 272, (1985), 155 - 16O.

[C/F] Cafarelli, L.A.; Friedman, A. Convexity of solutions of semilinear elliptic equations, Duke Math. J. 52 (1985), 431-456.

[C/S] Cafarelli, L.A.; Spruck, J. Convexity properties of solutions to some classical variational problems, Comm. Partial Differential Equations 7 (1982), 1337-1379.

[D/K/H] Davidovič, Ju.S.; Korenbljum, B.I; Hacet, I. A property of logarithmically concave functions, Soviet Math. Dokl. 10 (1969), 477 - 480.

[H] Hajek, B. Mean stochastic comparison of diffusions, Z. Wahrsch. Verw. Gebiete 68 (1985), 315-329.

[I/W] Ikeda, N.; Watanabe, S. A comparison theorem for solutions of stochastic differential equations and its applications, Osaka J. Math. 14 (1977), 619-633.

[Ko] Korevaar, N.J. Convex solutions to nonlinear elliptic and parabolic boundary value problems, Indiana Univ. Math. J. 32 (1983), 6o3 - 614.

[K] Kröger, P. Vergleichssätze für Diffunsionsprozesse, Thesis Erlangen (1986).

[K1] Kröger, P. Comparison de diffusions, C.R. Acad. Sci. Paris Sér.I 3o5 (1987), 89-92.

[K/S] Krylov, N.V. Safonov, M.V. A certain property of solutions of parabolic equations with measurable coefficients, Math. USSR-Izv. 16 (1981), 151-164.

[L] Lions, P.L. Two geometrical properties of solutions of semilinear problems. Applicable Anal. 12 (1981), 267-272.

[N] Nickel, K. Gestaltaussagen über Lösungen parabolischer Differentialgleichungen, J. Reine Angew. Math. 211 (1962), 78-94.

[P] Phillips, R.S. Perturbation theory for semigroups of linear operators, Trans. Amer. Math. Soc. 74 (1953),199-221.

[Sk] Skorochod, A.V. Existence and uniqueness of solutions to stochastic diffusion equations (russian),Sibirsk Mat.Zh. 2 (1961), 129-137.

[S/V] Stroock,D.W. Varadhan, S.R.S. Multidimensional diffusion processes, Springer Berlin 1979.

[W] Walter, W. Differential and integral inequalities, Springer Berlin 1970.

[Y] Yamada, T. On a comparison theorem for solutions of stochastic differential equations and its applications, J.Math. Kyoto Univ. 13 (1973), 497-512.

[Y1] Yamada, T. On the non-confluent property of solutions of one-dimensional stochastic differential equations, Stochastics 17 (1986), 111-124.

[Y/O] Yamada, T.; Ogura, Y. On the strong comparison theorem for solutions of stochastic differential equations, Z. Wahrsch. Verw. Gebiete 56 (1985), 3-19.

PERFORMANCE AND ROBUSTNESS IN ADAPTIVE CONTROL OF LINEAR STOCHASTIC SYSTEMS

P. R. Kumar
Department of Electrical and Computer Engineering
and
Coordinated Science Laboratory
University of Illinois
1101 W. Springfield Avenue
Urbana, Illinois 61801/USA

ABSTRACT

We provide an account of some recent results concerning performance and robustness of adaptive controllers for linear stochastic systems. By performance is meant such properties as self-tuning and self-optimization when the system under control satisfies some assumed properties such as being of known order, with a certain spectrum for the noisy disturbance, etc. Under the topic of robustness we analyze the consequences when such idealized properties fail to be met.

I. INTRODUCTION

Over the last few years much progress has been made on the problems of analyzing the **performance** properties of adaptive controllers for linear stochastic systems, as well as analyzing their **robustness** to modelling errors. In this paper we provide an account of these results. Proofs are omitted; they may be found in the cited references.

By **performance**, we mean how closely the adaptive controller in closed-loop with the unknown system comes to optimizing a certain performance index, when the unknown system under control satisfies some assumptions such as, for example, being of known order, etc. Under the topic of **robustness** we examine the consequences when the known system fails to meet such idealized assumptions.

II. SELF-TUNING TRACKERS

Consider the linear stochastic system:

$$y(t) = \sum_{i=1}^{p} a_i y(t-i) + \sum_{i=1}^{q} b_i u(t-i) + w(t) + \sum_{i=1}^{s} c_i w(t-i) \tag{1}$$

where y is the output, u is the input, and $\{w(t)\}$ is a zero-mean "white" noise process, or more precisely, if F_t is the σ-algebra generated by $\{w(s): s \leq t\}$, then

i) $E[w(t) \mid F_{t-1}] = 0$ a.s.

ii) $E[w^2(t) \mid F_{t-1}] = \sigma^2$ a.s.

iii) For some $\delta > 0$, $\sup_t E[|w(t)|^{2+\delta} | F_{t-1}] < +\infty$ a.s.

If $y^*(t)$ is a desired reference trajectory which we want the output $y(t)$ to track as closely as possible, then we can postulate a cost function of the form

$$\lim_{t\to\infty} E[(y(t)-y^*(t))^2] ,$$

or alternatively,

$$\lim_{N\to+\infty} \frac{1}{N} \sum_{t=1}^{N} (y(t)-y^*(t))^2 \tag{2}$$

which we seek to minimize.

The Minimum Variance Control Laws for Tracking and Regulation

Let us first consider the situation where the parameters $(a_1, \ldots, a_p, b_1, \ldots, b_q, c_1, \ldots, c_s)$ are known.

We shall assume that $b_1 \neq 0$, i.e. the system has a unit time delay. Then it is clear that the control law:

$$u(t-1) = -\frac{1}{b_1}[-y^*(t) + \sum_{i=1}^{p} a_i y(t-i) + \sum_{i=2}^{q} b_i u(t-i) + \sum_{i=1}^{s} c_i w(t-i)] \tag{3}$$

will result in

$$y(t) = y^*(t) + w(t) \quad \text{for all } t . \tag{4}$$

Clearly, this results in,

$$E[(y(t)-y^*(t))^2] = \sigma^2$$

as well as,

$$\lim_{N\to\infty} \frac{1}{N} \sum_{t=1}^{N} y(t) - y^*(t))^2 = \sigma^2 \quad \text{a.s.}$$

which is the best possible, since σ^2 is the lowest possible value of either of the cost criteria above.

However, in order to implement the control law (3), one needs to have access to the values of $w(t-1),\ldots,w(t-s)$ at time t-1. This is not available to the controller since one should note that the white noise process $\{w(t)\}$ is merely a mathematical device used to model the "colored" additive disturbance $w(t) + \sum_{i=1}^{s} c_i w(t-i)$ entering into the system at time t.

Note however, that **under optimal control**, the relationship (4) will be satisfied, and so one will have

$$w(t-i) = y(t-i) - y^*(t-i) .$$

This suggests that in place of (3), one should use the control law,

$$u(t-1) = -\frac{1}{b_1}[-y^*(t) - \sum_{i=1}^{s} c_i y^*(t-i) + \sum_{i=1}^{p} (a_i+c_i)y(t-i) + \sum_{i=2}^{q} b_i u(t-i)] \tag{5}$$

which is obtained by replacing w(t-i) for i=1,...,s in (3) by $y(t-i)-y^*(t-i)$. (Just for simplicity we assume throughout that $s \leq p$.)

It can be shown, see Kumar and Varaiya [1], that if the **implementable** control law (5) is used, then one will have

$$\lim_{N\to+\infty} \frac{1}{N} \sum_{t=1}^{N} (y(t)-y^*(t))^2 = \sigma^2 \quad \text{a.s.} \tag{6}$$

provided that:

All the roots of the polynomial $1 + c_1z+...+c_sz^s$ are outside the closed–unit disk. (7)

Now note that by the Spectral Factorization Theorem, when modeling a disturbance v(t) as a linear combination of {w(t)} one can choose spectral factors to lie within or outside the unit disk. Thus (7) is without too much loss of generality, since only the case where some roots lie exactly on the unit circle is excluded. Thus, (7) is a reasonable assumption which we henceforth make, and we shall call (5) as the **minimum variable control** law, since it minimizes the cost criterion (2).

Note however, that since the cost criterion (2) does not include a positive definite weighting on control energy, it could happen that

$$\lim_{t\to\infty} u^2(t) = +\infty$$

under a minimum variance control law, as the following example shows.

Example

Consider $y(t)=u(t-1)-2u(t-2)+w(t)$ with $y^*(t)\equiv 0$. Then the minimum variance control law is

$$u(t-1) = 2u(t-2)$$

which results in a divergent u(t).

Such situation is avoided if we make the following assumption:

All the roots of the polynomial $b_1 + b_2z+...+b_qz^{q1}$ are outside the closed–unit disk. (8)

This assumption is called a **minimum-phase** assumption, and will henceforth be made.

The Adaptive Control Law

Let us define:

$$\phi(t-1) := (y(t-1),...,y(t-p),u(t-1),...,u(t-q),-y^*(t),...,-y^*(t-s))^T , \tag{9}$$

and

$$\theta^0 := (a_1+c_1, \ldots, a_p+c_p, \ldots, b_1, \ldots, b_q, 1, c_1, \ldots, c_s)^T . \tag{10}$$

Then it is easy to see that the minimum variance control law (5) can be rewritten as:

$$\phi^T(t-1)\theta^0 = 0 . \tag{11}$$

Moreover, if the system is under optimum control, then one also has,

$$y(t) - y^*(t) = \phi^T(t-1)\theta^0 + w(t) .$$

Let us now consider the situation where the parameters $(a_1, \ldots, a_p, b_1, \ldots, b_q, c_1, \ldots, c_s)$, or equivalently, the **parameter vector** θ^0, is unknown.

Note that

$$\nabla_\theta(y(t)-y^*(t)-\phi^T(t-1)\theta)^2 = -2\phi(t-1)[y(t)-y^*(t)-\phi^T(t-1)\theta] .$$

Thus $\phi(t-1)[y(t)-y^*(t)-\phi^T(t-1)\theta]$ is the **direction** in which the squared error of fit $(y(t)-y^*(t)-\phi^T(t-1)\theta)^2$ is reduced. This suggests the so-called **Stochastic Gradient Algorithm**;

$$\hat{\theta}(t) = \hat{\theta}(t-1) + \frac{\mu}{r(t-1)} \phi(t-1)[y(t)-y^*(t)-\phi^T(t-1)\theta] \tag{13}$$

$$r(t) = \sum_{i=0}^{t} \phi^T(i)\phi(i) \tag{14}$$

to recursively estimate the unknown parameter vector θ^0 by $\hat{\theta}(t)$, where $\frac{\mu}{r(t-1)}$ controls the step-size.

Once the estimate $\hat{\theta}(t-1)$ for θ^0 is made at time t-1, then one can use the control law

$$\phi^T(t-1)\hat{\theta}(t-1) = 0 \tag{15}$$

which mimics (11), since θ^0 is unknown. This is called a **certainty-equivalent** control law. It should be noted that the relationship (15) is actually an implicit specification of how u(t-1) is to be chosen, in exactly the same way as (11) is an implicit specification of the choice of u(t-1) chosen according to (5).

The adaptively controller is thus specified by (13-15) once initial conditions $\hat{\theta}(0)$ and $r(0)>0$ are chosen.

The Results

We now arrive at the critical questions concerning the **performance** of the adaptive controller. We shall say that the adaptive controller (13-15) is **self-optimizing** if when applied to the system (1) it results in (6), i.e. it achieves the minimum cost. We shall say that the adaptive controller is **self-tuning** if

$$\lim_{t\to\infty} \hat{\theta}(t) = \gamma\theta^0 \qquad \text{for } \gamma\neq 0 ,$$

since this would imply that the control law (15) is asymptotically the same as the optimal control law (11). Lastly, we shall say that the parameter estimates are **strongly consistent** if

$$\lim \hat{\theta}(t) = \theta^0 \qquad \text{a.s.} ,$$

i.e. if the parameter estimates converge to the true values. It should be noted that:

Strong Consistency $\not\Rightarrow$ Self–Tuning $\not\Rightarrow$ Self–Optimality.

In order to establish such results, it has been discovered that the property:

$$\mathrm{Re}[1+c_1e^{j\omega}+c_2e^{2j\omega}+\ldots+c_se^{sj\omega}-\frac{\mu}{2}] > 0 \ \text{ for } 0\leq\omega\leq 2\pi \tag{16}$$

plays a crucial role. This is called a **Positive-Real** condition, and its usefulness arises from the following fact. Let us define $C(q^{-1}):=1+c_1q^{-1}+\ldots+c_sq^{-s}$ where q^{-1} represents the backward-shift operator. Then if

$$z(t) = C(q^{-1})v(t) ,$$

i.e. $z(t)$ is the output of a system with transfer function $C(q^{-1})$, when $v(t)$ is the input, then there exists a δ such that

$$\sum_{i=0}^{t} z(i)v(i) \geq \delta \qquad \text{for all } t .$$

Such Positive Real Transfer functions play an important role in the analysis of passive electrical networks. In our context it turns out that

$$\phi^T(t)\theta^0 = C(q^{-1})[E(y(t+1)-y^*(t+1)|F_t] .$$

and so the Positive Real property can be usefully employed to establish certain inequalities needed in the analysis.

The following Theorem has been proved in Kumar and Praly [2].

Theorem

The above adaptive control law is **self-optimizing.**

In order to prove a self-tuning result, some conditions have to be imposed on the reference trajectory $\{y^*(t)\}$. We shall say that **$y^*(t)$ is exciting of order n** if

There exist integers M and T, and an $\epsilon > 0$ such that

$$\sum_{k=t+1}^{t+M} \begin{bmatrix} y^*(k-1) \\ \vdots \\ y^*(k-n) \end{bmatrix} [y^*(k-1),\ldots,y^*(k-n)] \geq \epsilon I \quad \text{for all } t \geq T .$$

This condition arises essentially from the fact that without enough "excitation" of the system it is not possible to identify it.

The following result is also from [2].

Theorem

If $\{y^*(t)\}$ is exciting of order (q+s), then

$$\lim_{t\to+\infty} \hat{\theta}(t) = \xi\theta^0 \qquad \text{a.s.} \tag{17}$$

where ξ is an a.s. finite, non-zero random variable.

Thus, under an excitation condition on $\{y^*(t)\}$, the adaptive controller is **self-tuning.**

Note that since θ^0 is defined by (10), the (p+q+1)-th component is 1, i.e.

$$\theta^0_{p+q+1} = 1.$$

Thus, from (17) it follows that

$$\lim_{t\to\infty} \hat{\theta}_{p+q+1}(t) = \xi \quad \text{a.s.}$$

Hence the value of the scaling factor is determined; therefore,

$$\lim_{t\to\infty} \frac{(\hat{\theta}_1(t),\hat{\theta}_2(t),\ldots,\hat{\theta}_{p+q}(t),\hat{\theta}_{p+q+2}(t),\ldots,\hat{\theta}_{p+q+s+1}(t)}{\hat{\theta}_{p+q+1}(t)}$$

$$= (a_1+c_1, \ldots, a_p+c_p, b_1, \ldots, b_q, c_1, \ldots, c_s) \text{ a.s.} ,$$

which means that even the parameters $(a_1, \ldots, a_p, b_1, \ldots, b_q, c_1, \ldots, c_s)$ can be identified. Thus the adaptive control law also results in asymptotic identification of the true parameters.

There is one crucial difference between the adaptive control law studied above and that proposed by Goodwin, Ramadge and Caines [3]. They have suggested using,

$$\phi^T(t-1) := (y(t-1),\ldots,y(t-p),u(t-1),\ldots,u(t-q),-y^*(t-1),\ldots,-y^*(t-s))$$

and

$$\theta^0 := (a_1, \ldots, a_p, b_1, \ldots, b_q, c_1, \ldots, c_s) .$$

Thus their θ^0 vector is of one less dimension since it does not include a 1. Similarly, their **regression vector** $\phi(t)$ is also of one less dimension, since it does not include $-y^*(t)$. Hence their Stochastic Gradient Algorithm is:

$$\hat{\theta}(t) = \hat{\theta}(t-1) + \frac{\mu}{r(t)} \phi(t-1)[y(t) - \phi^T(t-1)\hat{\theta}(t-1)] \tag{18}$$

$$r(t) = r(t-1) + \phi^T(t)\phi(t) \tag{19}$$

and the certainty equivalent control law is:

$$\phi^T(t-1)\hat{\theta}(t-1) = y^*(t) . \tag{20}$$

The difference between the right hand sides of (20) and (15), as well as the difference between (18) and (13) are worth noting.

Goodwin, Ramadge and Caines [3] have proved the self-optimality of the adaptive control law (18-20). Recently, Lin and Kumar [4] have established some self-tuning properties also. We will not detail these results here.

The Regulation Problem

It is worth noting that if one considers the special case of **regulation**, where

$$y^*(t) \equiv 0 ,$$

then the last (s+1) components of $\phi(t-1)$ in (9) play no role in the control law (11). Thus one can eliminate them and define

$$\phi^T(t-1) := (y(t-1),\ldots,y(t-p),u(t-1),\ldots,u(t-q))$$

with $\theta^0 := (a_1+c_1, \ldots, a_p+c_p, b_1, \ldots, b_q)^T$. Then we obtain the following adaptive control law:

$$\hat{\theta}(t) = \hat{\theta}(t-1) + \frac{\mu}{r(t-1)} \phi(t-1)y(t)$$

$$r(t) = r(t-1) + \phi^T(t)\phi(t)$$

with the certainty equivalent control,

$$\phi^T(t-1)\hat{\theta}(t-1) = 0 .$$

This adaptive control law has been proved to be self-optimizing, i.e.

$$\lim_{N\to+\infty} \frac{1}{N} \sum_{t=1}^{N} y^2(t) = \theta^2 \qquad \text{a.s.}$$

by Goodwin, Ramadge and Caines [3], while Becker, Kumar and Wei [5] have proved that it is also self-optimizing, i.e.

$$\lim_{t\to+\infty} \hat{\theta}(t) = \xi\theta^0 \qquad \text{a.s.}$$

where ξ is a.s. finite and nonzero.

The Linear Model Following Problem

Above, we have obtained self-tuning adaptive control laws when:
either $\{y^*(t)\}$ is exciting of order (q+s),
or $y^*(t) \equiv 0$, i.e. it is exciting of order 0.
In the latter case we have also reduced the dimension of the regression vector $\phi(t)$ by eliminating some unnecessary variables.

In many practical problems the excitation in the reference trajectory falls in between the two extremes above. For example, an important application is the **set-point** problem, where

$$y^*(t) \equiv y^* \neq 0 \ \ \text{(i.e. a constant) .,}$$

In this case, $\{y^*(t)\}$ is exciting of order 1 only. The question then arises as to whether we can eliminate some unnecessary variables from the regression vector $\phi(t)$ and obtain self-tuning adaptive control laws.

Let us consider a reference trajectory $\{y^*(t)\}$ which satisfies

$$y^*(t) + h_1 y^*(t-1) + \ldots + h_m y^*(t-m) \equiv 0 ,$$

i.e. it is the solution of linear homogeneous difference equation of order m. Without loss of generality, we can assume that m is the smallest possible, i.e. there is no linear homogeneous difference of smaller order than in which $\{y^*(t)\}$ satisfies. The implication of this assumption is that all the "modes" of the equation are present in $\{y^*(t)\}$.

We shall make one further assumption that all the roots of the polynomial

$$H(z) := 1+h_1z+\ldots+h_mz^m$$

are exactly on the unit circle in the complex plane, and there are no repeated roots.

This assumption is also reasonable, since roots inside the open unit disk correspond to modes that are unbounded, while roots outside the closed unit disk correspond to modes that decay exponentially to 0, and hence vanish asymptotically.

It is a consequence of these assumptions that $\{y^*(t)\}$ is exciting of order m. We will consider the case where $m<s$.

It is worth noting that in the set-point problem, one has $y^*(t)=y^*(t-1)$ for all t. Thus $H(z)=1-z$, which has a single root at $z=1$.

Let us now see how to reduce the dimension of the regression vector $\phi(t)$. It should be noted that the minimum variance control is give by (5), which can also be written as (11), when $\phi(t)$ and θ^0 are specified by (9) and (10), respectively.

Note that one rewrite the control law (5) as:

$$u(t-1) = -\frac{1}{b_1}[-C(q^{-1})y^*(t)+\sum_{i=1}^{p}(a_i+c_i)y(t-i)+\sum_{i=2}^{q}b_iu(t-i)]\,. \tag{12}$$

Let us now divide the polynomial C(z) by H(z). This gives a quotient F(z) and a remainder G(z), where

$$C(z) = F(z)H(z)+G(z)$$

with,

$$G(z) = \sum_{i=0}^{m-1} g_iz^i$$

and

$$F(z) = \sum_{i=0}^{s-m} f_iz^i\,.$$

Now noting that $H(q^{-1})y^*(t)\equiv 0$, we obtain,

$$\begin{aligned} C(q^{-1})y^*(t) &= F(q^{-1})H(q^{-1})y^*(t)+G(q^{-1})y^*(t) \\ &= G(q^{-1})y^*(t)\,. \end{aligned}$$

Hence we can rewrite (5) as,

$$u(t-1) = -\frac{1}{b_1}[-G(q^{-1})y^*(t) + \sum_{i=1}^{p}(a_i+c_i)y(t-i) + \sum_{i=2}^{q} b_i u(t-i)] .$$

Thus if one defines θ^0 and $\phi(t-1)$ by:

$$\theta^0 := (a_1+c_1, \ldots, a_p+c_p, b_1, \ldots, b_q, g_0, \ldots, g_{m-1})^T$$

and

$$\phi^T(t-1) := (y(t-1),\ldots,y(t-p),u(t-1),\ldots,u(t-q),-y^*(t),\ldots,-y^*(t-m+1)) ,$$

then the optimal control law can be written as,

$$\phi^T(t-1)\theta^0 = 0 .$$

Thus we have reduced the dimension of $\phi(t)$ by (s-m+1).

We now have the following adaptive control law:

$$\hat{\theta}(t) = \hat{\theta}(t-1) + \frac{\mu}{r(t-1)}\phi(t-1)(y(t)-y^*(t))$$
$$r(t) = r(t-1) + \phi^T(t)\phi(t)$$
$$\phi^T(t-1)\hat{\theta}(t-1) = 0 .$$

The following results have been proved in Kumar and Praly [2].

Theorem: Self-tuning tracking for the linear model following problem

i) $\lim_{N\to\infty} \frac{1}{N}\sum_{t=1}^{N}(y(t)-y^*(t))^2 = \sigma^2$ a.s.

ii) $\lim_{t\to+\infty} \hat{\theta}(t) = \xi\theta^0$ a.s. where ξ is finite and non-zero a.s.

III. A ROBUST ADAPTIVE CONTROLLER

In the previous section we have exhibited adaptive controllers which are self-optimizing and self-tuning for systems which satisfy the following properties:

i) the system is a linear stochastic system whose orders are given by the integers p, q and s,

ii) it is strictly minimum phase,

iii) $b_1 \neq 0$,

iv) a positive real condition is satisfied by C(z).

In practical applications these "idealized" conditions are subject to violation. Thus one would like to design an adaptive controller which performs **optimally** when such ideal conditions are satisfied, but also preserves **some stability** of the overall closed-loop system when some of the conditions are violated to some degree, i.e. it is **robust.**

To study the robustness of the adaptive controller, it is useful to consider a topology on the space of linear systems, and then show that stability in an appropriate sense is preserved in an "open" neighborhood of ideal systems.

We shall consider the **graph topology** on the space of linear systems, introduced by Vidyasagar [6]. This is the weakest topology on the space of linear systems which satisfies the following two properties:

i) If a plant is stabilized by a linear dynamic feedback controller, then stability is preserved in an open neighborhood of the plant, and

ii) The frequency response of the closed loop system (with the "sup" norm) is a continuous function of the plant.

Our goal therefore is to design an adaptive controller which is self-optimizing for ideal systems which satisfy conditions of the type i)-iv) listed above, while also preserving stability in an appropriate sense in a graph-topological neighborhood of such systems.

The following adaptive controller has been considered in Praly, Lin and Kumar [7].

Let n_R, n_s and n_c be three integers chosen *a priori* which fix the dimension of the adaptive controller. We shall also choose a "nominal" parameter vector $\theta^c \in R^{n_R+n_s+n_c+2}$, a radius K which reflects our uncertainty about the plant, and a nominal value d which represents the **delay.** Finally, we choose two positive numbers $0<\lambda_0<\lambda_1$, which will be used to bound the eigenvalues of a certain "covariance" matrix, as well as a constant $\sigma_0>0$ which we believe is a lower bound on the high frequency gain, and another constant $\rho>0$ which will be used in the design.

Having committed ourselves to these prior choices, the adaptive controller is defined recursively by the following (cumbersome!) equations:

$$\phi^T(t) := (u(t),\dots,u(t-n_s),y(t),\dots,y(t-n_R),y^*(t+d-1),\dots,y^*(t+d-n_c))$$

$$r(t) := r(t-1)+\max(\rho,\phi^T(t-d)\phi(t-d))$$

$$\bar{\phi}(t-d) := \frac{\phi(t-d)}{r^{1/2}(t)}$$

$$g(t) := \frac{1}{1+\bar{\phi}^T(t-d)F(t-d)\bar{\phi}(t-d)}$$

$$e(t) := y(t)-\phi^T(t-d)\theta(t-d)$$

$$\bar{e}(t) := \frac{e(t)}{r^{1/2}(t)}$$

$$F^1(t) := F(t-d)-g(t)F(t-d)\bar{\phi}(t-d)\bar{\phi}^T(t-d)F(t-d)$$

$$F(t) := [1-\frac{\lambda_0}{\lambda_0}]F^1(t)+\lambda_0 I \quad (\text{with } \lambda_0 I \le F(0) \le \lambda_1 I) \tag{21}$$

$$\theta^1(t) := \theta(t-d) + g(t)F(t-d)\bar{\phi}(t-d)\bar{e}(t)$$

$$\theta^2(t) := \theta^1 + \max(0,\sigma_0 - s_0^1(t)) \frac{F_{.1}(t)}{F_{11}(t)} \tag{22}$$

where $s_0^1(t)$:= first element of the vector $\theta^1(t)$
$F_{.1}(t)$:= first column of the matrix F(t)
$F_{11}(t)$:= (1,1)-th element of F(t)

$$\theta(t) := \theta^c + (\theta^2(t) - \theta^c) \min(1, \frac{K\lambda_1}{\lambda_0 \|\theta^2(t) - \theta^c\|}) . \tag{23}$$

Finally the control input u(t) is specified by:

$$\theta^T(t)\phi(t) = y^*(t+d) .$$

Essentially, the above algorithm is a least-squares type algorithm to estimate the parameters of the unknown system, followed by a certainty-equivalent control law. The least-squares algorithm has several modifications, however. First, the covariance matrix is modified at each step through (21) in such a way that its eigenvalues are maintained between λ_0 and λ_1. Second, the estimate of the high frequency gain, which is the first component of $\theta(t)$, is kept above σ_0, by (22). Finally, the parameter estimates are kept in a sphere of bounded radius centered at θ^c through (23).

All the above modifications are well motivated, but it would be of interest to know whether they are indeed necessary.

The following self-optimization result is proved in Praly, Lin and Kumar [7].

Theorem: Self-Optimality in Ideal Case

If the true parameter vector θ^0 is indeed inside the region maintained by the adaptive controller, if n_R, n_s and n_c are compatible with the true orders of the system, and if

$$\sup_{0 \le \omega \le 2\pi} |C(e^{i\omega}) - 1| < \frac{1}{\sqrt{1+\lambda_1}}$$

then the system is self-optimizing.

Regarding robustness, we have the following [7].

Theorem: Robustness with Respect to the Graph Topology

There is an open neighborhood of ideal systems as above, where one has

$$\sup_T \frac{1}{T} \sum_{t=1}^{T} \{[y(t)-y^*(t)]^2+u^2(t)\} < +\infty .$$

It should be noted that we have not said much about continuity of the performance with respect to the graph topology.

ACKNOWLEDGEMENT

This research has been supported in part by the U.S. Army Research Office under Contract DAAL 03-88-K-0046, and in part by the Joint Services Electronics Program under Contract N00014-85-C-0149.

IV. REFERENCES

1. P. R. Kumar and P. P. Varaiya, *Stochastic Systems: Estimation, Identification and Adaptive Control*, Prentice-Hall, 1986.
2. P. R. Kumar and L. Praly, "Self-Tuning Trackers," *SIAM Journal on Control and Optimization*, vol. 25, no. 4, pp. 1053-1071, July 1987.
3. G. Goodwin, P. Ramadge and P. E. Caines, "Discrete-Time Stochastic Adaptive Control," *SIAM Journal on Control and Optimization*, vol. 19, pp. 829-853, 1981.
4. S.-F. Lin and P. R. Kumar, "Parameter Convergence in the Goodman-Ramadge-Caines and a Modified Least Squares Adaptive Control Algorithms," University of Illinois, 1988.
5. A. Becker, P. R. Kumar and C. Z. Wei, "Adaptive Control with the Stochastic Approximation Algorithm: Geometry and Convergence," *IEEE Transactions on Automatic Control*, vol. AC-30, pp. 330-338, 1985.
6. M. Vidyasagar, "The Graph Metric for Unstable Plants and Robustness Estimates for Feedback Stability," *IEEE Transactions on Automatic Control*, vol. AC-29, pp. 403-418, May 1984.
7. L. Praly, S.-F. Lin and P. R. Kumar, "A Robust Adaptive Minimum Variance Controller," to appear in *SIAM Journal on Control and Optimization*.

SINGULAR PERTURBATIONS FOR STOCHASTIC CONTROL

H. J. Kushner
Division of Applied Mathematics
Brown University
Providence, Rhode Island 02912

1. INTRODUCTION

Singular perturbations problems for stochastic systems have received attention mainly via methods of partial differential equations [6], [7]. Here we adopt a "weak convergence" based method and obtain many results more efficiently, and often under weaker conditions. Extensions to the wide bandwidth noise problem and to filtering problems will be dealt with elsewhere.

2. SINGULAR CONTROL ON THE INTERVAL [0,T]

We work with the singularity perturbed diffusion model

$$dx^{\varepsilon} = dt \int G(x^{\varepsilon},z^{\varepsilon},\alpha)m_t^{\varepsilon}(d\alpha) + \sigma(x^{\varepsilon},z^{\varepsilon})dw_1 \tag{2.1}$$

$$\varepsilon\, dz^{\varepsilon} = H(x^{\varepsilon},z^{\varepsilon})dt + \sqrt{\varepsilon}v(x^{\varepsilon},z^{\varepsilon})dw_2, \tag{2.2}$$

where the $w_i(\cdot)$ are mutually independent standard vector valued Wiener processes, with respect to some filtration $\{F_t\}$, and $m^{\varepsilon}(\cdot)$ is an admissible relaxed stochastic control. (See below for the definition.) The cost criterion is

$$E_x^m \int_0^T\int k(x^{\varepsilon}(s),z^{\varepsilon}(s),\alpha)m_s^{\varepsilon}(d\alpha)ds + E_x^m g(x^{\varepsilon}(T)) \equiv V^{\varepsilon}(x,m^{\varepsilon}). \tag{2.3}$$

We write the V^{ε} as a function of m, although it depends on the set $(m^{\varepsilon},w_1,w_2)$.

Definition. The set of control values is a compact set U. Let $\mathcal{U}$ denote the Borel σ-algebra on U. We say that $m(\cdot)$ is an *admissible relaxed control* [1], [2], if it is a measure on the Borel sets of $U \times [0,T]$ with the properties: (a) $m(U \times [0,t]) \equiv t$ for all $t \leq T$; (b) for $A \in \mathcal{U}$, $m(A \times [0,\cdot])$ is F_t-progressively measurable. For such $m(\cdot)$, there is a *derivative* $m_t(\cdot)$ which is a measure on $\mathcal{U}$ for each t and $m_.(A)$ is F_t-progressively measurable for each $A \in \mathcal{U}$ and for Borel A, B, $m(A \times B) = \int m_s(A)I_{\{s\in B\}}ds$.

Admissible $m(\cdot)$ is said to be a *relaxed feedback control* if there is $\hat{m}_x(\cdot)$, a measure on $\mathcal{U}$ for each x and such that $\hat{m}_.(A)$ is Borel measurable for each $A \in \mathcal{U}$ and $m_t(\cdot) = \hat{m}_{x(t)}(\cdot)$ for almost all ω, t.

We require the following conditions.

A2.1

$$G(x,z,\alpha) = G_0(x,z) + G_1(x,\alpha)$$

$$k(x,z,\alpha) = k_0(x,z) + k_1(x,\alpha),$$

where $\sigma(\cdot)$, $G_i(\cdot)$ *are continuous and are Lipschitz continuous and have a linear growth in* x, *all uniformly in* z, α. *The* $k_i(\cdot)$ *and* $g(\cdot)$ *are continuous and have a polynomial growth in* x, *uniformly in* z, α.

A2.2. $H(\cdot)$ *and* $v(\cdot)$ *are continuous and have a linear growth and are Lipschitz continuous in* z, *uniformly in* x.

A2.3. *For the controls used below., there is a function* $0 \le \hat{g}(z) \to \infty$ *as* $|z| \to \infty$ *such that*

$$\sup_{\epsilon, t \le T} \frac{1}{\Delta} \int_t^{t+\Delta} E\hat{g}(z^\epsilon(s))ds < \infty$$

for some sequence $\Delta \to 0$ *such that* $\epsilon/\Delta \to 0$.

<u>The Averaged Problem</u>. Define $z_0^\varepsilon(t) = z^\varepsilon(\varepsilon t)$, $x_0^\varepsilon(t) = x^\varepsilon(\varepsilon t)$. Then

$$\begin{aligned} dx_0^\varepsilon &= dt\ \varepsilon \int G(x_0^\varepsilon, z_0^\varepsilon, \alpha) m_t^\varepsilon(d\alpha) + \sqrt{\varepsilon}\sigma(x_0^\varepsilon, z_0^\varepsilon) dw_{01}, \\ dz_0^\varepsilon &= H(x_0^\varepsilon, z_0^\varepsilon)dt + v(x_0^\varepsilon, z_0^\varepsilon) dw_{02}, \end{aligned} \tag{2.4}$$

where the $w_{0i}(\cdot)$ are mutually independent vector valued standard Wiener processes (in fact $w_{0i}(t) = w_i(\varepsilon t)/\sqrt{\varepsilon}$). Define the fixed-x process $z_0(\cdot|x)$ (written simply as $z_0(\cdot)$ when the value of x is obvious) by

$$dz_0 = H(x,z_0)dt + v(x,z_0)dw_{02}. \tag{2.5}$$

Let A_x^0 denote the weak infinitesimal operator of $z_0(\cdot|x)$ and A^ε that of $(x^\varepsilon(\cdot), z^\varepsilon(\cdot))$. We need the further assumption:

A2.4. *For each* x, $z_0(\cdot|x)$ *has a unique invariant measure* $\mu_x(\cdot)$. *Also there is a continuous matrix valued function* $\bar{\sigma}(\cdot)$ *such that* $\bar{\sigma}(x)\bar{\sigma}'(x) = \int \sigma(x,z)\sigma'(x,z)\mu_x(dz) \equiv \bar{a}(x)$.

The second part of (A2.4) is convenient but can be dispensed with.

Define the *averaged system* by $\bar{G}_0(x) = \int G_0(x,z)\mu_x(dz)$, $\bar{k}_0(x) = \int k(x,z)\mu_x(dz)$, $\bar{G} = \bar{G}_0 + G_1$, $\bar{k} = \bar{k}_0 + k_1$ and

$$dx = dt \int \bar{G}(x,\alpha)m_t(d\alpha) + \bar{\sigma}(x)dw \tag{2.6}$$

$$V(x,m) = E_x^m \int_0^T \int \bar{k}(x(s),\alpha)m_s(d\alpha)ds + E_x^m g(x\ (T)), \tag{2.7}$$

where $m(\cdot)$ is admissible with respect to the standard vector valued Wiener process $w(\cdot)$. Let $\bar{A}^\alpha$ denote the weak infinitesimal operator of (2.6), with control fixed at α, and $\bar{A}^m$ that associated with relaxed control $m(\cdot)$. Finally, we assume

A2.5. *For each admissible relaxed control,* (2.6) *has a unique weak sense solution.*

Remark. By uniqueness, $\mu_x(\cdot)$ is (weakly) continuous in x, but $\bar{G}$ is not necessarily Lipschitz continuous in x. Nevertheless, all moments of the solution $x(\cdot)$ exist. We wish to have a well defined "averaged" problem such as (2.6), which is the limit in an appropriate sense of $x^\varepsilon(\cdot)$ for *any* weakly convergent sequence $\{m^\varepsilon(\cdot)\}$. If $m_t^\varepsilon(\cdot)$ "oscillates too much" for small ε, and there is a direct interaction between $z^\varepsilon(\cdot)$ and $m^\varepsilon(\cdot)$, then the *form* of the limit problem can be quite sensitive to the particular *form* chosen for $m^\varepsilon(\cdot)$, and we would have different limit "types" for different classes of $m^\varepsilon(\cdot)$. This problem is avoided by the "separation of variables" imposed in (A2.1).

The following lemma will be helpful in characterizing certain measures as invariant measures. It is proved by a slight extension of [8], Prop. 9.2.

Lemma 2.1. *Let* $z(\cdot)$ *be a diffusion process with weak infinitesimal operator* A, *and suppose that the coefficients have linear growth in* z *as* $|z| \to \infty$, *and let the solution be unique for each initial condition. Suppose that* $\int Af(x)\mu(dx) = 0$ *for all smooth* $f(\cdot)$ *with compact support. Then* $\mu(\cdot)$ *is an invariant measure for* $x(\cdot)$.

In the weak convergence arguments below, we use the *Skorohod topology* on the processes $x^\varepsilon(\cdot)$, $z_0^\varepsilon(\cdot)$, etc., and the *weak topology* on the $\{m^\varepsilon(\cdot)\}$.

Theorem 2.2. *Assume* (A2.1) - (A2.5), *and let* $m^\varepsilon(\cdot)$ *be admissible relaxed controls. Then* $\{x^\varepsilon(\cdot),m^\varepsilon(\cdot)\}$ *is tight. Let* $(x(\cdot),m(\cdot))$ *denote the limit of a weakly convergent subsequence (indexed by* ε_n*). Then there is a standard Wiener process* $w(\cdot)$ *such that* $m(\cdot)$ *is admissible with respect to* $w(\cdot)$ *and* $(x(\cdot),m(\cdot),w(\cdot))$ *satisfy* (2.6). *Also*

$$V^\varepsilon(x,m^\varepsilon) \to V(x,m). \tag{2.8}$$

Proof. Part 1. The tightness is obvious. We need only characterize the limit and prove (2.8). Let $f(\cdot)$ be a smooth real valued function of x with compact support, and let $h(\cdot)$ be a real valued, bounded and continuous function of its arguments. Define $(m,\phi_i)_t = \int_0^t \int \phi_i(s,\alpha)m(d\alpha ds)$ for $\phi_i(\cdot)$ real valued, continuous and with compact support. For given t, τ, we let $t_i \le t \le t + \tau$, $i \le q$. Then

$$A^\varepsilon f(x) = f_x'(x) \int G(x,z,\alpha)m_t^\varepsilon(d\alpha) + \text{trace}\,\frac{f_{xx}(x)}{2}\,\sigma(x,z)\sigma'(x,z),$$

and

$$Eh(x^\varepsilon(t_i),(m^\varepsilon,\phi_j)_{t_i},\ i \le q,\ j \le p)\Big[f(x^\varepsilon(t+\tau)) - f(x^\varepsilon(t)) - \int_t^{t+\tau} A^\varepsilon f(x^\varepsilon(s))ds\Big] = 0. \tag{2.9}$$

Suppose that we can show that

$$\Big|\int_t^{t+\tau} A^\varepsilon f(x^\varepsilon(s))ds - \int_t^{t+\tau}\int \bar{A}^\alpha f(x^\varepsilon(s))m_s^\varepsilon(d\alpha)ds\Big| \to 0 \tag{2.10}$$

in probability, and let ε index a weakly convergent subsequence with limit $(x(\cdot), m(\cdot))$. Then we would have

$$Eh(x(t_i),(m,\phi_j)_{t_i}, i \le q, j \le p)\left[f(x(t+\tau)) - f(x(t)) - \int_t^{t+\tau}\int \bar{A}^{\alpha} f(x(s))m_s(d\alpha)ds\right] = 0. \tag{2.11}$$

The expression (2.10) and the arbitrariness of $h(\cdot)$, $f(\cdot)$, $\phi_i(\cdot)$, t_i, q, p, t and τ, implies that $x(\cdot)$ solves the martingale problem for the averaged system (2.6) under the relaxed control $m(\cdot)$. Thus, there is a standard Wiener process $w(\cdot)$ such that $m(\cdot)$ is admissible with respect to $w(\cdot)$, and (2.6) holds.

Part 2. Proof of (2.10). By the weak convergence (and the Skorohod representation used, so that we can assume w.p.1 convergence in the appropriate topology)

$$\int_t^{t+\tau}\int G_0(x^{\varepsilon}(s),\alpha)m_s^{\varepsilon}(d\alpha)ds \to \int_t^{t+\tau}\int G_0(x(s),\alpha)m_s(d\alpha)ds$$

$$\int_0^T\int k_0(x^{\varepsilon}(s),\alpha)m_s^{\varepsilon}(d\alpha)ds \to \int_0^T\int k_0(x(s),\alpha)m_s(d\alpha)ds \tag{2.12}$$

We next show that

$$\frac{1}{\Delta}\int_{t_1}^{t_1+\Delta} f_x'(x^{\varepsilon}(s))G_1(x^{\varepsilon}(s),z^{\varepsilon}(s))ds \xrightarrow{P} f_x'(x(t_1))\bar{G}_1(x(t_1)) \tag{2.13}$$

uniformly in $t_1 \le T$ as $\varepsilon \to 0$, $\Delta \to 0$, $\varepsilon/\Delta \to 0$. This convergence implies that the terms in (2.10) involving $f_x(\cdot)$ converge as desired.

By changing the time scale and using the tightness of $\{x^{\varepsilon}(\cdot)\}$, the left hand side of (2.13) can be written as

$$\frac{\varepsilon}{\Delta}\int_{t_1/\varepsilon}^{(t_1+\Delta)/\varepsilon} f_x'(x^{\varepsilon}(t_1))G_1(x^{\varepsilon}(t_1),z_0^{\varepsilon}(s))ds + \delta^{\varepsilon,\Delta}(t_1), \tag{2.14}$$

where $\delta^{\varepsilon,\Delta} \xrightarrow{P} 0$ uniformly in $t_1 \le T$ as $\varepsilon \to 0$, $\Delta \to 0$, $\varepsilon/\Delta \to 0$.

Define the sample occupation measure

$$P_{t_1}^{\varepsilon,\Delta}(dz) = \frac{\varepsilon}{\Delta}\int_{t_1/\varepsilon}^{(t_1+\Delta)/\varepsilon} I_{\{z_0^{\varepsilon}(s)\in dz\}}ds.$$

By (A2.3), $\{P_{t_1}^{\varepsilon,\Delta}(\cdot),\varepsilon,\Delta\}$ is a tight sequence of measure valued random variables. Let (ε_n,Δ_n) index a weakly convergent subsequence with limit denoted by $P_{t_1}(\cdot)$. We will show below that (w.p.1)

$$P_{t_1}(dz) = \mu_{x(t)}(dz). \tag{2.15}$$

Equation (2.15), the representation of the main term of (2.14) as $\int f_x'(x^{\varepsilon}(t_1))G_1(x^{\varepsilon}(t_1),z)\cdot P_{t_1}^{\varepsilon,\Delta}(dz)$ and the weak convergence yield (2.13).

Part 3. Let $g(\cdot)$ be a smooth function with compact support. For each t_1, the process $(t \geq t_1)$

$$M_g^\varepsilon(t) = g(z^\varepsilon(t)) - g(z^\varepsilon(t_1)) - \int_{t_1}^{t} A^\varepsilon g(z^\varepsilon(s))ds$$

is a martingale. We can also write

$$M_g^\varepsilon(t) = g(z^\varepsilon(t/\varepsilon)) - g(z_0^\varepsilon(t_1/\varepsilon)) - \int_{t_1/\varepsilon}^{t/\varepsilon} A_0^\varepsilon g(z_0^\varepsilon(s))ds, \tag{2.16}$$

where A_0^ε is the infinitesimal operator of (2.4). By the martingale property and boundedness of $g(\cdot)$ and $A_0^\varepsilon g(z)$, we have $\varepsilon M_g^\varepsilon(t_1 + \Delta)/\Delta \to 0$ in probability (uniformly in t_1) as $\varepsilon \to 0$, $\Delta \to 0$, $\varepsilon/\Delta \to 0$.

By (A2.1) and the tightness of $\{x^\varepsilon(\cdot)\}$, we can replace $x^\varepsilon(\cdot)$ in $A_0^\varepsilon g(z_0^\varepsilon(s))$ by $x^\varepsilon(t_1)$ and still get the same limits of $\varepsilon M_g^\varepsilon(t_1 + \Delta)/\Delta$. Doing this and using the representation

$$\frac{\varepsilon}{\Delta}\int_{t_1/\varepsilon}^{(t_1+\Delta)/\varepsilon} A^0_{x^\varepsilon(t_1)} g(z_0^\varepsilon(s))ds = \int A^0_{x^\varepsilon(t_1)} g(z)P_{t_1}^{\varepsilon,\Delta}(dz),$$

we have that as $\varepsilon_n \to 0$, $\Delta_n \to 0$, $\varepsilon_n/\Delta_n \to 0$,

$$\int A^0_{x(t_1)} g(y)P_{t_1}(dy) = 0 \tag{2.17}$$

w.p.1. Thus (2.17) holds w.p.1 for all smooth $g(\cdot)$ with compact support. This yields that (for a.a.ω) $P_{t_1}(\cdot)$ must be an invariant measure for the process $z_0(\cdot|x(t_1))$. By uniqueness $P_{t_1}(\cdot) = \mu_{x(t_1)}(\cdot)$ w.p.1 irrespective of the subsequence $\{\varepsilon_n,\Delta_n\}$.

Part 4. By Part 2, (2.13) holds. An identical proof yields the averaging of the second order term in $A^\varepsilon f(x^\varepsilon(s))$ to those of $\bar{A}^\alpha f(x^\varepsilon(s))$ (the control doesn't actually appear in the second order terms). Thus (2.10) holds. An identical argument together with the uniform integrability of

$$\left\{\int_0^T \int k(x^\varepsilon(s),z^\varepsilon(s),\alpha)m_s^\varepsilon(d\alpha)ds,\quad \varepsilon > 0\right\}$$

yields (2.8). Q.E.D.

Convergence of $V^\varepsilon(x)$ to $V(x)$. Let $(\bar{m}(\cdot),w(\cdot))$ denote an optimal admissible pair for the averaged system (2.6), (2.7). I.e., $V(x,\bar{m}) = V(x)$.

Theorem 2.3. *Assume* (A2.1) *to* (A2.5). *Then*

$$V^\varepsilon(x) \to V(x).$$

Proof. Let $\{m^\varepsilon(\cdot),w_1(\cdot),w_2(\cdot)\}$ be admissible for (2.1)-(2.2). By Theorem 2.1, $\underline{\lim}_\varepsilon V^\varepsilon(x,m^\varepsilon) \geq V(x)$. Thus, we need only show that for each $\delta > 0$, there is a sequence $\tilde{m}^\varepsilon(\cdot)$ such that ($\bar{m}$ is the optimal control)

$$\lim_{\varepsilon} V^\varepsilon(x,\tilde{m}^\varepsilon) \le V(x) + \delta = V(x,\bar{m}) + \delta \tag{2.18}$$

We outline two approaches to getting $\tilde{m}^\varepsilon(\cdot)$. In [2] (where $k(\cdot)$, $G(\cdot)$ and $\sigma(\cdot)$ were bounded), it was shown that for each $\delta > 0$ there is a $\Delta > 0$ and a sequence of U-valued continuous functions $u_i(\cdot)$ such that the control taking values $u(t) = u_i(x(t))$ on $[i\Delta, i\Delta + \Delta)$ yields a $\delta/2$-optimal control for (2.6), (2.7). The same idea works here due to the boundedness of $\bar{\sigma}(\cdot)$, and the linear growth conditions in x. Now, apply $u_i(\cdot)$ to (2.1), (2.2) by using $u^\varepsilon(t) = u_\varepsilon(x^\varepsilon(t))$ on $[i\Delta, i\Delta + \Delta)$. We have $V^\varepsilon(x,u^\varepsilon) \to V(x,u)$, hence (2.18).

For an alternative point of view, we do the special case where $\bar{\sigma}(\cdot)$ is (uniformly) invertable - (but the idea can be extended to work in general).

Recall that the solution to (2.6) is (weakly) unique, given $(m(\cdot),w(\cdot))$. Thus, if $\{m^n(\cdot),w(\cdot)\} \Rightarrow (m(\cdot),w(\cdot))$, we have $V(x,m^n) \to V(x,m)$. Fix $\delta > 0$. We can then approximate the optimal control $\bar{m}(\cdot)$ by an admissible $m^\delta(\cdot)$ which takes only finitely many values and is constant on the intervals $[i\Delta, i\Delta + \Delta)$, for small $\Delta > 0$ and such that $V(x,\bar{m}) \ge V(x,m^\delta) - \delta/2$. Define

$$w^\varepsilon(t) = \int_0^t \bar{\sigma}^{-1}(x^\varepsilon(s))\sigma(x^\varepsilon(s),z^\varepsilon(s))dw_1(s).$$

Then $\{w^\varepsilon(\cdot)\}$ is tight and the weak limits are standard Wiener processes. In fact, whatever (admissible with respect to $w_1(\cdot)$) $m^{\delta,\varepsilon}(\cdot)$ are, the limits of the sequence $\{x^\varepsilon(\cdot),m^{\delta,\varepsilon}(\cdot),w^\varepsilon(\cdot)\}$ satisfy (2.6).

We can define $m^{\delta,\varepsilon}(\cdot)$ such that it is admissible with respect to $(w_1(\cdot),w_2(\cdot))$ and $\{m^{\delta,\varepsilon}(\cdot),w^\varepsilon(\cdot)\} \Rightarrow (m^\delta(\cdot),w(\cdot))$. Now, we have $\{x^\varepsilon(\cdot),m^{\delta,\varepsilon}(\cdot),w^\varepsilon(\cdot)\} \Rightarrow (x(\cdot),m^\delta(\cdot),w(\cdot))$ satisfying (2.6), and also $V^\varepsilon(x,m^{\delta,\varepsilon}) \to V(x,m^\delta)$. Q.E.D.

On the Condition (A2.3). Conditions such as (A2.3) can often be verified by Liapunov function methods. The idea is essentially the same as used in the deterministic case [3], [4], and we do one special case for illustrative purposes only. We use the system

$$\begin{aligned} dx^\varepsilon &= (A_0x^\varepsilon + A_1y^\varepsilon + Bu)dt + \sigma\, dw_1 \\ \varepsilon\, dy^\varepsilon &= (-H_0x^\varepsilon + H_1y^\varepsilon)dt + \sqrt{\varepsilon}v\, dw_2, \end{aligned} \tag{2.19}$$

where σ and v are bounded and the matrices H_1 and $\tilde{A}_0 = [A_0 - A_1H_1^{-1}H_0]$ are stable. We now show that

$$\sup_{\substack{t<\infty \\ \varepsilon>0}} (E|y^\varepsilon(t)|^2 + E|x^\varepsilon(t)|^2) < \infty.$$

Owing to the linearity and the boundedness of U, we need only work with the deterministic and uncontrolled system

$$\dot{x}^\varepsilon = A_0x^\varepsilon + A_1y^\varepsilon = \tilde{A}_0x^\varepsilon + A_1(y^\varepsilon - H_1^{-1}H_0x^\varepsilon)$$
$$\dot{y}^\varepsilon = H_1(y^\varepsilon - H_1^{-1}H_0x^\varepsilon)/\varepsilon. \tag{2.20}$$

Note that $y = H_1^{-1}H_0x$ is a stable point of the second system in (2.20), for fixed x, and that $\tilde{A}_0$ is obtained by centering y about the stable point.

Let $y'Qy$ and $x'Px$ ($Q > 0$, $P > 0$, symmetric) be Liapunov functions for $\dot{y} = H_1y$ and $\dot{x} = \tilde{A}_0x$, resp. Then set $W(x,y) = x'Px + (y - H_1^{-1}H_0x)'Q(y - H_1^{-1}H_0x)$.

We have (for $x = x^\varepsilon$, $y = y^\varepsilon$)

$$\dot{W}(x,y) = 2[x'P(\tilde{A}_0x + A_1(y - H_1^{-1}H_0x)] + 2(y - H_1^{-1}H_0x)'Q[H_1(y - H_1^{-1}H_0x)/\varepsilon$$
$$- H_1^{-1}H_0(\tilde{A}_0x + A_1(y - H_1^{-1}H_0x))].$$

From this it follows that for some $a > 0$, $b > 0$

$$\dot{W}(x,y) \leq -\frac{a}{\varepsilon}|y - H_1^{-1}H_0x|^2 - b|x|^2.$$

The "stochastic" result follows from this.

3. AVERAGE COST PER UNIT TIME

We now extend the results of the last section to the interval $[0,\infty)$. For admissible $m(\cdot)$, define the costs

$$\gamma_T^\varepsilon(m^\varepsilon) = \frac{1}{T}\int_0^T \int k(x^\varepsilon(s), z^\varepsilon(s), \alpha)m_s^\varepsilon(d\alpha)ds$$
$$\gamma^\varepsilon(m^\varepsilon) = \overline{\lim_T}\, E^m\gamma_T^\varepsilon(m^\varepsilon) \tag{3.1}$$

$$\gamma_T(m) = \frac{1}{T}\int_0^T \int \bar{k}(x(s),\alpha)m_s(d\alpha)ds$$
$$\gamma(m) = \overline{\lim_T}\, E^m\gamma_T(m). \tag{3.2}$$

Note that $\gamma_T^\varepsilon(m^\varepsilon)$ and $\gamma_T(m)$ are pathwise averages - there is no expectation taken.

We assume

A3.1. *For each $\delta > 0$, there is a δ-optimal m^ε such that the sequence $\{x^\varepsilon(n)$, $n = 0,1, \ldots$, small $\varepsilon > 0$, m^ε used$\}$ is tight.*

If the state space is not compact, then conditions such as (A3.1) are generally verified by a stability argument, as done for the linear case in the last part of Section 2.

A3.2. *For each $\delta > 0$,* (A2.3) *holds for $T = \infty$, and the controls of* (A3.1).

A3.3. *For each admissible feedback relaxed control,* (2.6) *has a unique weak sense solution and is a Feller process with a unique invariant measure.*

Define the occupation measures

$$P^{\varepsilon,T}(dxd\alpha) = \frac{1}{T}\int_0^T I_{\{x^\varepsilon(s)\in dx\}} m_s^\varepsilon(d\alpha)ds.$$

Such occupation measures were first used in [9] for a problem concerning existence of an optimal ergodic control for a diffusion process.

Theorem 3.1. *Assume (A2.1), (A2.2), (A2.4), (A2.5), (A3.3), but with $k(\cdot)$ bounded. Let $\{m^\varepsilon(\cdot)\}$ denote a sequence of admissible relaxed controls such that (A3.1), (A3.2) holds. Then $\{P^{\varepsilon,T}(\cdot)\}$ is tight. If (ε_n,T_n) indexes a weakly convergent subsequence, then there is a relaxed feedback $m_x(\cdot)$ for (2.6) with associated invariant measure $v(\cdot)$ such that*

$$\gamma_{T_n}^{\varepsilon_n}(m^\varepsilon) \xrightarrow{P} \gamma(m_x) = \int \bar{k}(x,\alpha)m_x(d\alpha)v(dx), \tag{3.3}$$

where the initial condition used in the definition of $\gamma(m_x)$ is the invariant one.

Proof. Condition (A3.1) and the conditions on $G(\cdot)$ and $\sigma(\cdot)$ in (A2.1) imply the tightness of

$$\{x^\varepsilon(t),\ \text{small } \varepsilon > 0,\ t < \infty\}\ , \tag{3.4}$$

hence the tightness of $\{P^{\varepsilon,T}(\cdot)\}$. The tightness (3.4) and the boundedness of $k(\cdot)$ imply that the limits of $\gamma_T^\varepsilon(m^\varepsilon)$ and those of (as $\varepsilon \to 0$, $T \to \infty$, $\Delta \to 0$)

$$\sum_0^{T/\Delta} \frac{1}{T}\int_{i\Delta}^{i\Delta+\Delta} k(x^\varepsilon(i\Delta),z^\varepsilon(s),\alpha)m_s^\varepsilon(d\alpha)ds \tag{3.5}$$

are the same w.p.1: If $\gamma_{T_n}^{\varepsilon_n}(m^\varepsilon) \to \tilde{\gamma}$ w.p.1, then so will the same subsequence of (3.5) and conversely.

By the tightness (3.4) and the proof of Theorem 2.1, the limits of (3.5) and those of

$$\frac{1}{T}\int_0^T \int \bar{k}(x^\varepsilon(s),\alpha)m_s^\varepsilon(d\alpha)ds = \int \bar{k}(x,\alpha)P^{\varepsilon,T}(dxd\alpha) \tag{3.6}$$

are the same. We now characterize the weak limits of $\{P^{\varepsilon,T}(\cdot)\}$.

We follow the scheme used in Theorem 2.1. Let $g(\cdot)$ be a smooth function with compact support. The process defined by

$$M_g^\varepsilon(t) = g(x^\varepsilon(t)) - g(x(0)) - \int_0^t A^\varepsilon g(x^\varepsilon(s))ds$$

is a martingale. Since $\sup_{\varepsilon,n} E[M_g^\varepsilon(n+1) - M_g^\varepsilon(n)]^2 < \infty$, we have $M_g(t)/t \xrightarrow{P} 0$ as $\varepsilon \to 0$, $t \to \infty$. Thus (w.p.1)

$$\begin{aligned}0 = \lim_{\varepsilon,T}\frac{1}{T}\int_0^T A^\varepsilon g(x^\varepsilon(s))ds = \lim_{\varepsilon,T}\frac{1}{T}\int_0^T ds\Big[g_x'(x^\varepsilon(s))\int G_0(x^\varepsilon(s),\alpha)m_s^\varepsilon(d\alpha) \\ + g_x'(x^\varepsilon(s))G_1(x^\varepsilon(s),z^\varepsilon(s)) + \frac{1}{2}\,\text{trace}\ g_{xx}(x^\varepsilon(s)) \cdot a(x^\varepsilon(s),z^\varepsilon(s))\Big].\end{aligned} \tag{3.7}$$

Similarly to what was done in the first part of the proof, the right hand side of (3.7) has the same limits (w.p.1) as has

$$\frac{1}{T}\int_0^T \Big[g_x'(x^\varepsilon(s)) \int G_0(x^\varepsilon(s),\alpha)m_s^\varepsilon(d\alpha) + \bar{G}_1(x^\varepsilon(s)) + \frac{1}{2}\,\text{trace}\, g_{xx}(x^\varepsilon(s))\cdot\bar{a}(x^\varepsilon(x))\Big]ds$$
$$= \frac{1}{T}\int_0^T \bar{A}^\alpha g(x^\varepsilon(s))m_t^\varepsilon(d\alpha)\,ds = \int \bar{A}^\alpha g(x)P^{\varepsilon,T}(dxd\alpha). \tag{3.8}$$

Let ε_n, T_n index a weakly convergent subsequence of $\{P^{\varepsilon,T}(\cdot)\}$, with limit $\tilde{P}(\cdot)$. Then by (3.7) and the equivalence of the limits of (3.7) and (3.8), we have (w.p.1)

$$0 = \int \bar{A}^\alpha g(x)\tilde{P}(dxd\alpha).$$

Now, factor $\tilde{P}(\cdot)$ in the form $\tilde{P}(dxd\alpha) = m_x(d\alpha)v(dx)$. We can choose $m_x(\cdot)$ such that is is an admissible relaxed feedback control (i.e., such that $m_x(A)$ is Borel measurable for each $A \in \mathcal{U}$). Then, w.p.1,

$$0 = \int \bar{A}^\alpha g(x)m_x(d\alpha)v(dx). \tag{3.9}$$

Hence, w.p.1, (3.9) holds for all smooth $g(\cdot)$ with compact support. Since $x(\cdot)$ is a continuous Feller process under $m_x(\cdot)$ by (A3.3), (3.9) implies that $v(\cdot)$ is an invariant measure for $x(\cdot)$ under $m_x(\cdot)$. Finally, taking limits in (3.6) (for ε, T replaced by ε_n, T_n, resp.) yields the limit

$$\int \bar{k}(x,\alpha)\tilde{P}(dxd\alpha) = \int k(x,z,\alpha)\mu_x(dy)m_x(d\alpha)v(dx) = \gamma(m_x)$$

as desired. Q.E.D.

We now examine the following questions. Let $\bar{m}^\varepsilon(\cdot)$ be optimal for (3.1), under (2.1), (2.2), in the sense of minimizing $E^m\gamma^\varepsilon(m)$, and let $\bar{m}(\cdot)$ be optimal for (3.2), under (2.6). Then does $\gamma_T^\varepsilon(\bar{m}^\varepsilon) \xrightarrow{P} \gamma(\bar{m})$ as $\varepsilon \to 0$, $T \to \infty$? Next, suppose that there is a continuous function $u^\delta(\cdot)$ which is δ-optimal in the sense that $\gamma(u^\delta) \le \gamma(m_x) + \delta$ for any feedback relaxed control, and where the initial conditions are the invariant ones. Then we wish to know whether $u^\delta(\cdot)$ is "nearly δ-optimal" for (2.1), (2.2).

Assume

A3.4. *For $\delta > 0$, there is a continuous δ-optimal feedback control $u^\delta(\cdot)$ for* (2.6), *under which* (A3.1), (A3.2) *hold.*

See [5] for criteria guaranteeing (A3.4).

<u>Theorem 3.2</u>. *Assume the conditions of Theorem* 3.1 *and* (A3.4). *Then*

$$\gamma_T^\varepsilon(u^\delta) \xrightarrow{P} \gamma(u^\delta) \quad \textit{as} \quad \varepsilon \to 0,\ T \to \infty \tag{3.10}$$

and

$$\underline{\lim}_{\varepsilon,T}\, P\{\gamma_T^\varepsilon(\bar{m}^\varepsilon) \ge \gamma(u^\delta) - 2\delta\} = 1. \tag{3.11}$$

Remark. The proof is essentially the same as that of Theorem 3.1. The limit measure associated with $u^{\delta}(\cdot)$ must be invariant, by the uniqueness in (A3.3).

The work was partially supported by contracts AFOSR 85-0315, NSF ECS-8505674 and ARO DAAL 03-86-K-0171.

BIBLIOGRAPHY

1. W. H. Fleming, M. Nisio, "On stochastic relaxed controls for partially observed diffusions," *Magoya Math. J.*, 93, 1984, 71-108.

2. H. J. Kushner, W. Runggaldier, "Nearly optimal state feedback controls for stochastic systems with wideband noise disturbances," *SIAM J. on Control and Optimization*, 25, 1987, 298-315.

3. A. Saberi, H. Khalil, "Quadratic-type Liapunov functions for singularly perturbed systems," *IEEE Trans. on Automatic Control*, AC-29, 1984, 542-550.

4. H. Khalil, "Stability analysis of singularly perturbed systems," Vol. 90, *Lect. Notes on Control and Inf. Sci.* (Ed.: Kokotovic, Bensoussan, Blankenship), Springer, Berlin, 1986, 357-373.

5. H. J. Kushner, "Necessary and sufficient conditions for optimality for the average cost per unit time problem with a diffusion model," *SIAM J. on Control and Optimization*, 16, 1977, 330-346.

6. A. Bensoussan, *Méthodes de Perturbations en Contrôle Optimal*, Dunod, Paris, 1988, to appear.

7. A. Bensoussan, G. Blankenship, "Singular perturbations in stochastic control," same volume as in [4], pp. 171-262.

8. S. N. Ethier, T. G. Kurtz, *Markov Processes; Characterization and Convergence*, Wiley, New York, 1986.

9. V. Borkar, M. K. Ghosh, "Ergodic control of multidimensional diffusions I: the existence results," *SIAM J. on Control and Optimization*, 28, 1988, 112-126.

EXTENDED STOCHASTIC LYAPUNOV FUNCTIONS AND RECURSIVE ALGORITHMS IN LINEAR STOCHASTIC SYSTEMS

Tze Leung Lai
Department of Statistics, Stanford University
Stanford, CA 94305, USA

1. Introduction

In many scientific and engineering applications, the problem of parameter estimation not only involves the classical concept of efficiency of the estimator θ_n based on a sample of n observations $Y_1, \ldots, Y_n$, but for on-line implementation the computational complexity of successively updating the estimator must also be considered. For example, a simple recursion of the form $\theta_n = g(n,\theta_{n-1},Y_n)$ requires much less storage and computational burden than an estimator of the form $\theta_n = g_n(Y_1,\ldots,Y_n)$ which needs the values of all the previous observations. In addition to the computational advantage for on-line identification, these recursive algorithms can also be easily tailored to track time varying parameters in dynamical systems (cf. [18]).

An important stochastic model in the time series and control systems literature is that governed by the linear difference equation

$$A(q^{-1})y_n = B(q^{-1})u_{n-d} + C(q^{-1})\varepsilon_n , \qquad (1)$$

where $\{y_n\}$, $\{u_n\}$ and $\{\varepsilon_n\}$ denote the output, input, and disturbance sequences, respectively, and

$$A(q^{-1}) = 1 + a_1q^{-1} +\ldots+ a_kq^{-k} , \quad B(q^{-1}) = b_0 +\ldots+ b_hq^{-h} ,$$
$$C(q^{-1}) = 1 + c_1q^{-1} +\ldots+ c_rq^{-r}$$

are polynomials in the unit delay operator q^{-1}. There is a large literature on recursive estimation of the parameter vector

$$\theta \triangleq (a_1,\ldots,a_k, b_0,\ldots,b_h, c_1,\ldots,c_r)' \qquad (2)$$

of (1), for which many recursive estimators have been proposed. The recent monographs by Ljung and Söderström [18] and by Caines [1] provide excellent unified overviews of the subject.

A basic problem concerning a recursive identification algorithm is whether, or under what conditions, it converges. The seminal papers by Ljung [16], [17] represent a pioneering effort to study the consistency problem of various recursive identification algorithms in parameter estimation for the linear stochastic system (1). In these papers, Ljung introduced the method of studying the convergence of an identification algorithm, defined recursively by

$$\theta_n = \theta_{n-1} + \rho_n R_n^{-1} \phi_n e_n(\theta_{n-1}) , \tag{3}$$

via the stability properties of an associated non-random ODE (ordinary differential equation). The ρ_n in (3) is a positive scalar ≤ 1 such that $\Sigma_1^\infty \rho_i = \infty$ and $\Sigma_1^\infty \rho_i^2 < \infty$. Typically, $\rho_n \sim 1/n$. In (3), $e_n(\theta_{n-1})$ denotes the prediction error of the one-step-ahead predictor of y_n, and ϕ_n is a $(k+r+h+1)$-dimensional vector satisfying the recursive relation

$$\phi_n = F(\theta_{n-1})\phi_{n-1} + G(\theta_{n-1})(y_{n-1},\ldots,y_{n-k}, u_{n-d},\ldots,u_{n-d-h}, e_{n-1}(\theta_{n-2}),\ldots,e_{n-r}(\theta_{n-r-1}))' , \tag{4}$$

where F and G are matrix functions. The R_n in (3) is a positive definite matrix defined recursively by either of the following:

$$R_n = R_{n-1} + \rho_n(\phi_n\phi_n' - R_{n-1}) , \tag{5a}$$

or

$$R_n = r_n I, \quad r_n = r_{n-1} + \rho_n(\phi_n'\phi_n - r_{n-1}) . \tag{5b}$$

The underlying heuristic idea in Ljung's approach is that if θ_n should converge to a limit θ^*, then one may approximate $e_i(\theta_{i-1})$ and ϕ_i in (3)-(5) by $e_i(\theta^*)$ and $\phi_i(\theta^*)$, which are stationary ergodic sequences under certain assumptions on the disturbances ε_n and the inputs u_n. This suggests that if R_n should also converge to a limit R^*, then one may approximate (3) by

$$(\theta_n - \theta_{n-1})/\rho_n \doteq R^{*-1} E[\phi_n(\theta^*)e_n(\theta^*)] \doteq R_n^{-1} h(\theta^*) , \tag{6}$$

where $h(\theta^*) = E[\phi_n(\theta^*)e_n(\theta^*)]$ for all n by stationarity. Let $t_n = \Sigma_1^n \rho_i (\to\infty)$, $\theta(t) = \theta_n$ at $t = t_n$, and define $\theta(t)$ by linear interpolation when $t_{n-1} < t < t_n$. Since $t_n - t_{n-1} = \rho_n \to 0$, the left hand side of (6) can be approximated by $d\theta/dt|_{t=t_n}$. This suggests that the ODE

$$d\theta/dt = R^{-1}(t)\, h(\theta(t)) \tag{7}$$

is a limiting version of the recursion (3) and that the constant solution (equilibrium) $\theta(t) \equiv \theta^*$ satisfies the equation $h(\theta^*) = 0$. Moreover, letting $H(\theta) = E\{\phi_n(\theta^*)\phi_n'(\theta^*)\}$, the same argument shows that (5a) and (5b) can be approximated by the ODEs

$$dR/dt = H(\theta(t)) - R , \tag{8a}$$

$$dr/dt = \mathrm{tr}\, H(\theta(t)) - r \quad \text{and} \quad R = rI , \tag{8b}$$

respectively. Hence the constant solution $R(t) \equiv R^*$ (or $r(t) \equiv \mathrm{tr}\, R^*$) of the ODE (8a) (or (8b)) satisfies the equation $R^* = H(\theta^*)$ (or $\mathrm{tr}\, R^* = \mathrm{tr}\, H(\theta^*)$).

The above argument shows that if the recursive scheme (3)-(5) should converge to a limit (θ^*, R^*) a.s., then the limit (θ^*, R^*) is an equilibrium of the ODE (7)-(8). Ljung showed further that under certain assumptions, a stability analysis of the ODE (7)-(8) can also characterize the set of limit points of the recursive scheme. Different sets of assumptions have been proposed to ensure the validity of this approach, and we refer the reader to Ljung and Söderström [18, Chapter 4] for the details. Although the basic heuristic ideas of this approach are quire simple and the stability analysis of the ODE (7)-(8) can often be handled elegantly by using Lyapunov functions, the technical details of the argument and the verification of the underlying assumptions are difficult and tedious and require "persistent excitation" and other restrictive conditions on the inputs as well as stability assumptions on system dynamics.

Consider the ODE $dx/dt = f(x)$, where $x(t) = (x_1(t),\ldots,x_m(t))'$ and $f = (f_1,\ldots,f_m)'$. In particular, for the ODE defined by (7) and (8b), $x' = (\theta',r)$. A Lyapunov function $V(x)$ associated with the ODE is a continuously differentiable nonnegative function such that

$$<f, \text{grad } V> \leq 0 . \tag{9}$$

From (9) and the ODE $dx/dt = f(x)$, it follows that

$$\frac{d}{dt} V(x(t)) = <f(x(t)), \text{grad } V(x(t))> \leq 0 . \tag{10}$$

Hence V is non-increasing along the trajectories of the ODE and the set

$$D \stackrel{\Delta}{=} \{x : <f(x), \text{grad } V(x)> = 0\} \tag{11}$$

is an invariant set of the ODE (i.e., any trajectory that starts in D remains there). Outside D, V is strictly decreasing, but since V is bounded below, it cannot continue to decrease indefinitely. Hence $x(t)$ must converge to D as $t \to \infty$, showing that D is a globally asymptotically stable invariant set. The construction of such functions V is a basic ingredient in Ljung's stability analysis. Usually one cannot, however, define such a function V on the whole space, but can only define it with (9) holding on some bounded connected subset S. In this case, Ljung's idea is to first choose S such that it is a domain of attraction of the invariant set D (i.e., any trajectory that starts in S converges to D as $t \to \infty$) and then to show that $P\{(\theta_n,R_n) \in S$ infinitely often$\} = 1$. Carrying out this idea, however, often involves substantial technical difficulties.

Instead of working with a Lyapunov function associated with the limiting deterministic ODE (7)-(8), an obvious alternative is to develop an analogue for the original stochastic recursive algorithm (3)-(5). This is the idea behind the "stochastic Lyapunov function" approach introduced by Moore and Ledwich [19] and Solo [24] for convergence analysis of recursive identification algorithms. This approach, which will be discussed in Section 3, has a substantial history in the stability theory of stochastic dynamical systems and the almost sure (a.s.) convergence theory of stochastic approximation schemes. Section 2 reviews the method of stochastic Lyapunov functions, the supermartingale property of these functions and related martingale convergence theorems. Section 3 extends this method by replacing the supermartingale property with another generalization of the fundamental property (10) of a Lyapunov function and by applying other martingale limit theorems (not restricted to convergence). Applications of this notion of "extended stochastic Lyapunov functions" to recursive identification, adaptive prediction and adaptive control will also be discussed in Sections 3 and 4.

2. Martingale convergence theorems and the method of stochastic Lyapunov functions in stochastic stability theory and stochastic approximation

Stochastic Lyapunov functions have been introduced as analogues of Lyapunov's auxiliary functions in the stability theory of deterministic ODEs to develop similar stability theorems for stochastic differential (or difference) equations and other Markov processes (cf. [5], [6]). Consider, for example, the stochastic differential equation (SDE)

$$dY(t) = b(t,Y(t))dt + \sum_{i=1}^{m} \sigma_i(t,Y(t))dw_i(t) , \tag{12}$$

where $Y(t)$, $b(t,y)$, $\sigma_i(t,y)$ are m-dimensional vectors and $w_1(t), \ldots, w_m(t)$ are independent Wiener processes. Assume that $b(t,y)$ and $\sigma_i(t,y)$ are uniformly Lipschitz continuous in y and that $b(t,0) \equiv 0$, $\sigma_i(t,0) \equiv 0$, so that $Y(t) \equiv 0$ is a constant solution of (12). This constant solution is said to be "stable with probability 1" if for any $s \geq 0$ and $\varepsilon > 0$,

$$\lim_{y \to 0} P\{\sup_{t \geq s} |Y^{s,y}(t)| \geq \varepsilon\} = 0 , \tag{13}$$

and is said to be "asymptotically stable with probability 1" if it is stable with probability 1 and for all $s \geq 0$ and y in some neighborhood of 0 ,

$$P\{\lim_{t\to\infty} Y^{s,y}(t) = 0\} = 1 , \tag{14}$$

where $Y^{s,y}(t)$, $t \geq s$, denotes the solution of (12) with initial condition $Y(s) = y$.

In the stability theory for a deterministic ODE $dx/dt = f(x)$, the second (or direct) method of Lyapunov uses the existence of Lyapunov functions satisfying certain conditions to show that an equilibrium is stable or asymptotically stable. A basic property of these functions is (10), which can be written as

$$D_0 V \leq 0 , \quad \text{where} \quad (D_0 g)(u) = \lim_{\delta \downarrow 0} \delta^{-1}\{g(x^u(\delta)) - g(u)\} , \tag{15}$$

and $x^u(t)$ denotes the solution of the ODE with initial condition $x^u(0) = u$. For the SDE (12) with time-varying coefficients, a Lyapunov-type function is a function of (t,y) and a natural analogue of the Lyapunov operator D_0 in (15) now takes the form

$$(Lg)(t,y) = \lim_{\delta \downarrow 0} \delta^{-1} E\{g(t+\delta, Y^{t,y}(t+\delta)) - g(t,y)\} . \tag{16}$$

Note that L is the infinitesimal generator of the space-time process $(t,Y(t))$. A nonnegative function V defined on the domain $[0,\infty) \times B$, where B is some neighborhood of the equilibrium 0 of (12), is called a "stochastic Lyapunov function" associated with the SDE if

$$LV \leq 0 \quad \text{on} \quad [0,\infty) \times B , \tag{17}$$

in analogy with (15). Like the stability theory of ODEs, the existence of such stochastic Lyapunov functions satisfying certain conditions implies that the equilibrium 0 of the SDE is stable or asymptotically stable with probability 1 (cf. [5]). A key tool in proving these results is that (17) implies that $\{V_t \triangleq V(\tau_B \wedge t,\ Y^{s,y}(\tau_B \wedge t)),\ t \geq s\}$ is a supermartingale for all $s \geq 0$ and $y \in B$, where $\tau_B = \inf\{t \geq s : Y^{s,y}(t) \notin B\}$. By Doob's martingale convergence theorem, the nonnegative supermartingale V_t converges a.s. as $t \to \infty$. In addition to the martingale convergence theorem, one can also apply martingale inequalities to obtain bounds for probabilities of the type (13).

Roughly speaking, the stochastic Lyapunov function method is to transform a given stochastic dynamical system (possibly multidimensional, e.g., (12)) into a one-dimensional nonnegative process V_t so that the dynamics of the original system induce a supermartingale structure on V_t. We next review the application of this method to prove the a.s. convergence of stochastic approximation schemes. Here an obvious candidate for a stochastic Lyapunov function does not yield a supermartingale but has an "almost supermartingale" property. To fix the ideas, consider the Robbins-Monro [21] scheme

$$u_{n+1} = u_n - \rho_n y_n \tag{18}$$

to find the root θ of an unknown regression function in the regression model

$$y_i = M(u_i) + \varepsilon_i \qquad (i = 1,2,\ldots)\ , \tag{19}$$

where ρ_n are positive constants such that $\Sigma_1^\infty \rho_n^2 < \infty$ and $\Sigma_1^\infty \rho_n = \infty$. The random disturbances ε_i in the regression model are assumed to form a martingale difference sequence with respect to an increasing sequence of σ-fields $\mathcal{F}_i$ such that $\sup_i E(\varepsilon_i^2 | \mathcal{F}_{i-1}) < \infty$ a.s., while the regression function M is assumed to satisfy

$$\begin{aligned} &M(\theta) = 0,\ \inf_{\varepsilon < |u-\theta| < 1/\varepsilon} (u-\theta)M(u) > 0 \quad \text{for all} \quad 0 < \varepsilon < 1\ , \\ &|M(u)| \leq c(|u-\theta|+1) \quad \text{for some} \quad c > 0 \quad \text{and all} \quad u\ . \end{aligned} \tag{20}$$

To prove that $u_n \to \theta$ a.s., or equivalently that $(u_n-\theta)^2 \to 0$ a.s., a natural candidate for a stochastic Lyapunov function in the stochastic dynamical system (18) is the quadratic function $V(u) = (u-\theta)^2$. The recursion (18) implies the following recursion for $V_n \triangleq V(u_{n+1})$:

$$\begin{aligned} V_n = (u_{n+1}-\theta)^2 = V_{n-1} &+ \rho_n^2 M^2(u_n) - 2\rho_n M(u_n)(u_n-\theta) \\ &-2\rho_n \varepsilon_n [u_n-\theta-\rho_n M(u_n)] + \rho_n^2 \varepsilon_n^2\ . \end{aligned} \tag{21}$$

Hence, by (20) and the fact that $E(\varepsilon_n | \mathcal{F}_{n-1}) = 0$,

$$E(V_n | \mathcal{F}_{n-1}) \leq (1+2c^2\rho_n^2)V_{n-1} + \rho_n^2\{2c^2+E(\varepsilon_n^2 | \mathcal{F}_{n-1})\} - 2\rho_n M(u_n)(u_n-\theta)\ , \tag{22}$$

and therefore

$$E(V_n | \mathcal{F}_{n-1}) \leq (1+\alpha_{n-1})V_{n-1} + \beta_{n-1} - \gamma_{n-1}\ , \text{ say,} \tag{23a}$$

where α_i, β_i, γ_i are nonnegative $\mathcal{F}_i$-measurable random variables such that

$$\sum_1^\infty \alpha_i + \sum_1^\infty \beta_i < \infty \quad \text{a.s.} \tag{23b}$$

Robbins and Siegmund [22] call a sequence $V_n \geq 0$ satisfying (23) a "nonnegative almost supermartingale", for which they prove the following convergence theorem:

$$V_n \text{ converges a.s. and } \sum_1^\infty \gamma_i < \infty \quad \text{a.s.} \tag{24}$$

The a.s. convergence of the Robbins-Monro scheme u_n to θ follows easily from (20) and from the a.s. convergence of $|u_n-\theta|\,(=V_{n-1}^{1/2})$ and that of $\Sigma_1^\infty \rho_i\, M(u_i)(u_i-\theta)\,(=\Sigma_1^\infty \gamma_{i-1})$.

From (23a), it follows that $U_n \triangleq V_n/\Pi_{i=1}^{n-1}(1+\alpha_i) - \Sigma_{i=1}^{n-1}\, \beta_i/\Pi_{j=1}^{i}(1+\alpha_j)$ is a supermartingale. Although U_n need not be nonnegative, it is bounded below by $-k$ on the event $\{\Sigma_1^\infty \beta_i \leq k\}$ and therefore U_n converges a.s. on $\{\Sigma_1^\infty \beta_i \leq k\}$ for every $k = 1, 2, \ldots$. Since $\Sigma_1^\infty \beta_i < \infty$ a.s., U_n converges a.s. Since $\Sigma_1^\infty \alpha_i < \infty$ a.s., it then follows that V_n also converges a.s. This is the basic idea behind the Robbins-Siegmund theorem.

Starting with (22), we can in fact transform the V_n into a nonnegative supermartingale if we further assume that $\sup_i E(\varepsilon_i^2|\mathcal{F}_{i-1}) \leq k$ (which has no loss of generality since we can apply the martingale convergence theorem locally to the events $\{\sup_i E(\varepsilon_i^2|\mathcal{F}_{i-1}) \leq k\}$, $k = 1, 2, \ldots,$ as in the preceding argument). Let

$$W_n = (V_n+1) \prod_{i=n+1}^{\infty} \{1+(2c^2+k)\rho_i^2\} \; . \tag{25}$$

From (22), it follows that W_n is a nonnegative supermartingale, so the a.s. convergence of W_n , and hence of V_n also, follows. This argument is due to Gladyshev [3].

The monograph by Nevel'son and Has'minskii [20] gives a systematic treatment of the a.s. convergence theory of other stochastic approximation schemes by using the method of stochastic Lyapunov functions.

3. Recursive identification and extended stochastic Lyapunov functions

As will be shown in the subsequent discussion, in applications to recursive identification and to adaptive prediction and control in the linear stochastic system (1), it provides much greater flexibility and convenience by not insisting that a stochastic Lyapunov-type function must always converge. Motivated by these applications, we generalize the Robbins-Siegmund theorem by dropping the condition $\Sigma_1^\infty \beta_i < \infty$ a.s. in the following.

Theorem 1. Let $\{\varepsilon_n, \mathcal{F}_n, n\geq 1\}$ be a martingale difference sequence such that $\sup_n E(\varepsilon_n^2|\mathcal{F}_{n-1}) < \infty$ a.s. Let α_n, β_n, γ_n, V_n be nonnegative $\mathcal{F}_n$-measurable random variables such that $\Sigma_1^\infty \alpha_n < \infty$ a.s., and let w_n be $\mathcal{F}_n$-measurable. Suppose that for $n \geq 2$,

$$V_n \le (1+\alpha_{n-1})V_{n-1} + \beta_n - \gamma_n + w_{n-1}\,\varepsilon_n \quad \text{a.s.} \tag{26}$$

(i) On the event $\{\Sigma_1^\infty E(\beta_i|\mathcal{F}_{i-1}) < \infty\}$, V_n converges a.s. and $\Sigma_1^\infty E(\gamma_i|\mathcal{F}_{i-1}) < \infty$ a.s.

(ii) For every $\delta > 0$,

$$\max(V_n, \Sigma_1^n \gamma_i) = O(\Sigma_1^n \beta_i + (\Sigma_1^{n-1} w_i^2)^{\frac{1}{2}+\delta}) \quad \text{a.s.}$$

Proof. (i) follows from Theorem 1 of Robbins and Siegmund [22]. To prove (ii), let $Q_n = V_n/\Pi_{i=1}^{n-1}(1+\alpha_i)$. Then by (26),

$$Q_n - Q_{n-1} + \tilde{\gamma}_n \le \tilde{\beta}_n + \tilde{w}_{n-1}\,\varepsilon_n \quad \text{a.s.},$$

where $\tilde{\gamma}_j = \gamma_j/\Pi_{i=1}^{j-1}(1+\alpha_i)$, $\tilde{\beta}_j = \beta_j/\Pi_{i=1}^{j-1}(1+\alpha_i)$, $\tilde{w}_j = w_j/\Pi_{i=1}^{j}(1+\alpha_i)$. Therefore

$$Q_N - Q_1 + \sum_{i=2}^N \tilde{\gamma}_i \le \sum_{i=2}^N \tilde{\beta}_i + \sum_{i=2}^N \tilde{w}_{i-1}\,\varepsilon_i .$$

The desired conclusion then follows by a truncation argument and the strong law for martingales, as in the proof of Lemma 2(iii) of [10], noting that $1 \le \Pi_{i=1}^\infty(1+\alpha_i) < \infty$ a.s. since $\Sigma_1^\infty \alpha_i < \infty$ a.s.

A stochastic Lyapunov function of a stochastic dynamical system (possibly vector-valued) is a sequence of nonnegative random variables that have the supermartingale property, or more generally, an almost supermartingale property in the sense of (23a) and (23b). An "extended stochastic Lyapunov function" drops the assumption $\Sigma_1^\infty \beta_i < \infty$ in (23b) but reinforces (23a), which provides an inequality for $E(V_n - V_{n-1}|\mathcal{F}_{n-1})$, into an inequality (26) for $V_n - V_{n-1}$ with an additional martingale difference term. To see the usefulness of this idea, we use Theorem 1 to derive the results of [10] and [12] on the strong consistency of the recursive least squares and extended least squares algorithms, and compare these results with those obtained by Moore and Ledwich [19] and Solo [24] using the a.s. convergence of a stochastic Lyapunov function.

As is well known, the linear stochastic system (1) can be written as a stochastic regression model

$$y_n = \theta' \psi_n + \varepsilon_n , \tag{27}$$

where θ is defined in (2) and

$$\psi_n = (-y_{n-1}, \ldots, -y_{n-k}, u_{n-d}, \ldots, u_{n-d-h}, \varepsilon_{n-1}, \ldots, \varepsilon_{n-r})' . \tag{28}$$

It is commonly assumed that the ε_n form a martingale difference sequence with respect to an increasing sequence of σ-fields $\{\mathcal{F}_n\}$ such that

$$\sup_n E(|\varepsilon_n|^\alpha|\mathcal{F}_{n-1}) < \infty \quad \text{a.s. for some } \alpha > 2 . \tag{29}$$

The input u_n at stage n is assumed to be $\mathcal{F}_n$-measurable (i.e., involving only the current and past observations $y_n, y_{n-1}, u_{n-1}, \ldots,$ but no future observations).

In the "white-noise" case $C(q^{-1}) = 1$, the regressors ψ_n are observable random vectors such that ψ_n is $\mathcal{F}_{n-1}$-measurable, and a commonly used estimator is that based on the least squares criterion. The least squares estimate $\theta_n \triangleq (\Sigma_1^n \psi_i \psi_i')^{-1} \Sigma_1^n \psi_i y_i$ based on observations up to stage n can be expressed in the recursive form

$$\theta_n = \theta_{n-1} + P_n \psi_n (y_n - \theta_{n-1}' \psi_n) , \tag{30}$$

where $P_n = (\Sigma_1^n \psi_i \psi_i')^{-1}$ satisfies the recursion

$$P_n^{-1} = P_{n-1}^{-1} + \psi_n \psi_n' . \tag{31}$$

Note that (30)-(31) is a special case of (3), (5a) with $\phi_n = \psi_n$, $e_n(\theta_{n-1}) = y_n - \theta_{n-1}'\psi_n$, $\rho_n = n^{-1}$ and $R_n = n^{-1}\Sigma_1^n \psi_i \psi_i' = n^{-1} P_n^{-1}$. In this special case, Ljung's ODE approach requires that

$$n^{-1} \sum_1^n \psi_i \psi_i' \ (= R_n) \to R^* \ \text{(positive definite) a.s.} , \tag{32}$$

and a natural Lyapunov function in the stability analysis of the limiting ODE (7)-(8a) is $V(\theta(t),R(t)) = (\theta(t)-\theta)' R(t)(\theta(t)-\theta)$, cf. [1, page 553], [16]. Instead of working with the Lyapunov function of the limiting ODE, an obvious alternative is to work directly with its analogue

$$(\theta_n - \theta)' R_n (\theta_n - \theta) = n^{-1} V_n , \text{ where } V_n = (\theta_n - \theta)' P_n^{-1} (\theta_n - \theta) . \tag{33}$$

From (30) and (31), we obtain the following recursive relation for V_n:

$$\begin{aligned} V_n = V_{n-1} &- [(\theta_{n-1}-\theta)'\psi_n]^2 (1-\psi_n' P_n \psi_n) + \psi_n' P_n \psi_n \varepsilon_n^2 \\ &+ 2[(\theta_{n-1}-\theta)'\psi_n](1-\psi_n' P_n \psi_n)\varepsilon_n . \end{aligned} \tag{34}$$

Hence V_n satisfies the inequality (26), which defines an extended stochastic Lyapunov function, with $\alpha_n \equiv 0$,

$$\begin{aligned} \beta_n &= \psi_n' P_n \psi_n \varepsilon_n^2, \quad w_{n-1} = 2[(\theta_{n-1}-\theta)'\psi_n](1-\psi_n' P_n \psi_n) , \\ \gamma_n &= [(\theta_{n-1}-\theta)'\psi_n]^2 (1-\psi_n' P_n \psi_n) \geq 0 , \end{aligned} \tag{35}$$

noting that ψ_n is $\mathcal{F}_{n-1}$-measurable and that $\psi_n' P_n \psi_n \leq 1$. Since $w_{n-1}^2 \leq 4\gamma_n$, it then follows from Theorem 1(ii) that

$$\max(V_n, \Sigma_1^n \gamma_i) = O(\Sigma_1^n \psi_i' P_i \psi_i \varepsilon_i^2) + O((\Sigma_1^n \gamma_i)^{2/3}) \quad \text{a.s.}$$

and therefore

$$\sum_1^n \gamma_i = O(\Sigma_1^n \psi_i' P_i \psi_i \varepsilon_i^2) \ \text{a.s.}, \quad V_n = O(\Sigma_1^n \psi_i' P_i \psi_i \varepsilon_i^2) \ \text{a.s.} \tag{36}$$

From (29) and Lemma 2 of [10], it follows that

$$\sum_1^n \psi_i' P_i \psi_i \varepsilon_i^2 = O(\Sigma_1^n \psi_i' P_i \psi_i) = O(\log \lambda_{max}(\Sigma_1^n \psi_i \psi_i')) \quad \text{a.s.} , \tag{37}$$

where $\lambda_{max}(A)$ and $\lambda_{min}(A)$ denote the largest and smallest eigenvalues of a symmetric matrix A. Since

$$V_n = \| P_n^{-1/2}(\theta_n - \theta)\|^2 \geq \|\theta_n - \theta\|^2 \lambda_{min}(P_n^{-1}) = \|\theta_n - \theta\|^2 \lambda_{min}(\Sigma_1^n \psi_i \psi_i') ,$$

(36) and (37) give the following result of Lai and Wei [10] on the strong consistency

of θ_n in stochastic regression models. Moreover, from the definition (35) of γ_n, (36) and (37) also imply the result of [7] on adaptive predictors using the least squares identification method.

Theorem 2. Suppose that in the regression model (27), $\{\varepsilon_n, \mathcal{F}_n, n \geq 1\}$ is a martingale difference sequence satisfying (29) and ψ_n is $\mathcal{F}_{n-1}$-measurable.

(i) On the event $\{\lambda_{min}(\Sigma_1^n \psi_i \psi_i') \to \infty$ and $\log \lambda_{max}(\Sigma_1^n \psi_i \psi_i') = o(\lambda_{min}(\Sigma_1^n \psi_i \psi_i'))\}$, $\theta_n \to \theta$ a.s.

(ii) Let $\tilde{y}_n = \theta' \psi_n$ be the optimal predictor of y_n when θ is known, and let $\hat{y}_n = \theta_{n-1}' \psi_n$ be the adaptive predictor of y_n using the least squares estimate θ_{n-1} to replace θ. Then for every $0 < \delta < 1$,

$$\sum_1^n (\hat{y}_i - \tilde{y}_i)^2 I_{\{\psi_i' (\Sigma_1^i \psi_t \psi_t')^{-1} \psi_i \leq \delta\}} = O(\log \lambda_{max}(\Sigma_1^n \psi_i \psi_i')) \text{ a.s.}$$

For the linear system (1), the regressors ψ_n, however, are not completely observable in the "colored-noise" case with $r \geq 1$ moving average parameters $c_1, \ldots, c_r$, since the components $\varepsilon_{n-1}, \ldots, \varepsilon_{n-r}$ of ψ_n are unobservable. Replacing the unobservable ε_i by their estimates $\hat{\varepsilon}_i$ in the recursions (30) and (31) leads to the "extended least squares" algorithm of the form

$$\begin{aligned} \theta_n &= \theta_{n-1} + P_n \phi_n (y_n - \theta_{n-1}' \phi_n), \text{ where} \\ \phi_n &= (-y_{n-1}, \ldots, -y_{n-k}, u_{n-d}, \ldots, u_{n-d-h}, \hat{\varepsilon}_{n-1}, \ldots, \hat{\varepsilon}_{n-r})' . \end{aligned} \tag{38}$$

The estimates $\hat{\varepsilon}_i$ of ε_i are given either by the residuals $\hat{\varepsilon}_i = y_i - \theta_i' \phi_i$, in which case (38) is called the AML method, or by the prediction errors $\hat{\varepsilon}_i = y_i - \theta_{i-1}' \phi_i$, in which case (38) is called the RML_1 method (cf. [1], [18]). Ljung's stability analysis of the associated ODE shows that the following positive real assumption is needed for the extended least squares algorithm to be strongly consistent:

$$\begin{aligned} &C(z) \triangleq 1 + c_1 z + \ldots + c_r z^r \text{ has all zeros outside the unit circle,} \\ &\text{and } \operatorname{Re}(1/C(z) - 1/2) > 0 \text{ for all } |z| = 1 . \end{aligned} \tag{39}$$

Under this assumption, Lai and Wei [12] showed that $V_n \triangleq (\theta_n - \theta)' P_n^{-1} (\theta_n - \theta)$ again satisfies an inequality of the type (26) for the AML algorithm, and that for the RML_1 algorithm, an inequality of this type holds on the event $\{\lim_{n\to\infty} \phi_n' P_n \phi_n = 0\}$. In this way they obtained the following analogue of Theorem 2 for the extended least squares method under the positive real assumption (39).

Theorem 3. Assume the positive real condition (39).

(i) For the AML algorithm θ_n defined by (38) with $\hat{\varepsilon}_i = y_i - \theta_i' \phi_i$,

$$\theta_n - \theta = O(\{\log \lambda_{max}(\Sigma_1^n \phi_i \phi_i')\}^{1/2} / \lambda_{min}^{1/2}(\Sigma_1^n \phi_i \phi_i')) \text{ a.s.} \tag{40}$$

Moreover, $\theta_n \to \theta$ a.s. if

$$\lambda_{min}(\Sigma_1^n \psi_i \psi_i') \to \infty \quad \text{and} \quad \log \lambda_{max}(\Sigma_1^n \psi_i \psi_i') = o(\lambda_{min}(\Sigma_1^n \psi_i \psi_i')) \quad \text{a.s.} \tag{41}$$

For every $0 < \delta < 1$,

$$\sum_1^n (\theta_{i-1}' \phi_i - \theta' \psi_i)^2 \, I_{\{\phi_i' P_i \phi_i \leq \delta\}} = O(\log \lambda_{max}(\Sigma_1^n \phi_i \phi_i')) \quad \text{a.s.} \tag{42}$$

On the event $\{\lambda_{max}(\Sigma_1^n \phi_i \phi_i') \to \infty\}$, $\lambda_{max}(\Sigma_1^n \phi_i \phi_i') \sim \lambda_{max}(\Sigma_1^n \psi_i \psi_i')$ a.s.

(ii) For the RML_1 algorithm θ_n defined by (38) with $\hat{\varepsilon}_i = y_i - \theta_{i-1}' \phi_i$, (40) and (42) hold on the event $\{\lim_{n\to\infty} \phi_n' P_n \phi_n = 0\}$.

Earlier, Solo [24] and Moore and Ledwich [19] established the strong consistency of the AML algorithm under (39) and the persistent excitation assumption (32), which is much stronger than (41). Instead of working directly with the recursive inequalities for V_n , they had to transform V_n into an almost supermartingale so that they could apply martingale convergence theorems. Thus, Solo introduced the stochastic Lyapunov function

$$W_n = n^{-1}(V_n + 2f_n g_n), \quad \text{where} \quad f_n = (\theta - \theta_n)' \phi_n, \quad g_n = (1/C(q^{-1}) - 1/2) f_n .$$

Theorem 1 enables one to bypass this kind of transformations, and working directly with V_n as in [12] actually gives sharper results.

The excitation condition (41) for the strong consistency of the least squares estimator in Theorem 2 or the AML algorithm in Theorem 3 is in some sense the weakest possible, cf. [10]. As shown in [11], such excitation conditions on the stochastic regressors ψ_i can be translated into corresponding conditions involving the inputs alone. To ensure such excitation conditions in an adaptive control setting that will be discussed in the next section, one can either introduce occasional white-noise probing inputs when the data show inadequate information to estimate θ by θ_n , as in [12] and [13], or use the Caines-Lafortune [2] idea of "continually disturbed controls" of the form

$$u_n = g_n(y_n, y_{n-1}, u_{n-1}, \ldots) + \eta_n ,$$

where $\{g_n\}$ represents a feedback control scheme and $\{\eta_n\}$ represents an exogenous white-noise process satisfying certain assumptions.

4. Adaptive prediction, adaptive control and other applications of the method of extended stochastic Lyapunov functions

To begin with, consider the Robbins-Monro scheme (18) for finding the root of the regression function M satisfying (20). As shown in Section 3, application of the stochastic Lyapunov function $V_n \triangleq (u_{n+1} - \theta)^2$ not only proves that $u_n \to \theta$ a.s. but also shows that $\Sigma_1^\infty \rho_n M(\theta_n)(u_n - \theta) < \infty$ a.s. In the sequel we shall assume that $M'(\theta) = \lambda$ exists and is positive. It then follows that $\Sigma_1^\infty \rho_n (u_n - \theta)^2 < \infty$ a.s. Assuming also that $\rho_n \downarrow 0$, this in turn implies by Kronecker's lemma that

$$\rho_n \sum_1^n (u_i - \theta)^2 \to 0 \quad \text{a.s.} \tag{43}$$

The consequence (43) of the method of stochastic Lyapunov functions has important implications on the adaptive control aspects of stochastic approximation. Motivated by the following kind of applications, Lai and Robbins [8], [9] studied the Robbins-Monro scheme (18) as a control algorithm in the regression model (19). Suppose that in (19), u_i is the dosage level of a drug given to the ith patient who turns up for treatment and that y_i is the response of the patient. Suppose also that the optimal response is y^*, which we shall assume to be 0 without loss of generality (replacing y_i by $y_i - y^*$ if necessary). To achieve this mean response $y^* = 0$, if the unique (by (20)) root θ of the equation $M(\theta) = 0$ were known, then the dosage levels should all be set at θ. Since θ is usually unknown, how can the dosage levels u_i be chosen so that they approach θ rapidly? More precisely, the control problem is choose the design levels $u_1, \ldots, u_n$ so as to minimize in some sense, at least in the long run as $n \to \infty$, the loss (or regret, due to ignorance of θ)

$$L_n \triangleq \lambda^2 \sum_1^n (u_i-\theta)^2 \quad (\sim \Sigma_1^n [M(u_i)-M(\theta)]^2) \;. \tag{44}$$

Note that by choosing $\rho_n \sim (cn)^{-1}$ $(c > 0)$ in the Robbins-Monro scheme (18), (43) implies that

$$n^{-1} L_n \to 0 \quad \text{a.s.} \tag{45}$$

Suppose that the ε_i are i.i.d. with mean 0 and variance $\sigma^2 > 0$. The Robbins-Monro scheme (18) with $\rho_n \sim (cn)^{-1}$ is asymptotically normal if $0 < c < 2\lambda$. In fact, $n^{\frac{1}{2}}(u_n - \theta)$ converges in distribution to $N(0, \sigma^2/\{c(2\lambda - c)\})$, and an asymptotically optimal choice of ρ_n is $\rho_n \sim (\lambda n)^{-1}$ (cf. [8]). For this asymptotically optimal choice, not only do we have

$$n^{\frac{1}{2}}(u_n-\theta) \stackrel{\mathcal{D}}{\to} N(0,\sigma^2/\lambda^2) \;, \tag{46}$$

but the regret L_n defined in (44) also has the logarithmic order

$$L_n \sim (\sigma^2/\lambda^2) \log n \quad \text{a.s.} \;, \tag{47}$$

as shown in [8]. The constant σ^2/λ^2 in (46) and (47) is in some sense minimal when the ε_i are normal and M is linear, in which case the stochastic approximation scheme u_n (with $\rho_n = (\lambda n)^{-1}$) is equal to the maximum likelihood estimator of θ at stage n (cf. [8], [9]).

Since $\lambda = M'(\theta)$ is usually unknown in practice, Lai and Robbins [8], [9] considered an adaptive stochastic approximation scheme (18) with $\rho_n = (n\lambda_n)^{-1}$, where λ_n is a consistent estimator of λ at stage n. It is shown that for such adaptive stochastic approximation schemes, (46) and (47) still hold. Wei [25] subsequently extended these results to multivariate adaptive stochastic approximation schemes for multidimensional regression functions.

There are analogous results for adaptive control of the linear stochastic system (1). The problem is to choose the inputs u_i so that $\Sigma_1^n (y_i - y_i^*)^2$ is minimized in some sense, at least in the long run as $n\to\infty$, for a given sequence of target values $\{y_i^*\}$ for the outputs. The special case $y_i^* \equiv 0$ corresponds to the "regulation problem". When the system parameters are known and $b_0 \neq 0$, the optimal controller chooses the inputs u_i at stage i so that $E(y_{i+d}|\mathcal{F}_i) = y_{i+d}^*$. To focus on the main ideas, we shall only consider the case of unit delay $d = 1$. The outputs of this optimal input sequence are given by $y_i = y_i^*+\varepsilon_i$. In view of this, we define the "regret" (due to ignorance of the system parameters) of an input sequence $\{u_i\}$ to be

$$L_n = \sum_1^n (y_i-y_i^*-\varepsilon_i)^2 = \sum_1^n \{E(y_i|\mathcal{F}_{i-1})-y_i^*\}^2 , \tag{48}$$

in analogy with (44). An input sequence is called "globally convergent" if (45) holds, cf. [4].

An important breakthrough in the long-standing problem of developing, for the stochastic system (1), on-line recursive control algorithms that are globally convergent is due to Goodwin, Ramadge and Caines [4]. Their idea is to use the "stochastic gradient method" for recursive estimation of the parameters:

$$\begin{aligned} \theta_n &= \theta_{n-1} + a\phi_n(y_n-\theta_{n-1}'\phi_n)/r_n , \\ r_n &= r_{n-1} + \phi_n'\phi_n , \end{aligned} \tag{49}$$

where $a > 0$, ϕ_n is the same as in (38) with $\hat{\varepsilon}_i = y_i-\theta_{i-1}'\phi_i$, and $\theta_0, \phi_0, r_0(>0)$ are arbitrary initial conditions. The inputs u_i are then determined by the equation

$$\theta_{n-1}'\phi_n = y_n^* . \tag{50}$$

Under certain assumptions including the positive real condition

$$\begin{aligned} &C(z) \triangleq 1 + c_1 z +\ldots+ c_r z^r \text{ has all zeros outside the unit circle,} \\ &\text{and } \operatorname{Re}(C(z)-a/2) > 0 \text{ for all } |z| = 1 , \end{aligned} \tag{51}$$

a stochastic Lyapunov function argument similar to that for the Robbins-Monro scheme presented above can be used in conjunction with an analysis of the system dynamics to show that (45) holds for the regret L_n, defined by (48), of this input sequence. Actually Goodwin et al. [4] reparametrized the model (1) and used the stochastic gradient algorithm to estimate the transformed parameters. However, as shown in [15], a modification of their argument can be used to handle the original model without reparametrization.

In analogy with the result (47) for adaptive stochastic approximation, it is natural to ask whether one can improve the global convergence result of Goodwin et al. [4] and have a logarithmic order for the regret L_n. To answer this question, Lai and Wei [12] used the AML algorithm (38) (in which $\hat{\varepsilon}_i = y_i-\theta_i'\phi_i$) for recursive estimation of the system parameters to construct recursive control algorithms whose regrets satisfy

$$L_n = O(\log n) \quad \text{a.s.} \; , \tag{52}$$

under certain assumptions including stability of the polynomials $A(z)$ and $B(z)$ and the positive real condition (39). A basic tool in their analysis is Theorem 3(i) on the strong consistency of the AML method and on the associated adaptive predictors (42). In particular, the desired order of $\log n$ in (52) follows from the order of $\log \lambda_{max}(\Sigma_1^n \phi_i \phi_i')$ in (42). As pointed out in Section 3, Theorem 3 can be obtained by applying Theorem 1 to the extended stochastic Lyapunov function $V_n \triangleq (\theta_n-\theta)' P_n^{-1}(\theta_n-\theta)$. Thus, while the method of stochastic Lyapunov functions has led to the global convergence result (45) of Goodwin et al. [4], the method of extended stochastic Lyapunov functions is also instrumental in the theory of adaptive controllers satisfying the much stronger property (52).

In the white-noise case $C(z) = 1$, the AML algorithm reduces to the least squares estimator (30). For the regulation problem (i.e., $y_i^* \equiv 0$) in this case, Lai and Wei [13] constructed adaptive controllers using recursive least squares for system identification such that

$$\limsup_{n\to\infty} L_n/\log n \leq (k+h) \limsup_{n\to\infty} E(\varepsilon_n^2 | \mathcal{F}_{n-1}) \; , \tag{53}$$

which is a stronger result than (40). The constant on the right hand side of (53) is in some sense best possible (cf. [7], [13]). In particular, the factor $k+h$ corresponds to the number of parameters to be estimated if b_0 were known, noting the linear constraint (50) induced by the inputs on the estimates of the $(k+h+1)$-dimensional parameter vector θ.

For the case of colored noise, although the extended least squares algorithm (38) has been called "approximate maximum likelihood" (AML), it does not arise from maximization of the log-likelihood function, as in the off-line (non-recursive) maximum likelihood estimator, when the ε_i are assumed to be normally distributed with mean 0 and variance σ^2. Using a linear approximation to the log-likelihood function, Söderström [23] derived a recursive estimator, commonly called RML_2, for the system parameters of (1). This method is recently refined in [14], where the underlying idea is to make use of an auxiliary recursive estimator based on instrumental variables and to linearize the log-likelihood function only in a neighborhood of the auxiliary estimator. Under certain conditions, the recursive estimator is shown in [14] to be asymptotically normal; moreover, the covariance matrix of the limiting normal distribution is minimal when the ε_i are i.i.d. $N(0,\sigma^2)$ random variables. The method of extended stochastic Lyapunov functions can also be used to provide an analogue of Theorem 2 for this estimator, as shown in [14].

Making use of this new recursive estimator, an adaptive control algorithm can be constructed for the linear stochastic system (1) whose regret satisfies

$$\limsup_{n\to\infty} L_n/\log n \leq (k+h+r+1) \limsup_{n\to\infty} E(\varepsilon_n^2 | \mathcal{F}_{n-1}) \; . \tag{54}$$

Moreover, for the regulation problem $(y_i^* \equiv 0)$, (54) can be further strengthened into

$$\limsup_{n\to\infty} L_n/\log n \le \{\max(k,r)+h\} \limsup_{n\to\infty} E(\varepsilon_n^2|\mathcal{F}_{n-1}) , \quad (55)$$

which is a generalization of (53) that corresponds to the case $r = 0$. The details are given in [15].

REFERENCES

[1] Caines, P. E. (1988). Linear Stochastic Systems. Wiley, New York.

[2] Caines, P. E. and Lafortune, S. (1984). Adaptive control with recursive identification for stochastic linear systems. IEEE Trans. Automat. Contr. AC-29 312-321.

[3] Gladyshev, E. G. (1965). On stochastic approximation. Theory Probability Appl. 10 275-278.

[4] Goodwin, G. C., Ramadge, P. J., and Caines, P. E. (1981). Discrete time stochastic adaptive control. SIAM J. Contr. Optimiz. 19 829-853.

[5] Has'minskii, R. Z. (1980). Stochastic Stability of Differential Equations. Sijthoff & Noordhoff, Alphen aan den Rijn, The Netherlands.

[6] Kushner, H. J. (1967). Stochastic Stability and Control. Academic Press, New York.

[7] Lai, T. L. (1986). Asymptotically efficient adaptive control in stochastic regression models. Adv. Appl. Math. 7 23-45.

[8] Lai, T. L. and Robbins, H. (1979). Adaptive design and stochastic approximation. Ann. Statist. 7 1196-1221.

[9] Lai, T. L. and Robbins, H. (1981). Consistency and asymptotic efficiency of slope estimates in stochastic approximation schemes. Z. Wahrsch. Verw. Gebiete 56 329-360.

[10] Lai, T. L. and Wei, C. Z. (1982). Least squares estimates in stochastic regression models with applications to identification and control of dynamic systems. Ann. Statist. 10 154-156.

[11] Lai, T. L. and Wei, C. Z. (1986). On the concept of excitation in least squares identification and adaptive control. Stochastics 16 227-254.

[12] Lai, T. L. and Wei, C. Z. (1986). Extended least squares and their applications to adaptive control and prediction in linear systems. IEEE Trans. Automat. Contr. AC-31 898-906.

[13] Lai, T. L. and Wei, C. Z. (1987). Asymptotically efficient self-tuning regulators. SIAM J. Contr. Optimiz. 25 466-481.

[14] Lai, T. L. and Ying, Z. (1988). Recursive estimation and adaptive prediction in stochastic regression models with moving average errors. Tech. Report, Dept. Statistics, Stanford Univ.

[15] Lai, T. L. and Ying, Z. (1988). Parallel recursive algorithms in adaptive control of linear stochastic systems. Tech. Report, Dept. Statistics, Stanford Univ.

[16] Ljung, L. (1977). Analysis of recursive stochastic algorithms. IEEE Trans. Automat. Contr. AC-22 551-575.

[17] Ljung, L. (1977). On positive real transfer functions and the convergence of some recursive schemes. IEEE Trans. Automat. Contr. AC-22 539-551.

[18] Ljung, L. and Söderström, T. (1983). Theory and Practice of Recursive Identification. MIT Press, Cambridge, Massachusetts.

[19] Moore, J. B. and Ledwich, G. (1979). Multivariable adaptive parameter and state estimators with convergence analysis. J. Austr. Math. Soc. Ser. B 21 176-197.

[20] Nevel'son, M. B. and Has'minskii, . (1973). Stochastic Approximation and Recursive Estimation. Providence, R.I.: Amer. Math. Soc. Transl.

[21] Robbins, H. and Monro, S. (1951). A stochastic approximation method. Ann. Math. Statist. 22 400-407.

[22] Robbins, H. and Siegmund, D. (1971). A convergence theorem for nonnegative almost supermartingales and some applications. In Optimizing Methods in Statistics (Ed. J. S. Rustagi), 233-257. Academic Press, New York.

[23] Söderström, T. (1973). An on-line algorithm for approximate maximum likelihood identification of linear dynamic systems. Tech. Report, Dept. Automatic Control, Lund Institute of Technology, Lund, Sweden.

[24] Solo, V. (1979). On the convergence of AML. IEEE Trans. Automat. Contr. **AC-24** 958-962.

[25] Wei, C. Z. (1987). Multivariate adaptive stochastic approximation. Ann. Statist. **15** 1115-1130.

CONSISTENCY SETS OF LEAST SQUARES ESTIMATES IN STOCHASTIC REGRESSION MODELS*

by

Alain Le Breton
Laboratory TIM3-USTMG-CNRS
BP.68,38402 Saint Martin d'Hères
FRANCE

and

Marek Musiela
School of Mathematics
The University of New South Wales
PO Box 1, Kensington, NSW, 2033
AUSTRALIA

Abstract

A stochastic linear regression model is investigated. Consistency sets of the least squares estimates are characterized in predictable terms. New sufficient conditions guaranteeing strong consistency of the estimates are derived.

1. Introduction

Consider the multiple regression model

(1) $$y_n = x_n^* \beta + \varepsilon_n \ , \ n = 1, 2, \ldots,$$

where y_n is the observed response corresponding to the design vector $x_n \in R^d$, $\beta \in R^d$ is an unknown parameter and ε_n are unobservable random errors. We shall assume that (ε_n) is a local martingale difference sequence with respect to an increasing sequence of σ -fields $(\mathcal{F}_n)$ i.e. $E(\varepsilon_n | \mathcal{F}_{n-1}) = 0$ and that (x_n) is $(\mathcal{F}_n)$ − predictable i.e. x_n, is $\mathcal{F}_{n-1}$ − measurable .

If the matrix

$$\Lambda_n = \sum_{k=1}^{n} x_k x_k^*$$

* Research supported by the Australian Research Grants Scheme

is nonsingular then the least squares estimate β_n of β based on the observations $y_1,\ldots,y_n$ is given by

$$\beta_n = \Lambda_n^{-1}\sum_{k=1}^{n} x_k y_k .$$

The asymptotic properties of β_n were recently studied by many authors. In particular Lai and Wei (1972) established the following results concerning the strong consistency of β_n.

Let $\lambda_{\min}(n)$ and $\lambda_{\max}(n)$ denote the minimum and maximum eigenvalues of Λ_n respectively. Moreover, let for $\gamma \geq 2$ (A_γ) stands for the condition:

$$(A_\gamma) \qquad \sup_n E(|\varepsilon_n|^\gamma|\mathcal{F}_{n-1}) < \infty \quad a.s. .$$

Let also for $\delta \geq 0$ (B_δ) be the condition:

$$(B_\delta) \qquad \lambda_{\min}(n) \to \infty \ a.s. \qquad \text{and} \qquad (\log\lambda_{\max}(n))^{1+\delta} = o(\lambda_{\min}(n)) \ a.s. .$$

Assume that in the regression model (1) either (A_2) and (B_δ) for some $\delta > 0$ or (A_γ) for some $\gamma > 2$ and (B_0) hold. Then

$$\beta_n \to \beta \quad a.s. .$$

Lai and Wei have also used the following example of Lai and Robbins (1981) to show that the condition (B_0) is in some sense weakest possible.

Example: Let $\varepsilon_1,\varepsilon_2,\ldots$ be *i.i.d.* random variables with $E\varepsilon_1 = 0$, $E\varepsilon_1^2 = 1$ and let $\mathcal{F}_n$ be the $\sigma-$ field generated by $\varepsilon_1,\ldots,\varepsilon_n$. Consider regression model (1) where $d = 2$, $x_n = (1\ a_n)^*$, $a_1 = 0$ and $a_n = \sum_{k=1}^{n-1} k^{-1}\varepsilon_k$, $n \geq 2$. It is clear that a_n converges $a.s.$ and that the limit a_∞ satisfies $Ea_\infty^2 = \sum_{k=1}^{\infty} k^{-2}$. Furthermore, with probability 1

$$n^{-1}\sum_{k=1}^{n} a_k \to a_\infty \qquad \text{and} \qquad n^{-1}\sum_{k=1}^{n} a_k^2 \to a_\infty^2$$

and because

$$\Lambda_n = \begin{bmatrix} n & \sum_{k=1}^n a_n \\ \sum_{k=1}^n a_n & \sum_{k=1}^n a_n^2 \end{bmatrix}$$

we have, with probability 1,

$$n^{-1}\,\mathrm{tr}\Lambda_n \to 1 + a_\infty^2 \qquad \text{and} \qquad n^{-2}\det\Lambda_n \to 0 .$$

Now since $(\mathrm{tr}\Lambda_n)^{-2}\det\Lambda_n \to 0$ $a.s.$ and

$$\lambda_{\max}(n) = \frac{\mathrm{tr}\Lambda_n}{2}(1 + (1 - 4\frac{\det\Lambda_n}{(\mathrm{tr}\Lambda_n)^2})^{1/2})$$

we obtain that

$$n^{-1}\lambda_{\max}(n) \to 1 + a_\infty^2 \quad a.s..$$

Moreover, since $a_{k+1} = \overline{a}_k + \overline{\varepsilon}_k$, where $\overline{a}_k$ denotes the arithmetic mean of k numbers $a_1, \ldots, a_k$, we can write

$$\det \Lambda_n = n\sum_{k=1}^{n}(a_k - \overline{a}_n)^2 = n\sum_{k=2}^{n}\frac{k-1}{k}(a_k - \overline{a}_{k-1})^2 = n\sum_{k=2}^{n}\frac{k-1}{k}\overline{\varepsilon}_{k-1}^2$$

and by Corollary 1 of Lai and Robbins (1979)

$$(n\log n)^{-1}\det \Lambda_n \to 1 \quad a.s..$$

Consequently, with probability 1,

$$(\log n)^{-1}\lambda_{\min}(n) = (n\log n)^{-1}\det \Lambda_n(n^{-1}\lambda_{\max}(n))^{-1} \to (1 + a_\infty^2)^{-1}$$

and therefore

$$\frac{\log \lambda_{\max}(n)}{\lambda_{\min}(n)} \to 1 + a_\infty^2 \quad a.s.$$

which shows that the condition (B_0) is only marginally violated. One can show however, that with probability 1

$$\beta_n - \beta \to (a_\infty - 1)^*$$

and hence the least squares estimate β_n is inconsistent.

The above-mentioned results indicate that it may be difficult to improve the conditions for consistency and also that a condition on the relative behaviour of $\lambda_{\max}(n)$ and $\lambda_{\min}(n)$ could be necessary. However, in Section 2, we show that sufficiently fast convergence of $\lambda_{\min}(n)$ to infinity (and no restriction on the behaviour of $\lambda_{\max}(n)$) guarantees the strong consistency of β_n.

2. Consistency sets of least squares estimates

In order to justify the above assertion the following theorem will be proved.

Theorem:

(i) Suppose that in the regression model (1) condition (A_2) holds. Then the least squares estimate β_n satisfies

$$\left\{\operatorname{tr}(I + \Lambda_n)^{-1} \to 0\,,\ \sum_{k=1}^{\infty}\operatorname{tr}(I + \Lambda_k)^{-1}x_k^*(I + \Lambda_k)^{-1}x_k < \infty\right\} \subset \{\beta_n \to \beta\} \quad a.s..$$

(ii) Suppose that in the regression model (1) condition (A_γ) holds for some $\gamma > 2$. Then β_n satisfies

$$\left\{\operatorname{tr}(I + \Lambda_n)^{-1}\sum_{k=1}^{n}x_k^*(I + \Lambda_k)^{-1}x_k \to 0\right\} \subset \{\beta_n \to \beta\} \quad a.s..$$

Proof: We use the notation Δ for the difference operator. Hence, if (a_n) is a sequence of vectors then $\Delta a_n = a_n - a_{n-1}$. Define

$$N_n = \sum_{k=1}^{n} x_k \varepsilon_k \ , \ n = 1,2,\dots \ , \ N_0 = 0 \ ; \ \Gamma_n = I + \Lambda_n \ , \ n = 1,2,\dots \ , \ \Gamma_0 = I \ ,$$

then for $n = 1,2,\dots$

$$\Delta(N_n^* \Gamma_n^{-1} N_n) = 2N_{n-1}^* \Gamma_n^{-1} \Delta N_n + (\Delta N_n)^* \Gamma_n^{-1} \Delta N_n + N_{n-1}^* (\Delta \Gamma_n^{-1}) N_{n-1} \ ,$$

$$\Delta \Gamma_n^{-1} = -\Gamma_n^{-1}(\Delta \Gamma_n)\Gamma_n^{-1} - \Gamma_n^{-1}(\Delta \Gamma_n)\Gamma_{n-1}^{-1}(\Delta \Gamma_n)\Gamma_n^{-1}$$

and consequently

$$N_n^* \Gamma_n^{-1} N_n = 2 \sum_{k=1}^{n} N_{k-1}^* \Gamma_k^{-1} \Delta N_k - \sum_{k=1}^{n} N_{k-1}^* \Gamma_k^{-1} (\Delta \Gamma_k) \Gamma_k^{-1} N_{k-1}$$

$$+ \sum_{k=1}^{n} (\Delta N_k)^* \Gamma_k^{-1} \Delta N_k - \sum_{k=1}^{n} N_{k-1}^* \Gamma_k^{-1} (\Delta \Gamma_k) \Gamma_{k-1}^{-1} (\Delta \Gamma_k) \Gamma_k^{-1} N_{k-1} \ .$$

Note that $m_n = \sum_{k=1}^{n} N_{k-1}^* \Gamma_k^{-1} \Delta N_k \ , \ n = 1,2,\dots, \ , \ m_0 = 0$ is a locally square integrable local martingale and that

$$< m >_n = \sum_{k=1}^{n} E\left((\Delta m_k)^2 | \mathcal{F}_{k-1}\right) = \sum_{k=1}^{n} N_{k-1}^* \Gamma_k^{-1} \Delta \Gamma_k \Gamma_k^{-1} N_{k-1} E(\varepsilon_k^2 | \mathcal{F}_{k-1}) \ .$$

Let $\xi = \sup_n E(\varepsilon_n^2 | \mathcal{F}_{n-1})$ which is well defined under (A_2). Then because m_n converges *a.s.* to a finite limit on the set $\{< m >_\infty < \infty\}$ we deduce that

$$\sup_n (m_n - (2\xi)^{-1} < m >_n) = \eta < \infty \ a.s.$$

and consequently that

$$N_n^* \Gamma_n^{-1} N_n \leq 2\eta + \sum_{k=1}^{n} (\Delta N_k)^* \Gamma_k^{-1} \Delta N_k \ .$$

Moreover, the Borel-Cantelli lemma (cf. Lépingle 1978), implies

$$\left\{ \sum_{k=1}^{\infty} \mathrm{tr} \Gamma_k^{-1} x_k^* \Gamma_k^{-1} x_k < \infty \right\} \subset \left\{ \sum_{k=1}^{\infty} \mathrm{tr} \Gamma_k^{-1} (\Delta N_k)^* \Gamma_k^{-1} \Delta N_k < \infty \right\} \ a.s.$$

and since

(2) $$N_n^* \Gamma_n^{-2} N_n \leq \ \mathrm{tr} \Gamma_n^{-1} (2\eta + \sum_{k=1}^{n} (\Delta N_k)^* \Gamma_k^{-1} \Delta N_k)$$

and Kronecker lemma gives

$$\left\{\mathrm{tr}\Gamma_n^{-1} \to 0\,,\ \sum_{k=1}^{\infty} \mathrm{tr}\Gamma_k^{-1} x_k^* \Gamma_k^{-1} x_k < \infty\right\} \subset \{\Gamma_n^{-1} N_n \to 0\}\,.$$

Now, because $\beta_n - \beta = (I + \Lambda_n^{-1})\Gamma_n^{-1} N_n$ we also have $\{\Gamma_n^{-1} N_n \to 0\} \subset \{\beta_n \to \beta\}$ *a.s.* and the first assertion follows.

Let for $n = 1, 2, \ldots, B_n = \sum_{k=1}^{n} x_k^* \Gamma_k^{-1} x_k \varepsilon_k^2$ and $\tilde{B}_m = \sum_{k=1}^{n} x_k^* \Gamma_k^{-1} x_k E(\varepsilon_k^2 | \mathcal{F}_{k-1})$.
Using (2) we can write

$$N_n^* \Gamma_n^{-2} N_n \leq 2\eta\ \mathrm{tr}\Gamma_n^{-1} + \mathrm{tr}\Gamma_n^{-1}(1 + \xi \sum_{k=1}^{n} x_k^* \Gamma_k^{-1} x_k)(1 + \tilde{B}_n)^{-1} B_n\,.$$

Consequently to complete the proof it is sufficient to show that, under $(A\gamma)$ for some $\gamma > 2$

$$\sup_n (1 + \tilde{B}_n)^{-1} B_n < \infty \ a.s.\,. \tag{3}$$

This fact follows from Lemma 2 of Lai and Wei (1982).

3. Remarks

(a) Here we assume that (A_2) holds. At first note that part (i) of the Theorem says that if $\mathrm{tr}(I + \Lambda_n)^{-1} \to 0$ *a.s.* and $\sum_{k=1}^{\infty} \mathrm{tr}(I + \Lambda_k)^{-1} x_k^* (I + \Lambda_k)^{-1} x_k < \infty$ *a.s.* then $\beta_n \to \beta$ *a.s.* that is nothing but Theorem 12.4 of Solo (1986). In fact, since the following inequality

$$x_n^*(I + \Lambda_n)^{-1} x_n \leq d \min(1, \log\det(I + \Lambda_n)(I + \Lambda_{n-1})^{-1}, \mathrm{tr}(I + \Lambda_n)^{-1} \|x_n\|^2) \tag{4}$$

holds, under (A_2) one has *a.s.*

$$\left\{\sum_{k=1}^{\infty} \mathrm{tr}(I + \Lambda_k)^{-1} < \infty\right\} \cup \left\{\sum_{k=1}^{\infty} \mathrm{tr}(I + \Lambda_k)^{-1} \log\det(I + \Lambda_k)(I + \Lambda_{k-1})^{-1} < \infty\right\} \subset \{\beta_n \to \beta\}\,.$$

Therefore if $\sum_{k=1}^{\infty} \mathrm{tr}(I + \Lambda_k)^{-1} \log\det(I + \Lambda_k)(I + \Lambda_{k-1})^{-1} < \infty$ *a.s.* then β_n is consistent *a.s.* (see Theorem 12.4 of Solo (1986)). Similarly if the series $\sum_{k=1}^{\infty} \mathrm{tr}(I + \Lambda_k)^{-1}$ converges *a.s.* then β_n is consistent *a.s.* . Finally if in addition to (A_2) the condition $\sup_n \|x_n\| < \infty$ *a.s.* holds, then β_n is consistent on the set where the series $\sum_{k=1}^{\infty} (\mathrm{tr}(I + \Lambda_k)^{-1})^2$ converges.

(b) Now we assume that (A_γ) holds for some $\gamma > 2$. Note that from the Theorem part (ii) and the inequality (4) we get

$$\{n^{-1}\lambda_{\min}(n) \to \infty\} \cup \{\lambda_{\min}(n) \to \infty\,,\ \log \lambda_{\max}(n) = o(\lambda_{\min}(n))\} \subset \{\beta_n \to \beta\} \ a.s.$$

and therefore β_n is consistent *a.s.* if (B_0) holds (what is nothing but the second part of Lai and Wei results) or if $\lambda_{\min}(n)$ tends *a.s.* to infinity faster than n. Finally if $\sup_n \|x_n\| < \infty$ *a.s.*, then β_n is consistent on the set where $\mathrm{tr}\,(I + \Lambda_n)^{-1} \sum_{k=1}^{n} \mathrm{tr}\,(I + \Lambda_k)^{-1}$ tends to zero.

(c) Note that the inclusion in the Theorem part (ii) is also true under (A_2) provided in addition $E\sup_n \varepsilon_n^2 < \infty$. Indeed since

$$E\sup_n \Delta B_n = E\sup_n x_n^* \Gamma_n^{-1} x_n \varepsilon_n^2 \leq dE\sup_n \varepsilon_n^2 < \infty$$

the inequality (3) follows from Theorem 1 of Lépingle (1978).

(d) More detailed study of laws of large numbers for vector valued martingales normalized by increasing sequences of matrices is carried out in Le Breton and Musiela (1987). The continuous time case is considered, applications to stochastic regression are given and other references are listed.

References

1. Lai, T.L. and Robbins, M. (1979): Adaptive design and stochastic approximation. *Ann. Statist.* **7**, 1196–1221.
2. Lai, T.L. and Robbins, M. (1981): Consistency and asymptotic efficiency of slope estimates in stochastic approximation schemes. *Z. Wahrsch. verw. Gebiete* **56**, 329–360.
3. Lai, T.L. and Wei, C.Z. (1982): Least squares estimates in stochastic regression models with applications to identification and control of dynamic systems. *Ann. Statist.* **10**, 1154–166.
4. Le Breton, A. and Musiela, M. (1987): Laws of large numbers for semimartingales with applications to stochastic regression. Preprint.
5. Lépingle, D. (1978): Sur le comportement asymptotique des martingales locales. Séminaire de Probabilités XII. *Lecture Notes in Mathematics* **649**, 148–161, Berlin, Springer.
6. Solo, V. (1986): Topics in advanced time series analysis. *Lectures in Probability and Statistics. Lecture Notes in Mathematics* **1215**, 165–328, Berlin, Springer.

CONSISTENCY OF ESTIMATORS IN CONTROLLED SYSTEMS

Petr Mandl
Faculty of Mathematics and Physics, Charles University
Sokolovská 83, 186 00 Prague 8, Czechoslovakia

1. Introduction

The paper deals with linear controlled systems satisfying

(1) $$dX_t = f(\alpha)X_t dt + gU_t dt + dW_t, \qquad t \geq 0.$$

X is assumed to be n-dimensional, $X_0 = x$. $f(\alpha)$ is an $n \times n$ matrix depending linearly on $\alpha = (\alpha^1, \ldots, \alpha^m) \in R^m$,

$$f(\alpha) = f_0 + \alpha^1 f_1 + \ldots + \alpha^m f_m,$$

where $f_0, f_1, \ldots, f_m$ are given matrices. $\{W_t\}$ is the n-dimensional Wiener process with incremental variance matrix $h \neq 0$, i.e.,

$$dW dW' = h dt.$$

We shall concentrate on the case of U_t one-dimensional. g is thus an n-dimensional vector.

The true value of the parameter α, say α_0, is supposed to be unknown to the controller. We shall investigate least squares estimates α^*_T of α_0 on the basis of $\{X_t, t \in [0,T]\}$ defined as follows. Let ℓ be a nonnegatively definite symmetric matrix. α^*_T is the minimizer of

(2) $$\int_0^T (\dot{X}_t - f(\alpha)X_t - gU_t)' \ell (\dot{X}_t - f(\alpha)X_t - gU_t) dt.$$

$\dot{X}_t$ denotes the derivative of X_t which does not exist in the rigorous sense. Despite of it, the system of equations for $\alpha^{*1}_T, \ldots, \alpha^{*m}_T$ obtained by equating the gradient of (2) to zero is well defined, and reads

$$\sum_j \int_0^T X' f_i' \ell f_j X dt \, \alpha^{*j}_T = \int_0^T X' f_i' \ell (dX - f_0 X dt - gU dt), \qquad i = 1, \ldots, m.$$

Inserting from (1) with $\alpha = \alpha_0$ we obtain from here

$$\sum_j \int_0^T X' f_i' \ell f_j X dt (\alpha^{*j}_T - \alpha^j_0) = \int_0^T X' f_i' \ell \, dW,$$

and hence,

(3) $\sum_{ij} \int_0^T X' f_i' \ell f_j X dt (\alpha^*{}_T^i - \alpha_0^i)(\alpha^*{}_T^j - \alpha_0^j) = \sum_i \int_0^T X' f_i' \ell dW (\alpha'^i_T - \alpha_0^i).$

Recently, control theory methods have been applied to study the convergence of α'_T to α_0 (see [1],[3]). Earlier results are found in [2], [4] .

2. Statement of results

We present a method to investigate the strong consistency of α'_T, $T > 0$, i.e. the validity of

(4) $\lim_{T\to\infty} \alpha'_T = \alpha_0$ a.s.

under the controls $\{U_t\}$ such that

(5) $U_t = K_t X_t$, $t \geq 0$,

where $\{K_t\}$ is nonanticipative,

(6) $|K_t| \leq \varkappa$, $t \geq 0$.

This is motivated by the analysis of self-tuning controls with

$$K_t = k(\alpha_t^*).$$

If the feedback gains $k(\alpha)$, $\alpha \in R^m$, depend continuously on α and are bounded, then the consistency of α_T^* under (5), (6) implies the self-tuning property.

The proper norming in (3) is the division by $\int_0^T |X|^2 dt$. It holds

$$\lim_{T\to\infty} \int_0^T X' f_i' \ell \, dW / \int_0^T |X|^2 dt = 0 \quad \text{a.s.},$$

since $\int_0^T |X|^2 dt = \infty$ under (5), (6) by Lemma 1. Hence it can be seen from (3) that (4) is implied by the following <u>persistent excitation property</u>

(7) $\liminf_{T\to\infty} \sum_{ij} \int_0^T X' f_i' \ell f_j X dt \, \mu^i \mu^j / \int_0^T |X|^2 dt \geq a|\mu|^2$, $\mu \in R^m$. a.s.,

where $a > 0$.

Write

$$\sum_{ij} \int_0^T X' f_i' \ell f_j X dt \, \mu^i \mu^j = \int_0^T X' q(\mu) X dt.$$

It is not difficult to derive the estimate

$$\left|\int_0^T X'q(\mu)X\,dt - \int_0^T X'q(\nu)X\,dt\right| / \int_0^T |X|^2 dt \le k \quad |\mu - \nu| .$$

Consequently, the proof of (7) can be reduced to the proof of

(8) $$\liminf_{T\to\infty} \int_0^T X'q(\mu_i)X\,dt / \int_0^T |X|^2 dt \quad \ge \delta > 0 \quad \text{a.s.},$$

$i=1,..,N$, where μ_i are to be such that the balls

$$\{ \nu : k\,|\nu - \mu_i| < \delta \}, i=1,\dots,N,$$

cover the unit sphere in R^m.

If the columns of the matrices $f_1,\dots,f_m$ are mutually orthogonal, then $q(\mu)$ can be written in the form

(9) $$q(\mu) = \sum_{j=1}^{m} (\mu^j)^2 q_j .$$

This is the case if $f(\alpha)$ contains at most one parameter in each row and each column. (9) reduces the problem to the verification of (8) for q_j, $j=1,\dots,m$.

Fix μ_i and set $q = q(\mu_i)$. We shall give a numerical algorithm for the verification of (8). The argument consists in applying the control theory methods to estimate

(10) $$\liminf_{T\to\infty} \int_0^T (X'qX+cU^2)dt / \int_0^T |X|^2 dt, \quad c > 0,$$

and in selecting c sufficiently small.

Since

$$q = \sum_{i=1}^{p} e_i e_i' ,$$

it suffices to consider the case

$$q = ee', \quad n > 1.$$

The asymptotic lower bound θ for (10) is obtained by solving the matrix Riccati equation ($f = f(\alpha_0)$, I unit matrix)

(11) $$wf + f'w - \frac{1}{c} wgg'w + ee' - \theta I = 0$$

with θ chosen so that

(12) $$\operatorname{trace}(wh) = 0.$$

There is an iteration procedure to solve (11), (12), which the specialists on controlled Markov chains would call Howard's iteration. Our approach is based on the expansions

(13) $$w = \sqrt{c}\,w_1 + c w_2 + o(c),$$

(14) $$\theta = \theta_1\sqrt{c} + O(c), \quad c \to 0.$$

Inserting (13), (14) into (11), (12) one obtains the equations

(15) $w_1 g = e, \quad \mathrm{trace}(hw_1) = 0,$

(16) $w_2 ge' + eg'w_2 + \theta_1 I = w_1 f + f'w_1, \quad \mathrm{trace}(hw_2) = 0.$

For n=2 (15) represents as many equations as unknowns. There is only the choice between $\pm e$. For $n > 2$ the additional number of equations is obtained from the conditions for the solvability of (16), the matrix of which does not have full rank.

Example. Let n=3, $e'=(1,0,0)$. (15) is then

$$\begin{aligned}
g_1 w_{11} + g_2 w_{21} \quad\quad + g_3 w_{31} \quad\quad\quad\quad\quad\quad &= 1\\
g_1 w_{21} + g_2 w_{22} \quad\quad\quad\quad + g_3 w_{32} \quad\quad\quad &= 0\\
+ g_1 w_{31} + g_2 w_{32} + g_3 w_{33} &= 0\\
h_{11}w_{11} + 2h_{21}w_{21} + h_{22}w_{22} + 2h_{31}w_{31} + 2h_{32}w_{32} + h_{33}w_{33} &= 0 .
\end{aligned}$$

On positions (2,2) and (3,3) the left-hand side of (16) equals θ_1, on position (3,2) it equals 0. This imposes on the right-hand side the conditions

$$\begin{aligned}
f_{12}w_{21} + f_{22}w_{22} - f_{13}w_{31} - (f_{32} - f_{23})w_{32} - f_{33}w_{33} &= 0,\\
f_{13}w_{21} + f_{23}w_{22} + f_{12}w_{31} + (f_{22} + f_{33})w_{32} + f_{32}w_{33} &= 0.
\end{aligned}$$

Algorithm. Solve (15), (16) with $\theta_1 > 0$, and decompose w_1 into

$$w_1 = w_1^+ - w_1^-,$$

where w_1^+, w_1^- are positively definite.

Proposition. Let $f - c^{-1}ge' - gg'w_2$ be a stable matrix for $c > 0$. If

(17) $g'w_1^+ g < 2g'w_1 g,$

then for c sufficiently small, $c \in (0, c_0]$,

(18) $$\liminf_{T\to\infty} \int_0^T ((e'x)^2 + cU^2)dt \Big/ \int_0^T |X|^2 dt \geq \frac{1}{2}\theta_1\sqrt{c} \quad \text{a.s.}$$

under arbitrary control $\{U_t\}$ fulfilling (5), (6).

An explicit choice of c for which (18) holds is made by verifying the inequalities (23), (29). From (6) and from (18) follows

$$\liminf_{T\to\infty} \int_0^T (e'x)^2 dt \Big/ \int_0^T |X|^2 dt \geq \frac{1}{2}\theta_1\sqrt{c} - \varkappa c,$$

which yields (8) for c small.

3. Proofs

First we present a lemma.

Lemma. Under arbitrary control fulfilling (5), (6)

(19) $$\liminf_{T\to\infty} \int_0^T |X|^2 dt/T \geq \frac{1}{2} \text{ trace } h/\sqrt{(|f|+\varkappa|g|)} \text{ a.s.}$$

Proof. The proof is based on the following inequality

(20) $$\int_0^T (|X|^2 + c|fX+gU|^2)dt - \sqrt{c}T \text{ trace } h + 2|Z_T|^2 \geq$$
$$\geq 2\sqrt{c}\int_0^T X'dW \text{ for } c > 0,$$

where

$$Z_T = \sqrt[4]{c}(X_0 \exp(-T/\sqrt{c}) + \int_0^T \exp(-(T-s)/\sqrt{c})dW_s).$$

To establish it set

$$V_t = fX_t + gU_t,$$

and write (1) as

(21) $$dX_t = -\frac{1}{\sqrt{c}}X_t \, dt + (V_t + \frac{1}{\sqrt{c}}X_t)dt + dW_t.$$

Expressing $\sqrt{c}\int_0^T d|X|^2$ by means of the Itô formula one obtains

(22) $$\int_0^T (|X|^2 + c|V|^2)dt + \sqrt{c}|X_T|^2 - \sqrt{c}|X_0|^2 - \sqrt{c}\, T \text{ trace } h -$$
$$- c\int_0^T |V + \frac{1}{\sqrt{c}}X|^2 dt = 2\sqrt{c}\int_0^T X'dW.$$

Further from (21),

$$\sqrt[4]{c}X_T = Z_T + \sqrt[4]{c}\int_0^T \exp(-(T-s)/\sqrt{c})(V_s + \frac{1}{\sqrt{c}}X_s)ds.$$

The square of the norm of the last term is estimated by

$$\sqrt{c}\int_0^T \exp(-2(T-s)/\sqrt{c})ds\, B_T \leq \frac{c}{2}B_T,$$

where

$$B_T = \int_0^T |V + \frac{1}{\sqrt{c}}X|^2 ds.$$

Consequently,

$$\sqrt{c}|X_T|^2 - cB_T \leq |Z_T|^2 + \sqrt{2c}|Z_T|\sqrt{B_T} - \frac{c}{2}B_T \leq 2|Z_T|^2.$$

Inserting from here into (22) we obtain (20).

Set

$$1 + c(|f| + \varkappa|g|) = a, \quad \sqrt{c} \text{ trace } h = b.$$

Representing the integral on the right-hand side of (20) by means of a random time change in a Wiener process $\{\mathcal{W}(t),\ t \geq 0\}$ we get

(23) $$a \int_0^T |X|^2 dt - bT + Z_T \geq \mathcal{W}(4c \int_0^T X'hX dt).$$

It holds for $\varepsilon > 0$

$$\sum_{n=1}^{\infty} P(\inf_{t \in [0,n/\varepsilon]} \mathcal{W}(t) \leq -\varepsilon n) = \sum_{n=1}^{\infty} 2(1-\Phi(\sqrt{n}\varepsilon^{3/2})).$$

Hence it is concluded that with probability 1 for n sufficiently large either

$$\mathcal{W}(4c \int_0^n X'hX dt) \geq -\varepsilon n \quad \text{or} \quad 4c|h| \int_0^n |X|^2 dt > n/\varepsilon .$$

In the first case it follows from (23) that

$$a \int_0^n |X|^2 dt + o(n) \geq (b-\varepsilon)n,$$

since $Z_n/n \to 0$ a.s. Letting ε be arbitrarily small we infere that

$$\liminf_{T\to\infty} \int_0^T |X|^2 dt/T \geq b/a \quad \text{a.s.},$$

which for $c=1/(|f| + \varkappa|g|)$ yields (20).

<u>Proof of the Proposition</u>. Set

$$w = \sqrt{c}w_1 + cw_2.$$

From (15),(16) it follows that

$$wf + f'w - \frac{1}{c}wgg'w + ee' - \theta_1\sqrt{c}I =$$

$$= c(-w_2f - f'w_2 + w_2gg'w_2).$$

Hence, for c sufficiently small and $\theta = \theta_1\sqrt{c}/2$, the following expression is nonnegatively definite,

(23) $$wf + f'w - \frac{1}{c}wgg'w + ee' - \theta I \geq 0.$$

For

$$k = -\frac{1}{c}g'w$$

it then holds

$$2X'wfX + (e'X)^2 - \theta|X|^2 \geq c(kX)^2.$$

Applying the Itô formula to $\int_0^T d(X'wX)$ and using trace(hw) = 0 we get

$$\int_0^T((e'X)^2+cU^2)dt - \theta \int_0^T |X|^2dt + X_T'wX_T - X_0'wX_0 \geq$$

$$\geq c \int_0^T(U^2+(kX)^2)dt + 2\int_0^T X'wgUdt + 2\int_0^T X'wdW =$$

$$= c \int_0^T(U-kX)^2dt + 2\int_0^T X'wdW.$$

Set

$$A_T = \int_0^T(U-kX)^2dt,$$

and consider the difference

$$X_T'wX_T - cA_T.$$

Write (1) as

(24) $dX_t=(f+gk)X_tdt + g(U_t-kX_t)dt + dW_t.$

According to the hypothesis

$$f + gk = f - \frac{1}{\sqrt{c}}ge' - gg'w_2$$

is a stable matrix. From (24) it follows

$$X_T=F(T)X_0 + \int_0^T F(T-s)g(U_s-kX_s)ds + \int_0^T F(T-s)dW_s$$

with

(25) $F(t) = \exp(t(f + gk)).$

Consequently, for

$$w^+ = \sqrt{c}w_1^+ + cw_2^+ ,$$

$$X_T'w^+X_T = |Z_T + \int_0^T \sqrt{w^+}F(T-s)g(U_s-kX_s)ds|^2,$$

where

$$Z_T = \sqrt{w^+}F(T)X_0 + \int_0^T\sqrt{w^+}F(T-s)dW_s.$$

It holds

$$|\int_0^T\sqrt{w^+}F(T-s)g(U_s-kX_s)ds|^2 \leq \int_0^T|\sqrt{w^+}F(T-s)g|^2dsA_T \leq \bar{v}A_T,$$

with

$$\bar{v} = g'vg, \qquad v = \int_0^\infty F(s)'w^+F(s)ds.$$

Note that in virtue of (25)

(26) $v(f+gk)+(f+gk)'v + w^+ = 0.$

If $\bar{v} < c$, it is concluded that

$$X_T'wX_T - cA_T \leq |Z_T|^2 + 2|Z_T|\sqrt{\bar{v}}\sqrt{A_T} + (\bar{v}-c)A_T \leq$$
$$\leq |Z_T|^2(1+\bar{v}/(c-\bar{v})).$$

To estimate $\bar{v}$ insert into (26)

$$k = -\frac{1}{c} g'(\sqrt{c}w_1 + cw_2), \quad w^+ = \sqrt{c}w_1^+ + cw_2^+ .$$

This yields

$$(27) \quad v = cv_2 + O(c^{3/2}),$$

where

$$-v_2gg'w_1 - w_1gg'v_2 + w_1^+ = 0,$$

and hence

$$2g'v_2gg'w_1g = g'w_1^+g.$$

Consequently, (17) implies

$$(28) \quad g'v_2g < 1,$$

and in virtue of (27)

$$(29) \quad \bar{v} < c$$

for c sufficiently small.

Putting together the inequalities obtained it seen that

$$(30) \quad \int_0^T((e'X)^2+cU^2)dt \geq \frac{1}{2}\sqrt{c}\theta_1 \int_0^T|X|^2dt - |X_0' wX_0| - |Z_T|^2(1+\bar{v}/(c-\bar{v}))+2\int_0^T X'wdX$$

for small c. It is not difficult to prove

$$\lim_{T\to\infty} |Z_T|^2/T = 0 \quad \text{a.s.}$$

Hence, (18) follows from (30) and from the Lemma.

References

[1] M. Boschková: Self-tuning control of stochastic linear systems in presence of drift. Kybernetika 24(1988), 347-362.

[2] T.E.Duncan and B.Pasik-Duncan: Adaptive control of continuous time linear systems. Preprint.University of Kansas 1986.

[3] P.Mandl,T.E.Duncan and B.Pasik-Duncan: On the consistency of a least squares identification procedure. Kybernetika 24(1988) 340-346.

[4] B.Pasik-Duncan: On Adaptive Control. (In Polish). Central School of Planning and Statistics Publishers. Warsaw 1986.

Stochastic controllability and stochastic Lyapunov functions with applications to adaptive and nonlinear systems

S.P. Meyn
Department of Systems Engineering
Research School of Physical Sciences
The Australian National University
Canberra, ACT, 2601 Australia

P.E. Caines
Department of Electrical Engineering
and The Canadian Institute for Advanced Research
McGill University, 3480 University Street
Montréal, P.Q., H3A 2A7 Canada

Abstract

Sufficient conditions are established under which the law of large numbers and related ergodic theorems hold for nonlinear stochastic systems operating under feedback. It is shown that these conditions hold whenever a moment condition is satisfied, which may be interpreted as a generalization of the martingale property.

If in addition a stochastic controllability condition holds, then it is shown that the underlying distributions governing the system converge to an invariant probability at a geometric rate.

These results are illustrated with general examples from linear, nonlinear, and adaptive control theory.

The key assumption used is that a Markov chain with stationary transition probabilities exists which serves as a state process for the closed loop system.

1 Introduction

In this paper we study the asymptotic behavior of discrete time nonlinear stochastic systems under feedback. Our principal assumption is that a Markov chain with stationary transition probabilities $\mathbf{\Phi}$ exists which may serve as a state process for the closed loop system.

In a large number of applications $\mathbf{\Phi}$ evolves on a subset $\mathbf{X} \subset \mathbb{R}^n$, and is generated by a nonlinear difference equation

$$\Phi_{k+1} = F(\Phi_k, w_{k+1}), \qquad k \in \mathbb{Z}_+, \tag{1}$$

where the disturbance $\mathbf{w}$ is an independent and identically distributed process on $\mathbb{R}^p$. Under the appropriate smoothness conditions on the function F and the distribution of $\mathbf{w}$, it has been shown in [Meyn and Caines, 1988] that the asymptotic behavior of $\mathbf{\Phi}$ is determined by invariant probabilities on $\mathbf{X}$ whenever a crude stability condition is satisfied, and the *weak stochastic controllability* condition holds. In this case there exist probabilities $\{\pi_x : x \in \mathbf{X}\}$, and *random* probabilities $\{\tilde{\pi}_x : x \in \mathbf{X}\}$, such that for every initial condition $\Phi_0 = x \in \mathbf{X}$,

$$\lim_{N\to\infty} \frac{1}{N} \sum_{k=1}^{N} f(\Phi_k) = \int f \, d\tilde{\pi}_x \qquad a.s.\ [\mathrm{P}_x] \tag{2}$$

$$\lim_{N\to\infty} \frac{1}{N} \sum_{k=1}^{N} \mathrm{E}_x[g(\Phi_k)] = \int g \, d\pi_x \tag{3}$$

for a large class of functions f and g on $\mathbf{X}$.

When (1) is viewed as a deterministic input/output system with input $\mathbf{w}$ and output $\mathbf{\Phi}$, weak stochastic controllability is equivalent to forward accessability (see [Jakubczyk and Sontag, 1988]).

In Section 3 we present some results for the deterministic system which will then be used to establish ergodic theorems for the original stochastic system.

The results of Section 3 will be used in Section 4 to establish a generalization of Doblin's condition for stochastic systems which possess uniformly bounded trajectories.

In the continuous time case, Φ is typically a diffusion process and the theory of stochastic Lyapunov functions has been a successful tool for assessing its stability properties (see [Kushner, 1967] and [Has'minskiĭ, 1980]). The idea is that if a positive function $V:\mathbf{X} \to \mathbb{R}_+$ exists such that $V(\Phi_k)$ decreases in an average sense, then under general conditions Φ will converge to a level set of the function V with probability one.

It is reasonable to expect that stochastic Lyapunov functions should be a useful tool in the discrete time case as well, and this is indeed the case (see for example [Kushner, 1967], [Solo, 1978] and [Goodwin Ramadge and Caines, 1981]). However when Φ is weakly stochastically controllable, the existence of a stochastic Lyapunov function is all but ruled out. In many cases (for example when V is a quadratic) the level sets of V are sets of Lebesgue measure zero. However, weak stochastic controllability implies that the set of limit points of the sequence

$$\{\Phi_k : k \in \mathbb{Z}_+\} \tag{4}$$

has nonempty interior with probability one for all initial conditions. In fact under the appropriate stability condition for almost every sample path the set of limit points of (4) is equal to the support of the random probability $\tilde{\pi}_x$, which always has nonempty interior under the weak stochastic controllability hypothesis.

In Section 5 we introduce an alternative stochastic Lyapunov function which is perfectly compatible with weak stochastic controllability, and which always exists for stable, linear systems even when an ordinary stochastic Lyapunov function does not exist. If Φ is weakly stochastically controllable, then the existence of this Lyapunov function may be used to prove that (3) holds at an exponential rate for a large class of functions g, and will allow us to establish generalizations of (2) and (3) even when the weak stochastic controllability condition is not satisfied.

In Section 6 these general results are illustrated using examples from nonlinear and adaptive control theory.

2 Preliminaries

Let $\mathbf{X}$ be an an open subset of $\mathbb{R}^n$. We let $\mathbf{C}$ denote the set of bounded and continuous functions $f:\mathbf{X} \to \mathbb{R}$, and $\mathcal{M}$ the set of probabilities on $\mathcal{B}(\mathbf{X})$, the Borel field on $\mathbf{X}$. A sequence $\{\mu_k : k \in \mathbb{Z}_+\} \subset \mathcal{M}$ of probabilities converges *weakly* to $\mu_\infty \in \mathcal{M}$ if

$$\lim_{k\to\infty} \int f\, d\mu_k = \int f\, d\mu_\infty \tag{5}$$

for all $f \in \mathbf{C}$, and this shall be denoted $\mu_k \xrightarrow{\text{weakly}} \mu_\infty$ as $k \to \infty$. It is well known (see [Billingsley, 1968]) that $\mathcal{M}$ is a metrizable topological space and that a subset $\mathcal{A} \subset \mathcal{M}$ is precompact if and only if it is tight, i.e. for all $\varepsilon > 0$ there exists a compact set $C \subset \mathbf{X}$ such that

$$\mu\{C\} \geq 1 - \varepsilon \qquad \mu \in \mathcal{A}.$$

A function $V:\mathbf{X} \to \mathbb{R}_+$ is called a *moment* if there exists a sequence of compact sets $C_n \uparrow \mathbf{X}$ such that

$$\lim_{n\to\infty} \inf_{x \in C_n^c} V(x) = \infty$$

where we adopt the convention that the infimum of a function over the empty set is infinity. It is easily verified that $\mathcal{A} \subset \mathcal{M}$ is tight if and only if a moment V exists such that

$$\sup_{\mu \in \mathcal{A}} \int V\, d\mu < \infty.$$

We let P denote a Feller Markov transition function on $(\mathbf{X}, \mathcal{B}(\mathbf{X}))$. That is, for all $x \in \mathbf{X}$, $A \in \mathcal{B}(\mathbf{X})$, and $f \in \mathbf{C}$,

$P(x, \cdot)$ is a probability on $\mathcal{B}(\mathbf{X})$

$P(\cdot, A)$ is $\mathcal{B}(\mathbf{X})$-measurable

$\int f(y)\, P(\cdot, dy)$ is continuous.

The k-fold iterates of P are defined inductively by $P^1 \triangleq P$, and

$$P^{k+1}(x, A) \triangleq \int P(x, dy) P^k(y, A),$$

and for $f \in \mathbf{C}$ and $\mu \in \mathcal{M}$ we use the standard notation

$$P^k f(\cdot) \triangleq \int P^k(\cdot, dy) f(y) \qquad \mu P^k(\cdot) \triangleq \int \mu(dx) P^k(x, \cdot).$$

We say $\mathbf{\Phi}$ is *irreducible* if the following condition is satisfied. The event $\{\mathbf{\Phi}$ enters $A\} \triangleq \{\bigcup_{k=0}^{\infty} \Phi_k \in A\}$, and the event $\{\mathbf{\Phi} \in A \text{ i.o.}\} \triangleq \{\bigcap_{N=0}^{\infty} \bigcup_{k=N}^{\infty} \Phi_k \in A\}$.

Irreducibility hypothesis. There exists a set $A \in \mathcal{B}(\mathbf{X})$, an integer n_0, a number $\lambda_0 > 0$, and a finite measure φ, such that

(i) $\mathrm{P}_x\{\mathbf{\Phi}$ enters $A\} > 0$ for all $x \in \mathbf{X}$;

(ii) $\sum_{k=1}^{n_0} \mathrm{P}_x\{\Phi_k \in E\} \geq \lambda_0 \varphi\{E\}$ for all $x \in A$ and $E \in \mathcal{B}(\mathbf{X})$.

When $\mathbf{\Phi}$ is irreducible, the set A used in the irreducibility hypothesis will be called *petite*, and the measure φ an *irreducibility measure*. $\mathbf{\Phi}$ is called *Harris recurrent* if the following condition is satisfied.

Recurrence hypothesis. $\mathbf{\Phi}$ satisfies the irreducibility hypothesis for some petite set A and irreducibility measure φ, and for every $x \in \mathbf{X}$,

$$\mathrm{P}_x\{\mathbf{\Phi} \text{ enters } A\} = 1.$$

A subset $B \subset \mathbf{X}$ is called *absorbing* if $P(x, B) = 1$ for every $x \in B$. If B is absorbing, then the Markov chain $\mathbf{\Phi}$ may be restricted to the set B, and B is called a *Harris set* if the restricted process is Harris recurrent.

Many of the important limit theorems for Markov chains require the existence of an invariant measure. That is, a σ-finite measure π on $\mathcal{B}(\mathbf{X})$ with the property

$$\pi\{A\} = \int \pi(dx) P(x, A) \qquad \text{for all } A \in \mathcal{B}(\mathbf{X}).$$

If the recurrence condition holds then it may be shown that an essentially unique invariant measure π exists. If the invariant measure is finite, then it may be normalized to a probability measure and in this case $\mathbf{\Phi}$ is called *positive Harris recurrent.*

If $\mathbf{\Phi}$ is irreducible, then there exists an integer $m \in \mathbb{Z}_+$ called the *period* of $\mathbf{\Phi}$, and a collection of sets $\{E_1, \ldots, E_m\}$ with the property that

$$P\mathbf{1}_{E_{i+1}} = \mathbf{1}_{E_i} \quad \text{and} \quad P^m \mathbf{1}_{E_i} = \mathbf{1}_{E_i}$$

for each $1 \leq i \leq m$. If $\mathbf{\Phi}$ is Harris recurrent with invariant measure π, then $\pi\{(\bigcup E_i)^c\} = 0$.

The following proposition shows that if Φ is positive Harris recurrent and *aperiodic* ($m = 1$), then its underlying distributions converge to the invariant probability for all initial conditions. If in addition, the distribution of the hitting time $\tau_A \triangleq \min\{k \geq 1 : \Phi_k \in A\}$ to the petite set A possesses geometrically decaying tails, uniformly for initial conditions lying in A, then the underlying distributions of Φ converge to π at a geometric rate.

Define the total variation norm $\|\mu - \nu\|$ for $\nu, \mu \in \mathcal{M}$ by

$$\|\mu - \nu\| \triangleq \sup \left| \int f \, d\mu - \int f \, d\nu \right|$$

where the supremum is taken over all Borel functions $f : \mathbf{X} \to [-1, 1]$.

Proposition 2.1. *Suppose that Φ is positive Harris recurrent and aperiodic with invariant probability π. Then,*

(i) *for each initial condition distribution $\mu \in \mathcal{M}$, and $f \in L^1(\mathbf{X}, \mathcal{B}(\mathbf{X}), \pi)$,*

$$\lim_{k \to \infty} \|\mu P^k - \pi\| = 0$$

$$\lim_{N \to \infty} \frac{1}{N} \sum_{k=1}^{N} f(\Phi_k) = \int f \, d\pi \qquad a.s. \ [\mathrm{P}_\mu]$$

(ii) *if in addition the set $A \in \mathcal{B}(\mathbf{X})$ defined in the recurrence hypothesis satisfies*

$$\sup_{x \in A} \mathrm{E}_x[r^{\tau_A}] < \infty \tag{6}$$

for some $r > 1$, then there exists $\rho < 1$, and an extended real valued function $M \in L^1(\mathbf{X}, \mathcal{B}(\mathbf{X}), \pi)$ such that for each $x \in \mathbf{X}$,

$$\|\delta_x P^k - \pi\| = \|P^k(x, \cdot) - \pi(\cdot)\| \leq M(x)\rho^k \qquad k \in \mathbb{Z}_+;$$

(iii) *if the conditions of (ii) hold, and an initial condition distribution $\mu_0 \in \mathcal{M}$ satisfies*

$$\mathrm{E}_{\mu_0}[r^{\tau_A}] < \infty,$$

then there exists $\rho < 1$ and $M < \infty$ such that

$$\|\mu_0 P^k - \pi\| \leq M\rho^k \qquad k \in \mathbb{Z}_+;$$

(iv) *Suppose that the conditions of (iii) hold, and $f : \mathbf{X} \to \mathbb{R}$ satisfies*

$$\sup_{k \in \mathbb{Z}_+} E_{\mu_0}[|f(\Phi_k)|^{1+\delta}] < \infty$$

for some $\delta > 0$. Then there exists $\rho < 1$ and $M < \infty$ such that

$$|E_{\mu_0}[f(\Phi_k)] - \int f \, d\pi| \leq M\rho^k \qquad k \in \mathbb{Z}_+.$$

□

For a proof of Proposition 2.1 (i) see [Nummelin, 1984] and [Athreya and Ney, 1980]. Results (ii) and (iii) may be found in [Nummelin and Tuominen, 1982], and result (iv) follows from (iii) and Hölder's inequality.

An aperiodic positive Harris recurrent Markov chain is sometimes called *ergodic*. If for all $x \in \mathbf{X}$ there exists $\rho(x) < 1$ and $M(x) < \infty$ such that

$$\|P^k(x, \cdot) - \pi(\cdot)\| \leq M\rho^k \qquad k \in \mathbb{Z}_+,$$

then $\mathbf{\Phi}$ will be called *geometrically ergodic*. This is weaker then the notion of geometric ergodicity introduced in [Tweedie, 1983], and stronger than that of [Nummelin, 1984]. In Section 5 we present sufficient conditions for geometric ergodicity using a stochastic Lyapunov function.

The following stability conditions will be shown to be closely connected to Harris (respectively positive Harris) recurrence:

Stability conditions.

S1 For each initial condition $x \in \mathbf{X}$ and each $\varepsilon > 0$, there exists a compact subset $C \subset \mathbf{X}$ such that

$$\mathrm{P}_x\{\mathbf{\Phi} \in C \ i.o.\} = \lim_{k\to\infty} \mathrm{P}_x\{\bigcup_{i=k}^{\infty}\{\Phi_i \in C\}\} \geq 1 - \varepsilon$$

S2 For each initial condition $x \in \mathbf{X}$ and each $\varepsilon > 0$ there exists a compact subset $C \subset \mathbf{X}$ such that

$$\liminf_{k\to\infty} \mathrm{P}_x\{\Phi_k \in C\} \geq 1 - \varepsilon.$$

It may be shown that if a moment V exists such that

$$\liminf_{k\to\infty} V(\Phi_k) < \infty \qquad a.s.\ [\mathrm{P}_x]$$

for each $x \in \mathbf{X}$ then condition S1 holds, and if

$$\limsup_{k\to\infty} \mathrm{E}_x[V(\Phi_k)] < \infty$$

for each $x \in \mathbf{X}$ then condition S2 is satisfied.

It is evident that condition S2 implies condition S1. In [Rosenblatt, 1971] a strengthening of condition S1 is used to establish the existence of a σ-finite invariant measure for Feller Markov chains. Condition S2 is called ***stability in probability*** in [Meyn and Caines, 1988], and is simply the tightness hypothesis of [Billingsley, 1968]. In [Beneš, 1968] a similar condition (among other assumptions) is used to establish the existence of an invariant probability for a continuous time Feller Markov process, and under the assumptions already made on $\mathbf{\Phi}$, condition S2 implies the existence of an invariant probability (see [Foguel, 1969] and [Meyn and Caines, 1988]).

The following result is taken from [Meyn, 1988a]. Similar results may also be found in [Tuominen and Tweedie, 1979].

Proposition 2.2. *Suppose that $\mathbf{\Phi}$ satisfies the irreducibility condition, and that the petite set A is open. Then $\mathbf{\Phi}$ is Harris recurrent if and only if condition S1 is satisfied, and $\mathbf{\Phi}$ is positive Harris recurrent if and only if condition S2 holds.*

□

In the following section we describe how the irreducibility condition may be established using ideas from dynamical systems theory.

3 Considerations from topological dynamics

In this section we study nonlinear, discrete time systems of of the form

$$x_{k+1} = F(x_k, u_{k+1}), \qquad k \in \mathbb{Z}_+, \tag{7}$$

where the state $\mathbf{x}$ evolves on $\mathbf{X}$ = an open subset of $\mathbb{R}^n$, the input $\mathbf{u}$ takes values in an open set $\mathcal{U} \subset \mathbb{R}^p$, and $F: \mathbf{X} \times \mathcal{U} \to \mathbf{X}$ is continuously differentiable (C^1).

Once we have established some basic properties of (7), we will replace the input $\mathbf{u}$ with and i.i.d. stochastic process $\mathbf{w}$ to obtain a Markov chain of the form (1).

For each time $k \in \mathbb{Z}_+$, the state x_k is a function of the initial condition x_0, and the control sequence $(u_1, \ldots, u_k)$ which we denote $F_{u_k \cdots u_1}(x_0)$. This function may be defined inductively by

$$F_{u_1}(x_0) = F(x_0, u_1) \qquad \text{and} \qquad F_{u_{k+1} \cdots u_1}(x_0) = F(F_{u_k \cdots u_1}(x_0), u_{k+1}).$$

For $E \subset \mathbf{X}$, $k \in \mathbb{Z}_+$, we define $A_+^k(E)$ to be the set of all states reachable from E at time k. That is,

$$A_+^k(E) \triangleq \left\{ F_{u_k \cdots u_1}(x) : x \in E, u_i \in \mathcal{U}, 1 \le i \le k \right\}.$$

The set $A_+(E)$ is defined to be the set of all states attainable from E at some time $k \geq 0$,

$$A_+(E) \triangleq E \cup \Big\{ \bigcup_{k=1}^{\infty} A_+^k(E) \Big\}.$$

Our objective here is to generalize standard definitions and results from dynamical systems theory as presented in, for example, [Bhatia and Szegö, 1970]. If $E \subset \mathbf{X}$ has the property that

$$A_+(E) \subset E$$

then E is called *invariant*. For example, for all $C \subset \mathbf{X}$, the set $A_+(C)$ is invariant, and since the closure, union, and intersection of invariant sets is invariant, the ω-limit set

$$\Omega_+(C) \triangleq \bigcap_{N=1}^{\infty} \overline{\Big\{ \bigcup_{k=N}^{\infty} A_+^k(C) \Big\}}$$

is also invariant.

A set $F \subset \mathbf{X}$ will be called *minimal* if it is (topologically) closed, invariant, and does not contain as a proper subset any closed invariant set.

In the stochastic framework, where the input $\mathbf{u}$ is replaced by an independent and identically distributed (i.i.d.) stochastic process, the minimal sets become the (possibly unstable) ergodic sets for the resulting Markov chain on $\mathbf{X}$, and every trajectory either enters a minimal set (and remains there), or converges to infinity.

3.1. Forward Accessible Systems.

Forward accessibility is the natural generalization of the notion of controllability from linear systems theory. It is useful because it ensures that the set of attainable states is not too small relative to the state space.

Definition The system (7) is called *forward accessible* if for each $x_0 \in \mathbf{X}$, the set $A_+(x_0)$ has non-empty interior. □

Necesary and sufficient conditions for forward accessibility based upon the theory of Lie algebras may be found in [Jakubczyk and Sontag, 1988]. We state here conditions involving a generalized controllability matrix.

For $y \in \mathbf{X}$ and a sequence $\{z_k : z_k \in \mathcal{U},\ k \in \mathbb{Z}_+\}$ let $\{\mathrm{A}_k,\ \mathrm{B}_k : k \in \mathbb{Z}_+\}$ denote the matrices

$$\mathrm{A}_k = \mathrm{A}_k(y, z_1, \dots, z_{k+1}) \triangleq \left[\frac{\partial F}{\partial x}\right]_{(F_{z_{k+1}\cdots z_1}(y))}$$

$$\mathrm{B}_k = \mathrm{B}_k(y, z_1, \dots, z_{k+1}) \triangleq \left[\frac{\partial F}{\partial z}\right]_{(F_{z_{k+1}\cdots z_1}(y))},$$

and let $\mathrm{C}_y^k = \mathrm{C}_y^k(z_1, \dots, z_k)$ denote the *generalized controllability matrix* (along the sequence z_1^k)

$$\mathrm{C}_y^k \triangleq [\mathrm{A}_{k-1} \cdots \mathrm{A}_1\mathrm{B}_0 | \mathrm{A}_{k-1} \cdots \mathrm{A}_2\mathrm{B}_1 | \ \cdots \ |\mathrm{A}_{k-1}\mathrm{B}_{k-2}|B_{k-1}] \tag{8}$$

We remark that if F is of the form

$$F(y, z) = \mathrm{A}y + \mathrm{B}z \tag{9}$$

then the generalized controllability matrix becomes the familiar controllability matrix

$$\left[\mathrm{A}^{T-1}\mathrm{B}|\mathrm{A}^{T-2}\mathrm{B}|\cdots|\mathrm{AB}|\mathrm{B}\right].$$

Note that all quantities in the matrix (8) are deterministic.

Proposition 3.1. *The system (7) is forward accessible if and only if for all initial conditions $x \in \mathbf{X}$, there exists $T \geq 1$ and $\lambda \in \mathcal{U}^T$ such that*

$$\operatorname{rank} C_x^T(\lambda) = n. \tag{10}$$

□

Because of lack of space, it is not possible to present here a complete description of the topological properties of forward accessible systems. One of the principle properties of interest is cyclicity. Given a minimal set F, it is possible to construct a unique (maximal) integer $\lambda \geq 1$, such that F may be decomposed as the union of disjoint closed sets $\{G_i : 1 \leq i \leq \lambda\}$ with the property that

$$A_+^1(G_i) \subset G_{i+1} \qquad (\text{mod } \lambda)$$

for all $i \in \{1, \dots, \lambda\}$. Hence regardless of the control $\mathbf{u}$, the sequence of sets $\{G_i\}$ form a periodic orbit in the sense that if $x_0 \in G_\lambda$, then $x_k \in G_k$ (mod λ) for all $k \in \mathbb{Z}_+$. This result may be compared to the Poincaré Bendixon Theorem for dynamical systems on $\mathbb{R}^2$.

For details of this general theory the reader is refered to [Meyn, 1988b]. We now specialize to the case where a unique minimal set F exists which is *aperiodic* (i.e. $\lambda = 1$). It will be shown that this is not a substantial restriction since it is implied by a general formulation of asymptotic controllability.

3.2. Irreducible and Indecomposable Systems.

It is difficult to find realistic examples in which more than one minimal set exists, and hence it is reasonable to specialize to the case where (7) possesses at most one minimal set.

We say that (7) is *indecomposable* if it does not contain two disjoint closed invariant sets. This is a necessary but not a sufficient condition for (7) to possess a unique minimal set. If (7) is indecomposable and also possesses a minimal set, then the system will be called *irreducible.* If this is the case, and if F denotes the unique minimal set, then it follows that

$$A_+(x) \cap F \neq \phi \quad \text{forall } x \in \mathbf{X}.$$

If this were not the case, then F and $\{x \in \mathbf{X} : A_+(x) \cap F = \phi\}$ would both be nonempty closed invariant sets, and this contradicts indecomposability. (One may show that $\{x \in \mathbf{X} : A_+(x) \cap F = \phi\}$ is a closed subset of $\mathbf{X}$ using the fact that F has non-empty interior.)

If (7) is irreducible, then it will be called *aperiodic* if its unique minimal set is aperiodic.

Suppose that a distinguished state $x^* \in \mathbf{X}$ exists such that

$$x^* \in \bigcap_{y \in \mathbf{X}} \overline{A_+(y)}.$$

Then it is easily verified that (7) is irreducible and that $\overline{A_+(x^*)}$ is the unique minimal set in $\mathbf{X}$. Now suppose that the following stronger condition holds. We will call (7) ***asymptotically controllable*** if a fixed state $x^* \in \mathbf{X}$ exists such that for every initial condition $y \in \mathbf{X}$, there exists a control sequence $(u_1, \ldots, u_k, \ldots)$ such that

$$\lim_{k \to \infty} F_{u_k \cdots u_1}(y) = x^*$$

This is in fact much weaker than the usual definition (see [Sontag, 1983]) but is sufficient for our purposes. If this condition holds, then a unique aperiodic minimal set exists. We collect these results together in the following proposition

Proposition 3.2. *If the system (7) is forward accessible and asymptotically controllable then it is irreducible and aperiodic.*

□

The following result is taken from [Meyn, 1988b]. We remark that a generalization of Lemma 3.1 is the fundamental property of forward accessible systems which allows the proof of the cyclicity result described in the previous section.

Lemma 3.1. *If (7) is forward accessible and $F \subset \mathbf{X}$ is an aperiodic minimal set, then there exists an open set $E \subset F$, and an integer $k_0 \in \mathbb{Z}_+$ with the property that*

$$A_+^k(e) \supset E \tag{11}$$

for all $e \in E$, and all $k \geq k_0$.

□

The following result is implied by Lemma 3.1.

Proposition 3.3. *Suppose that (7) is forward accessible, irreducible and aperiodic with minimal set F. Let $K \subset \mathbf{X}$ be compact, and $U \subset \mathbf{X}$ be an open set for which $U \cap F \neq \phi$. Then there exists an integer $N_0 = N_0(K, U) \in \mathbb{Z}_+$ for which*

$$A_+^k(x) \cap U \neq \phi$$

for all $x \in K$, and all $k \geq N_0$.

□

3.3. Applications to stochastic systems.

In this section we continue our investigation of forward accessible systems of the form (7). However, we now suppose that the input u is a stochastic process which we interpret as a disturbance.

To avoid confusion between the deterministic and stochastic frameworks, we will let $\mathbf{\Phi}$ denote the state process, and $\mathbf{w}$ denote the input (or disturbance process) so that (7) becomes (1). We stress that the definitions of invariant and minimal sets, attainability, etc. remain unchanged in this section.

> *Throughout the remainder of this paper we assume that the system (1) is forward accessible, irreducible, and aperiodic.*

Generalizations of the results below may be established when the irreducibility condition is removed. The details may be found in the [Meyn, 1988b] and [Meyn and Caines, 1988].

We henceforth assume that w and Φ_0 are defined on a probability space $(\Omega, \mathcal{F}, P_{\Phi_0})$, the disturbance $w = \{w_k : k \in \mathbb{Z}_+\}$ is independent and identically distributed (i.i.d.) and independent of the initial condition Φ_0, and we also make an assumption on the distribution μ_w of w_k, $k \in \mathbb{Z}_+$, which fits the assumption used in the previous section requiring the input to be constrained to lie in the open set $\mathcal{U}$. We assume:

> *The probability μ_w possesses a density p_w which is lower semi continuous. The open control set $\mathcal{U}$ used in the definition of forward accessibility will be taken to be $\mathcal{U} \triangleq \{z \in \mathcal{U} : p_w(z) > 0\}$.*

When this condition holds, forward accessibility is equivalent to the weak stochastic controllability hypothesis of [Meyn and Caines, 1988].

Because the function F is continuous and the disturbance w is i.i.d., the state process Φ becomes a Feller Markov chain with state space $\mathbf{X}$, and Markov transition function P defined for $x \in \mathbf{X}$ and $A \in \mathcal{B}(\mathbf{X})$ by

$$P(x, A) = \mu_w\{z \in \mathcal{U} : F(x, z) \in A\}.$$

Recall that a subset $A \subset \mathbf{X}$ is called *absorbing* if $P(x, A) = 1$ for all $x \in A$. Using the hypothesis on the distribution of the disturbance process, it is easily shown that

(i) If U is open and $x \in \mathbf{X}$, then $A_+(x) \cap U \neq \phi$ if and only if $P^k(x, U) > 0$ for some $k \in \mathbb{Z}_+$.
This implies

(ii) If the set A is (topologically) closed, then it is absorbing if and only if it is invariant;

We will use these facts repeatedly below.

Under the present conditions it is shown in [Meyn and Caines, 1988] that there exists (U, φ, k_0, β) such that U is an open subset of $\mathbf{X}$ with $U \subset F$, φ is a probability on $\mathcal{B}(\mathbf{X})$, $k_0 \in \mathbb{Z}_+$, $\beta > 0$, and for all $y \in U$ and $B \in \mathcal{B}(\mathbf{X})$,

$$P^{k_0}(y, B) \geq \beta\varphi\{B\}.$$

In fact φ may be taken to be the uniform distribution on a bounded open subset of $\mathbf{X}$. By Proposition 3.3 it may be shown that

$$P_y\{\Phi \text{ enters } U\} > 0$$

for all $y \in \mathbf{X}$. It follows that Φ is (stochastically) irreducible under the present conditions, and that the open set U is a petite set with φ the associated irreducibility measure.

Instability results.

Here we show that with probability one, the trajectory Φ either enters and forms a dense subset of the minimal set F, or "converges to infinity". To make this notion precise, let $K_n \subset \mathbf{X}$, $n \in \mathbb{Z}_+$, denote the compact set

$$K_n = \{x \in \mathbf{X} : |x| \leq n \quad \text{and} \quad \inf_{y \in \mathbb{R}^n \setminus \mathbf{X}} |x - y| \geq 1/n\}$$

We define the event $\{\Phi \to \infty\} \in \mathcal{F}$ by

$$\{\Phi \to \infty\} \triangleq \bigcap_{n=1}^{\infty} \{\Phi \in K_n \ i.o.\}^c$$

Proposition 3.4. *For each initial condition $\Phi_0 = x \in \mathbf{X}$, and each open set $V \subset \mathbf{X}$ satisfying $V \cap F \neq \phi$,*

$$P_x\Big\{\{\Phi \text{ enters } V \ i.o.\} \cup \{\Phi \to \infty\}\Big\} = 1.$$

Proposition 3.3 implies that under the conditions of this section, the support of an invariant measure is equal to the unique minimal set F. This may be seen as a generalization of a result from dynamical system theory which states that the support of an ergodic invariant probability is minimal (see exercise 7, page 67 of [Brown, 1976]).

Proof of Proposition 3.4.
Fix $n \in \mathbb{Z}_+$, and let $K_n \subset \mathbf{X}$ denote the compact set defined above. It may be shown that the function $P^k(\cdot, V)$ is lower semi-continuous ([Cogburn, 1975]). Hence by by Proposition 3.3, there exists $N_0 \in \mathbb{Z}_+$ such that

$$\inf_{x \in K_n} P^{N_0}(x, V) > 0.$$

By Proposition 5.1 of [Orey, 1971] this implies that for every initial condition $y \in \mathbf{X}$,

$$\mathrm{P}_y\{\{\Phi \in K_n \ i.o.\}^c \cup \{\Phi \in V \ i.o.\}\} = 1.$$

Since $n \in \mathbb{Z}_+$ is arbitrary, this completes the proof. □

4 Lagrange stability and Doblin's condiiton.

If (1) is linear, it is easily verified that it is forward accessible if and only if it is controllable in the usual sense, and if this is the case and (1) is stable, then it is possible to show by direct calculation that the distributions governing the system converge at a geometric rate for all initial conditions.

In this section we show that this is also the case for the class of Markov chains studied in the previous section whenever the system is Lagrange stable; i.e. whenever the trajectories of (7) lie in a compact set for each initial condition.

The following condition requires the stochastic process Φ to be uniformly bounded for each initial condition, and will be invoked throughout this section:

> *For each $x \in \mathbf{X}$, the closure of the set of attainable states $\overline{A_+(x)}$ is a compact subset of* $\mathbf{X}$.

This condition will hold in the case where $\mathbf{X} = \mathbb{R}^n$, the open set $\mathcal{U}$ supporting p_w is bounded, and the system (1) is BIBO stable in the usual sense. Observe that this is much stronger than the stability property S2.

A set $D \in \mathcal{B}(\mathbf{X})$ will be called a *D-set* if the following hypothesis from [Doob, 1953] is satisfied:

> **Condition D** The set $D \in \mathcal{B}(\mathbf{X})$ is absorbing, and a fixed probability φ on $\mathcal{B}(\mathbf{X})$, an integer $k_0 \in \mathbb{Z}_+$, and ε, $\delta > 0$ exist such that for all $A \in \mathcal{B}(\mathbf{X})$,
>
> $$\varphi\{A\} \leq \varepsilon \Longrightarrow \sup_{x \in D} P^{k_0}(x, A) \leq 1 - \delta$$

We may now present the main result of this section:

Proposition 4.1. *Under the stated conditions, for each $x \in \mathbf{X}$, the set $\overline{A_+(x)}$ is a D-set.*

Proof.
Fix $x \in \mathbf{X}$, and let (U, φ, k_0, β) be a quadruple satisfying the conditions of the previous section with $U \subset F$ open. As in the proof of Proposition 3.4, there exists $n_0 \in \mathbb{Z}_+$ such that

$$\inf_{y \in \overline{A_+(x)}} P^{n_0}(y, U) > 0,$$

and hence we may find $\alpha > 0$ such that

$$P^{n_0+k_0}(y, A) \geq \alpha\varphi\{A\}$$

for all $y \in \overline{A_+(x)}$ and $A \in \mathcal{B}(\mathbf{X})$. In particular, this shows that the set $\overline{A_+(x)}$ is petite.

Hence if $\varphi\{A\} \leq 1/2$, then for all $y \in \overline{A_+(x)}$,

$$1 - P^{n_0+k_0}(y, A) = P^{n_0+k_0}(y, A^c) \geq \alpha\varphi\{A^c\} \geq \alpha/2.$$

Hence condition D is satisfying with $\varepsilon \triangleq 1/2$ and $\delta = \alpha/2$. □

The following result shows that under the conditions of this section, the Markov chain $\mathbf{\Phi}$ is geometrically ergodic. Let $f: \mathbf{X} \to \mathbb{R}$ be a measurable function which is uniformly bounded on compact sets (for example, take f continuous) and let

$$\|f\|_\infty^x \triangleq \sup_{y \in \overline{A_+(x)}} |f(y)| < \infty.$$

Proposition 4.2. *Under the conditions of this section, a unique invariant probability π exists, and the following limit theorems hold:*

(i) *For each $x \in \mathbf{X}$, there exists $\rho < 1$, $M < \infty$ such that for every function $f: \mathbf{X} \to \mathbb{R}$ which is uniformly bounded on compact sets,*

$$\sup\Big\{\Big|\mathrm{E}_y[f(\Phi_k)] - \int f\, d\pi\Big| : y \in \overline{A_+(x)}\Big\} \leq M\, \|f\|_\infty^x \rho^k, \qquad k \in \mathbb{Z}_+;$$

(ii) *For all positive measurable functions $f: \mathbf{X} \to \mathbb{R}_+$, and alternatively, for all $f \in L^1(\mathbf{X}, \mathcal{B}(\mathbf{X}), \pi)$,*

$$\lim_{n\to\infty} \frac{1}{n}\sum_{k=1}^n f(\Phi_k) = \int f\, d\pi \qquad a.s.\ [\mathrm{P}_x]$$

for each initial condition $\Phi_0 = x \in \mathbf{X}$.

(iii) *Define $\sigma_f^2 \geq 0$ by*

$$\sigma_f^2 \triangleq \lim_{n\to\infty} \mathrm{E}_x\Big[\Big(\frac{1}{\sqrt{n}}\sum_{k=1}^n f(\Phi_k) - \mathrm{E}_x[f(\Phi_k)]\Big)^2\Big].$$

This limit always exists, is independent of $\Phi_0 = x \in \mathbf{X}$, and if $\sigma_f^2 > 0$ then for all initial conditions $x \in \mathbf{X}$, and all $\lambda \in \mathbb{R}$,

$$\lim_{n\to\infty} \mathrm{P}_x\Big\{\frac{1}{\sqrt{n}}\sum_{k=1}^n (f(\Phi_k) - \mathrm{E}_x[f(\Phi_k)]) \leq \lambda\Big\} = \frac{1}{\sqrt{2\pi\sigma_f^2}} e^{-\lambda^2/2\sigma_f^2}$$

Proof.
At most one invariant probability exists since $\mathbf{\Phi}$ is (stochastically) irreducible. By Proposition 2.2 there is a unique invariant probability π.

Result (i) follows from Proposition 4.1, and the hypothesis that the unique minimal set is aperiodic.

Result (ii) follows from Theorem 6.1 of Chapter 5 of [Doob, 1953], and (iii) is a direct consequence of Theorem 7.5 of the same chapter. □

5 Stochastic Lyapunov functions

In this section we show how stochastic Lyapunov functions may be used to verify the geometric ergodicity conditions of Proposition 2.1.

Let $V: \mathbf{X} \to \mathbb{R}_+$ be a positive measurable function. The following drift condition will be invoked throughout this section

For some $0 < \lambda < 1$, $K > 0$, and all $x \in \mathbf{X}$,

$$PV(x) \leq \lambda V(x) + K. \tag{12}$$

In terms of the stochastic process $\mathbf{\Phi}$ this condition may be expressed

$$\mathrm{E}[V(\Phi_{k+1}) \mid \mathcal{F}_k] \leq \lambda V(\Phi_k) + K$$

where $\mathcal{F}_k$ is the sigma algebra generated by past and present values of $\mathbf{\Phi}$:

$$\mathcal{F}_k \triangleq \sigma\{\Phi_0, \ldots, \Phi_k\}. \tag{13}$$

There is a great deal of motivation for introducing this condition. First of all, it is a natural generalization of the martingale property: In the degenerate case where $\lambda = 1$ and $K = 0$ (12) becomes

$$\mathrm{E}[V(\Phi_{k+1}) \mid \mathcal{F}_k] \leq V(\Phi_k). \tag{14}$$

Hence in this case $(V(\Phi_k), \mathcal{F}_k)$ is a supermartingale, and by the martingale convergence theorem (see [Doob, 1953]) for each initial condition $x \in \mathbf{X}$ there exists a random variable $V_\infty = V_\infty(x)$ such that

$$\lim_{k \to \infty} V(\Phi_k) = V_\infty \qquad a.s.\ [\mathrm{P}_x],$$

and furthermore $\mathrm{E}_x[V_\infty] \leq V(x) < \infty$.

A positive function $V: \mathbf{X} \to \mathbb{R}_+$ sayisfying (14) is called a *super harmonic* function in [Nummelin, 1984] and [Revuz, 1975], and a *stochastic Lyapunov function* in the stochastic systems theory literature.

Suppose that $\mathbf{\Phi}$ is a Markov chain generated by a forward accessible (weakly stochastically controllable) stochastic system of the form studied in Section 3.3. Then by Proposition 3.4 there are two possibilities. Either $\mathbf{\Phi} \to \infty$ or $\mathbf{\Phi}$ enters some minimal set F and visits all points in a dense subset of F. This implies that in the interesting case where $\mathrm{P}\{\mathbf{\Phi} \to \infty\} \neq 1$, a continuous Lyapunov function V takes on a constant value on each minimal set. It appears that in cases where the control set $\mathcal{U}$ is large, the corresponding minimal sets will usually be large subsets of the state space, and hence constructing a useful stochastic Lyapunov function for such a process would not be an easy task. On the other hand, the test function (12) does not have these negative properties, and is extremely useful when $\mathbf{\Phi}$ satisfies the irreducibility condition.

Proposition 5.1. *Suppose that V is a continuous moment satisfying (12), and $\mathbf{\Phi}$ is a Feller Markov chain. If the irreducibility condition holds for an open set $A \subset \mathbf{X}$, and if $\mathbf{\Phi}$ is aperiodic, then $\mathbf{\Phi}$ is geometrically ergodic.*

□

Similar results may be found in [Nummelin, 1984] and [Nummelin and Tuominen, 1982]. The principal difference between our result and these lies in our consideration of the topology of the state space. Our result also differs in the form of the test function V.

A connected result may also be found in the dissertation [Chan, 1986]. However, this result relies on an exponential stability hypothesis on the deterministic dynamical system obtained when the disturbance $\mathbf{w}$ is set equal to zero, and a global Lipschitz condition on the function F.

The original inspiration for such test function methods for establishing positive recurrence lies in the paper [Foster, 1953].

To prove Proposition 5.1 we closely follow [Nummelin and Tuominen, 1982]. The following result is an extension of Theorem 3.1 of that paper, and also may be seen as a sort of converse to a theorem of [Kushner, 1967] where a lyapunov function of the form (12) is used to estimate the probability that Φ will leave a given compact set. For $t \in \mathbb{R}$, let $C_t \triangleq \{x \in \mathbf{X} : V(x) \leq t\}$.

Lemma 5.1. *Suppose that $V: \mathbf{X} \to \mathbb{R}_+$ is a positive measurable function satisfying (12), and let $r \in \mathbb{R}_+$ satisfy $1 > r^{-1} > \lambda$. Then for each $t > K/(r^{-1} - \lambda)$, there exists a constant $K_1 > 0$ such that for every $x \in \mathbf{X}$,*

$$\mathrm{E}_x[r^{\tau_{C_t}}] \leq K_1(V(x) + 1).$$

Proof.

By (12) we have for every $x \in \mathbf{X}$,

$$rPV(x) \leq V(x) - ((1 - \lambda r)V(x) - rK). \tag{15}$$

Let $U(x) \triangleq (1-\lambda r)V(x) - rK$ for $x \in \mathbf{X}$, $\varepsilon \triangleq (1-\lambda r)t - rK > 0$, and observe that by the conditions of the lemma,

$$\mathbf{1}_{C_t^c}(x)\, U(x) \geq \varepsilon \mathbf{1}_{C_t^c}(x) \qquad \text{for all } x \in \mathbf{X}. \tag{16}$$

Equation (15) implies that for all $x \in \mathbf{X}$,

$$rP\mathbf{1}_{C_t^c}V(x) \leq V(x) - U(x) \tag{17}$$

where the operator "$rP\mathbf{1}_{C_t^c}$" is defined for a positive function $f: \mathbf{X} \to \mathbb{R}_+$ by

$$rP\mathbf{1}_{C_t^c}f(x) \triangleq r\int_{C_t^c} P(x, dy)f(y).$$

Applying this operator to both sides of (17) gives

$$\begin{aligned}\left(rP\mathbf{1}_{C_t^c}\right)^2 V &\leq V - U - rP\mathbf{1}_{C_t^c}U \\ &\leq V - U - \varepsilon rP\mathbf{1}_{C_t^c}1,\end{aligned}$$

and by induction we have for all $x \in \mathbf{X}$,

$$\varepsilon \sum_{k=1}^{\infty}\left(rP\mathbf{1}_{C_t^c}\right)^k 1\,(x) \leq V(x) - U(x). \tag{18}$$

The left hand side of this equation may be transformed as follows:

$$\begin{aligned}\sum_{k=1}^{\infty}\left(rP\mathbf{1}_{C_t^c}\right)^k 1\,(x) &= \sum_{k=1}^{\infty} r^k \mathrm{P}_x\{\tau_{C_t} > k\} \\ &= \sum_{k=1}^{\infty} r^k \sum_{i=k+1}^{\infty} \mathrm{P}_x\{\tau_{C_t} = i\} \\ &= \sum_{i=2}^{\infty} \mathrm{P}_x\{\tau_{C_t} = i\}\Big(\sum_{k=1}^{i-1} r^k\Big) \\ &= \frac{1}{r-1}\sum_{i=2}^{\infty} \mathrm{P}_x\{\tau_{C_t} = i\} r^i \\ &\quad - \frac{r}{r-1}\mathrm{P}_x\{\tau_{C_t} > 1\} \\ &\geq \frac{1}{r-1}\Big(\mathrm{E}_x[r^{\tau_{C_t}}] - r\mathrm{P}_x\{\tau_{C_t} \leq 1\}\Big) - \frac{r}{r-1}\mathrm{P}_x\{\tau_{C_t} > 1\} \\ &\geq \frac{1}{r-1}\mathrm{E}_x[r^{\tau_{C_t}}] - \frac{r}{r-1}.\end{aligned}$$

This together with (18) and the definitions of U and ε proves the lemma.

$\square$

Proof of Proposition 5.1.

The existence of a moment satisfying (12) implies that

$$\limsup_{k\to\infty} \mathrm{E}_x[V(\Phi_k)] \leq \frac{K}{1-\lambda}$$

for all initial conditions, and hence also the stability condtion S2. By Proposition 2.2, Φ is positive Harris recurrent.

Since V is a moment and is continuous, for each $t \in \mathbb{R}_+$, the set C_t defined above Lemma 5.1. is compact. By Lemma 2.2 of [Meyn,1988a] we may choose t so large that C_t is petite, and so the proof follows from Proposition 2.1. □

In the following result we show that condition (12) is an extremely useful property even when the irreducibility condition is not satisfied. For a proof see [Meyn, 1988a].

Proposition 5.2. *Suppose that V is a continuous moment satisfying (12), Φ is a Feller Markov chain, and exactly one invariant probability π exists. Define the occupation probabilities*

$$\tilde{\mu}_N\{A\} \triangleq \frac{1}{N}\sum_{k=1}^{N} \mathbf{1}_{\{\Phi_k \in A\}} \qquad N \in \mathbb{Z}_+,\ A \in \mathcal{B}(\mathbf{X}).$$

Then for each initial condition $x \in \mathbf{X}$,

$$\tilde{\mu}_k \xrightarrow{\text{weakly}} \pi \quad \text{as } k \to \infty \qquad a.s.\ [\mathrm{P}_x]$$

□

We now show how a moment satisfying (12) may be constructed. In the stable linear case (9) with $\mathbf{w} = 0$ such a construction was first carried out in [Kalman and Bertram, 1960]. Corollary 3.2* of that paper implies that there exists a positive definite matrix M with $I \leq M \leq mI$ for some $m \geq 1$, and

$$|Ax|_M^2 \leq \lambda |x|_M^2 \tag{19}$$

where $|y|_M^2 \triangleq y^\top M y$ for $y \in \mathbb{R}^n$, and $\lambda < 1$. In fact we can take

$$M \triangleq I + \sum_{i=1}^{\infty} A^{\top i} A^i.$$

In the case where $\mathbf{w} \neq 0$, suppose that the i.i.d. process $\mathbf{w}$ defined in (9) satisfies $\mathrm{E}[w_0^2] < \infty$, but is otherwise arbitrary, and let V be the moment on $\mathbb{R}^n$ defined by

$$V(x) \triangleq x^\top M x, \qquad x \in \mathbf{X}.$$

Then for each $x \in \mathbf{X}$ we have

$$PV(x) = x^\top A^\top M A x + \mathrm{E}[w_0^\top B^\top M B w_0] \leq \lambda V(x) + m|B|^2 \mathrm{E}[w_0^2],$$

which shows that the function V satisfies (12)

In the nonlinear case (1) with $\mathbf{X} = \mathbb{R}^n$ and $F: \mathbf{X} \times \mathbb{R}^p \to \mathbf{X}$ sufficiently smooth, it is possible to generalize this construction (see [Chan, 1986] for the details). However this requires extremely restrictive stability conditions. The idea is to look at the *freely evolving system*

$$d_{k+1} = F^{k+1}(x) \triangleq F(d_k, 0) \qquad k \in \mathbb{Z}_+$$

with $d_0 = x \in \mathbf{X}$ given. If there exists a fixed $0 < \lambda < 1$ and $K > 0$ such that

$$|d_k| \leq K\lambda^k|x|$$

for all $k \in \mathbb{Z}_+$ and $x \in \mathbf{X}$, then for fixed $\alpha \in (\lambda, 1)$ we may define the moment V by

$$V(x) \triangleq \sup_{k \geq 0} \alpha^{-k}|F^k(x)|,$$

where $F^0(x) \triangleq x$. Under the appropriate conditions, V will satisfy condition (12).

Among the difficulties with this approach is that to apply the Lyapunov function for the freely evolving system to the original system, a global Lipschitz condition on V is needed. Presently, the only way to obtain such a condition is by imposing a global Lipschitz condition on F, and this is ruled out in many examples.

However, in all of the examples previously studied (see [Guo and Meyn, 1988], [Meyn, 1988a] and the examples in the following section) a function satisfying (12) may be shown to exist by a reasonably simple calculation even though no global Lipschitz condition is satisfied.

6 Examples

In this section we consider a number of examples to illustrate how the results above may be applied to specific control problems. We assume that the Markov chain $\mathbf{\Phi}$ has the form studied in Section 3.3. For all k, $\Phi_k \in \mathbf{X}$, an n-dimensional manifold, $w_k \in \mathbb{R}^p$, and $F: \mathbf{X} \times \mathbb{R}^p \to \mathbf{X}$ is continuously differentiable (C^1).

We further assume that $(\Phi_0, \mathbf{w})$ are random variables on the probability space $(\Omega, \mathcal{F}, \mathrm{P}_{\Phi_0})$, Φ_0 is independent of $\mathbf{w}$, and that $\mathbf{w}$ is an independent and identically distributed (i.i.d) process.

We require the distribution μ_w of the random variables w_k, $k \in \mathbb{Z}_+$, to possess a density p_w which is lower semi-continuous, and we also require that $p_w(0) > 0$. This property will be used to obtain asymptotic controllability: In both of the examples to follow, we will show that when the disturbance $\mathbf{w}$ is set equal to the constant value 0, the resulting trajectory $\mathbf{\Phi}$ converges to zero. This is stronger than asymptotic controllability since the control required does not depend on the initial conditions of the system.

We may now proceed with the first example:

6.0. Linear Systems

We begin with the most elementary class of Markov chains: Those generated by a linear state space model of the form

$$\Phi_{k+1} = A\Phi_k + Bw_{k+1} \tag{20}$$

with $\mathbf{\Phi}$ evolving on $\mathbf{X} = \mathbb{R}^n$, $\mathbf{w}$ a Gaussian i.i.d. stochastic process evolving on $\mathbb{R}^p$, $w_k \sim N(0, I)$, $k \in \mathbb{Z}_+$, and A and B respectively $n \times n$ and $n \times p$ matrices.

We further assume that the eigenvalues of the matrix A lie within the unit circle in $\mathbb{C}$, so that (20) is bounded in probability.

This example is particularly easy because we can actually compute the k-step Markov transition probabilities. For a given initial condition $\Phi_0 = x \in \mathbf{X}$ we have for all $k \in \mathbb{Z}_+$,

$$\Phi_k \sim P^k(x, \cdot\,) = N(A^k x, \sum_{i=0}^{k-1} A^i BB^\top A^{i\top}), \tag{21}$$

and it is evident that for each initial condition, the resulting trajectory $\{P^k(x, \cdot\,) : k \in \mathbb{Z}_+\}$ converges weakly to the invariant probability π which in the present example is Gaussian $N(0, V)$ where V is the solution to the Lyapunov equation

$$V = AVA^\top + BB^\top.$$

A proof that the law of large numbers

$$\lim_{N\to\infty} \frac{1}{N} \sum_{k=1}^{N} f(\Phi_k) = \int f\, d\pi \tag{22}$$

holds for some class of functions $f: \mathbf{X} \to \mathbb{R}$ may be established using Propositions 5.1 and 5.2. Instead, we give here a direct proof for a special case. The general case is treated using the techniques of the subsequent example.

We henceforth suppose that the pair (A, B) is controllable, so that (20) is weakly stochastically controllable. In this case the covariance matrix V is full rank, and hence the invariant probability π is mutually absolutely continuous with respect to Lebesgue measure on $\mathbb{R}^n$. Likewise, for each $x \in \mathbf{X}$ the distribution of $\mathbf{\Phi}$ at time n is mutually absolutely continuous with respect to Lebesgue measure. By a theorem of [Doob, 1953] the limit (22) holds for all $f \in L^1(\mathbf{X}, \mathcal{B}(\mathbf{X}), \pi)$, and all initial condition distributions μ_0 which are absolutely continuous with respect to π. Conditioning on Φ_n, we obtain for an arbitrary initial condition $x \in \mathbf{X}$,

$$\mathrm{P}_x\Big\{\lim_{N\to\infty} \frac{1}{N} \sum_{k=1}^{N} f(\Phi_k) = \int f\, d\pi\Big\} = \int P^n(x, dy) \mathrm{P}_y\Big\{\lim_{N\to\infty} \frac{1}{N-n} \sum_{k=1}^{N-n} f(\Phi_k) = \int f\, d\pi\Big\} = 1.$$

In the case where the pair (A, B) is not controllable, this technique breaks down. In general, the probabilities $\{P^k(x, \cdot\,) : k \in \mathbb{Z}_+\}$ and the invariant probability π will be mutually singular, and hence conditioning gains nothing. The law of large numbers does not necessarily hold for arbitrary functions $f \in L^1$, and $\mathbf{\Phi}$ is not necessarily geometrically ergodic. However, using Proposition 5.2 it is possible to establish the law of large numbers for the class of continuous functions in $L^{1+\delta}(\mathbf{X}, \mathcal{B}(\mathbf{X}), \pi)$ with $\delta > 0$.

In [Caines, 1988] this result is used to present an explicit analysis of a linear Gaussian system under time invariant control

$$\begin{aligned} x_{k+1} &= A x_k + B u_k + w_k \\ y_k &= C x_k + D u_k + v_k \end{aligned}$$

where the control $u_k = K y_k$, $k \in \mathbb{Z}_+$, and K denotes the steady state Kalman gain.

6.1. Nonlinear Control

Here we consider a linear single input single output stochastic state space system with nonlinear feedback control law

$$u_k \triangleq -\varphi(y_k), \qquad \text{for all } k \in \mathbb{Z}_+, \tag{23}$$

where the function $\varphi \in C^1$. We assume that $\varphi(0) = 0$, and to simplify the analysis we also take $\frac{d\varphi}{dt}(0) \neq 0$.

The closed loop system equations are

$$\begin{aligned} x_{k+1} &= A x_k - b\varphi(c^{\mathsf{T}} x_k + \zeta_{k+1}) + G\xi_{k+1}, \\ y_k &= c^{\mathsf{T}} x_k + \zeta_{k+1}, \qquad k \in \mathbb{Z}_+, \end{aligned} \tag{24}$$

and it is easily seen that if $\mathbf{w} \triangleq \binom{\xi}{\zeta}$ satisfies the conditions given at the beginning of this section, then $\mathbf{x}$ is a Feller Markov chain of the form (1) with state space $\mathbb{R}^n$.

In fact, $\mathbf{\Phi} \triangleq \binom{\mathbf{x}}{\mathbf{y}}$ will also be a Markov chain under the appropriate conditions whose state space $\mathbf{X} \triangleq \mathbb{R}^{n+1}$. However, we may show that almost any result of interest obtainable for the process $\mathbf{x}$ will carry over to the joint process $\mathbf{\Phi}$, and so in the proof of Proposition 6.1 below we will restrict our attention to the simpler Markov chain.

This is a popular example in nonlinear systems theory (see e.g. [Zames, 1966], [Popov, 1973] and [Safonov, 1980]) and is ideal for illustrating the general results presented in the previous sections.

The following stability and controllability conditions will be needed below. We say that the control φ defined in equation (23) lies in the sector (α, r) (see [Safonov, 1980]) if for all $x \in \mathbb{R}$,

$$|\varphi(x) - \alpha x| \leq r|x|.$$

For a positive definite $n \times n$ matrix Q, a vector $z \in \mathbb{R}^n$, and an $n \times n$ matrix F we let

$$|z|_Q^2 = z^\top Q z \qquad \text{and} \quad |F|_Q^2 = \sup_{z \neq 0} \frac{|Fz|_Q^2}{|z|_Q^2}.$$

NC1 $\mathrm{E}[|w_0|^{2+\delta}] < \infty$ for some $\delta > 0$:

NC2 The control law φ lies in the sector (α, r), and for some positive definite $n \times n$ matrix Q,

$$\lambda \triangleq |(A - \alpha b c^\top)|_Q + r|b|_Q\,|c|_{Q^{-1}} < 1.$$

NC3 The pair $(A, [G \mid b])$ is controllable;

NC4 The distribution μ_w of w_0 satisfies the conditions introduced at the beginning of this section.

Let P denote the Markov transition function on $\mathbf{X} = \mathbb{R}^{n+1}$ for the joint process $\mathbf{\Phi}$, and let $\tilde{\mu}_k$, $k \in \mathbb{Z}_+$, denote the occupation probabilities defined in Proposition 5.2. The functions x, u and y on $\mathbf{X}$ are defined so that

$$x_k = x(\Phi_k) \quad u_k = u(\Phi_k) \quad y_k = y(\Phi_k) \qquad k \in \mathbb{Z}_+.$$

Our objective in this section is to prove

Proposition 6.1. *Suppose that conditions NC1, NC2, and NC4 hold for (24). Then a unique invariant proability π exists, and the following limits hold for each initial condition $x = \binom{x_0}{y_0} \in \mathbb{R}^{n+1}$:*

$$P^k(x, \cdot) \xrightarrow{\text{weakly}} \pi \tag{25}$$

$$\tilde{\mu}_k \xrightarrow{\text{weakly}} \pi \qquad \text{as } k \to \infty \quad a.s.\ [\mathrm{P}_x] \tag{26}$$

$$\lim_{N\to\infty} \frac{1}{N} \sum_{k=1}^{N} x_k^2 + y_k^2 + u_k^2 = \int x^2 + y^2 + u^2\, d\pi \qquad a.s.\ [\mathrm{P}_x] \tag{27}$$

$$\lim_{k\to\infty} \mathrm{E}_x[x_k^2 + y_k^2 + u_k^2] = \int x^2 + y^2 + u^2\, d\pi. \tag{28}$$

If in addition NC3 holds, then $\mathbf{\Phi}$ is geometrically ergodic, and (28) holds at a geometric rate. □

Proposition 6.1 will be proved in several steps. We first present necessary and sufficient conditions for the system generating the Markov chain $\mathbf{x}$ to be forward accessible.

Controllability

The generalized controllability matrix associated with $\mathbf{x}$ is defined for an initial condition $x \in \mathbb{R}^n$ by

$$C_x^T = [A_{T-1} \cdots A_1 B_0 | A_{T-1} \cdots A_2 B_1 | \cdots\cdots | A_{T-1} B_{T-2} | B_{T-1}] \tag{29}$$

where, letting $\alpha_k \triangleq \frac{d\varphi}{dt}(y_k)$,

$$A_k \triangleq \left[\frac{\partial F}{\partial x}\right]_{(x_k,w_{k+1})} = A - \alpha_k b c^{\mathsf{T}},$$

and

$$B_k \triangleq \left[\frac{\partial F}{\partial z}\right]_{(x_k,w_{k+1})} = [G| - \alpha_k b]$$

$$\text{for all } k \in \mathbb{Z}_+. \tag{30}$$

Observe that the generalized controllability matrix C_x^T is a function of the random variables $\{y_k : 0 \le k \le T-1\}$, and hence is also random. By Proposition 3.1 the system generating $\mathbf{x}$ is forward accessible (i.e. $\mathbf{x}$ is weakly stochastically controllable) if and only if for each $x \in \mathbb{R}^n$, there exists $T \in \mathbb{Z}_+$ such that the matrix C_x^T has rank n with positive probability.

The following lemma greatly simplifies the computation of the rank of the matrix C_x^T. For an $n \times m$ matrix H let co-ker(H) denote the n-dimensional vector space

$$\text{co-ker}(H) \triangleq \left\{x \in \mathbb{R}^n : x^{\mathsf{T}} H = 0\right\}.$$

For a proof of Lemmas 6.1 and 6.2 see [Meyn, 1988a].

Lemma 6.1. *The generalized controllability matrix C_x^T satisfies*

$$\text{co-ker}(C_x^T) = \text{co-ker}\left(\left[A^{T-1}[G|\alpha_0 b] \mid \cdots \mid [A[G|\alpha_{T-2}b] \mid [G|\alpha_{T-1}b]\right]\right) \tag{31}$$

and hence $\mathbf{x}$ is weakly stochastically controllable under conditions NC3 and NC4 if $\frac{d\varphi}{dt}(0) \neq 0$. □

Stability

We now show that a moment on $\mathbb{R}^n$ exists which satisfies (12). Let Q be the matrix defined in NC2, and let $V(\cdot) \triangleq |\cdot|_Q$.

Lemma 6.2. *Suppose that conditions NC1 and NC2 are satisfied. Then*

(i) *$\mathbf{x}$ is asymptotically controllable with $x^* = 0$;*

(ii) *the moment V satisfies (12);*

(iii) *for all initial conditions $\Phi_0 = \binom{x_0}{y_0} \in \mathbf{X}$,*

$$\sup_{k \in \mathbb{Z}_+} \mathrm{E}_{\Phi_0}[|x_k|^{2+\delta} + |u_k|^{2+\delta} + |y_k|^{2+\delta}] < \infty$$

$$\limsup_{N\to\infty} \int |x|^{2+\delta} + |u|^{2+\delta} + |y|^{2+\delta}\, d\tilde{\mu}_N$$

$$\triangleq \limsup_{N\to\infty} \frac{1}{N}\sum_{k=1}^{N}(|x_k|^{2+\delta} + |u_k|^{2+\delta} + |y_k|^{2+\delta}) < \infty \qquad a.s.\ [\mathrm{P}_{\Phi_0}]$$

where $\delta > 0$ is the constant used in condition NC1.

□

We may now prove Proposition 6.1:

Proof of Proposition 6.1.

We first suppose that conditions NC1-NC4 are satisfied. If this is the case then by Lemmas 6.1 and 6.2 the conditions of Proposition 5.1 are satisfied and hence $\mathbf{x}$ is geometrically ergodic. This implies that the joint process $\mathbf{\Phi} = \binom{\mathbf{x}}{\mathbf{y}}$ is also geometrically ergodic since $\mathbf{y}$ is virtually a function of $\mathbf{x}$.

Result (27) follows from this fact together with Proposition 2.1 (i). To show that the convergence result (28) holds at a geometric rate, apply Proposition 2.1 (iv).

We now relax condition NC3. If the pair $(A, [G|b])$ is not controllable, then $\mathbf{x}$ may be decomposed into controllable and uncontrollable parts using a similarity transformation M where

$$\begin{aligned} MAM^{-1} &= \begin{bmatrix} A_{11} & A_{12} \\ 0 & A_{22} \end{bmatrix} \\ M[G|b] &= \begin{bmatrix} G_1 & b_1 \\ 0 & 0 \end{bmatrix} \end{aligned}$$

and $(A_{11}, [G_1|b_1])$ is controllable.

Letting $\binom{x_k^c}{x_k^0} \triangleq Mx_k$ and $c_1^{\mathsf{T}} = c^{\mathsf{T}} M^{-1}$ it follows that

$$\begin{aligned} x_{k+1}^c &= A_{11}x_k^c + A_{12}x_k^0 + b_1\varphi(c_1^{\mathsf{T}}\begin{pmatrix} x_k^c \\ x_k^0 \end{pmatrix} + \zeta_{k+1}) + G_1 w_{k+1} \\ x_{k+1}^0 &= A_{22}x_k^0. \end{aligned}$$

If $x_0^0 = 0$ then $x_k^0 = 0$ for all $k \in \mathbb{Z}_+$, and in this case $\binom{\mathbf{x}^c}{\mathbf{y}}$ becomes a Markov chain for which the analysis above is valid.

By stability and asymptotic controllability, for all initial conditions $x \in \mathbf{X}$, $x_k^0 \to 0$ as $k \to 0$, and it follows that there exists a unique invariant probability π for $\mathbf{\Phi}$ under which $\mathrm{P}_\pi\{\mathbf{x}^0 \equiv 0\} = 1$. By Proposition 5.2 and Lemma 6.2 (iii) this implies that (26) and (27) hold. For a proof that (25) and (28) hold see [Meyn, 1988a].

□

6.2. Stochastic Adaptive Control

Consider the single input single output random parameter system model

$$y_{k+1} = \theta_k y_k + u_k + v_{k+1} \qquad k \in \mathbb{Z}_+ \tag{32}$$

where the parameter process $\boldsymbol{\theta}$ is the output of the $AR1$ model

$$\theta_{k+1} = \alpha\theta_k + e_{k+1} \qquad k \in \mathbb{Z}_+ \tag{33}$$

and $\alpha \in (-1, 1)$.

The parameter process $\boldsymbol{\theta}$ is assumed to be unknown, but is estimated by the gradient algorithm

$$\widehat{\theta}_{k+1} = \alpha\widehat{\theta}_k + \alpha\frac{y_k(y_{k+1} - \widehat{\theta}_k y_k - u_k)}{1 + y_k^2}. \tag{34}$$

This is a simplified version of the example analysed in [Meyn and Caines, 1987] where $\widehat{\theta}$ is a version of the conditional expectation

$$\widehat{\theta}_k = E[\theta_k \mid \sigma\{y_0, \ldots, y_k\}].$$

In the present example however, the parameter estimates have no simple interpretation.

Applying the certainty equivalence control law

$$u_k = -\widehat{\theta}_k y_k, \qquad k \in \mathbb{Z}_+ \tag{35}$$

and defining $\widetilde{\theta}_k \triangleq \theta_k - \widehat{\theta}_k$, the closed loop system equations become

$$\Phi_{k+1} \triangleq \begin{pmatrix} y_{k+1} \\ \widetilde{\theta}_{k+1} \end{pmatrix} = \begin{pmatrix} \widetilde{\theta}_k y_k + v_{k+1} \\ \alpha \frac{\widetilde{\theta}_k - y_k v_{k+1}}{1+y_k^2} + e_{k+1} \end{pmatrix}, \qquad k \in \mathbb{Z}_+. \tag{36}$$

It is evident that, under the appropriate conditions on the process $\mathbf{w} \triangleq \binom{\mathbf{v}}{\mathbf{e}}$, the state process $\mathbf{\Phi}$ is a Feller Markov chain of the form (1) with state space $\mathbf{X} \triangleq \mathbb{R}^2$

We henceforth assume that $\mathbf{w}$ satisfies conditions imposed at the beginning of this section, that $\mathbf{v}$ and $\mathbf{e}$ are independent, and that the following additional assumptions hold for some $\delta > 0$:

$$\mathrm{E}[w_1] = \begin{pmatrix} 0 \\ 0 \end{pmatrix}, \qquad \mathrm{E}[|w_1|^{4+\delta}] < \infty, \qquad \text{and} \quad \mathrm{E}[|e_1|^{2+\delta}] < 1. \tag{37}$$

These conditions imply that $\sigma_e^2 \triangleq \mathrm{E}[|e_1|^2] < 1$, $\sigma_v^2 \triangleq \mathrm{E}[|v_1|^2] < \infty$, and $\gamma_v^4 \triangleq \mathrm{E}[|v_1|^4] < \infty$.

The state process $\mathbf{\Phi}$ is weakly stochastically controllable, and is asymptotically controllable with $d^* = 0$. Using the results of Section 3 we may conclude that the irreducibility condition holds for an open petite set $A \subset \mathbf{X}$ and that $\mathbf{\Phi}$ is aperiodic. Our next task is to find a moment satisfying (12) so that we may apply Proposition 5.1.

Let $y\colon \mathbf{X} \to \mathbb{R}$ and $\widetilde{\theta}\colon \mathbf{X} \to \mathbb{R}$ denote the coordinate variables on $\mathbf{X}$ so that

$$y_k = y(\Phi_k) \qquad \widetilde{\theta}_k = \widetilde{\theta}(\Phi_k) \qquad \text{for all } k \in \mathbb{Z}_+,$$

and define the test function V on $\mathbf{X}$ by

$$V(y, \widetilde{\theta}) = \widetilde{\theta}^4 + \varepsilon_0 \widetilde{\theta}^2 y^2 + \varepsilon_0^2 y^2 \tag{38}$$

where $\varepsilon_0 > 0$ is a small constant which will be specified below.

Letting P denote the Markov transition function for $\mathbf{\Phi}$ we have by (36),

$$P y^2 = \widetilde{\theta}^2 y^2 + \sigma_v^2. \tag{39}$$

This is far from (12), but applying the operator P to the function $\widetilde{\theta}^2 y^2$ gives

$$\begin{aligned} P\widetilde{\theta}^2 y^2 &= \mathrm{E}\Big[\Big(\frac{\alpha\widetilde{\theta} - \alpha y v_1}{1+y^2} + e_1\Big)^2 (\widetilde{\theta} y + v_1)^2\Big] \\ &= \sigma_e^2 \widetilde{\theta}^2 y^2 + \sigma_e^2 \sigma_v^2 \\ &\quad + \Big(\frac{\alpha}{1+y^2}\Big)^2 \mathrm{E}[(\widetilde{\theta} - y v_1)^2 (\widetilde{\theta} y + v_1)^2] \\ &= \sigma_e^2 \widetilde{\theta}^2 y^2 + \sigma_e^2 \sigma_v^2 \\ &\quad + \Big(\frac{\alpha}{1+y^2}\Big)^2 \Big[\widetilde{\theta}^4 y^2 + \widetilde{\theta}^2 \sigma_v^2 - 4\widetilde{\theta}^2 y^2 \sigma_v^2 + \widetilde{\theta}^2 y^4 \sigma_v^2 + y^2 \gamma_v^4\Big], \end{aligned}$$

and hence we may find a constant $K_1 < \infty$ such that

$$P\widetilde{\theta}^2 y^2 \le \sigma_e^2 \widetilde{\theta}^2 y^2 + K_1(\widetilde{\theta}^4 + \widetilde{\theta}^2 + 1). \tag{40}$$

From (36) it is easy to show that for some constant $K_2 > 0$

$$P\widetilde{\theta}^4 \le \alpha^4 \widetilde{\theta}^4 + K_2(\widetilde{\theta}^2 + 1). \tag{41}$$

Combining equations (39 - 41) we may find a constant $K_3 < \infty$, such that for all $0 < \varepsilon < 1$,

$$\begin{aligned} P(\tilde{\theta}^4 + \varepsilon\tilde{\theta}^2 y^2 + \varepsilon^2 y^2) &\le (\alpha^4 + \varepsilon K_3)\tilde{\theta}^4 + (\sigma_e^2 + \varepsilon)\varepsilon\tilde{\theta}^2 y^2 + K_3(\tilde{\theta}^2 + 1) \\ &\le (\alpha^4 + 2\varepsilon K_3)\tilde{\theta}^4 + (\sigma_e^2 + \varepsilon)\varepsilon\tilde{\theta}^2 y^2 + 2K_3/\varepsilon \end{aligned} \tag{42}$$

where the second inequality follows from the estimate $\tilde{\theta}^2 \le \varepsilon\tilde{\theta}^4 + 1/\varepsilon$. Fix $1 > \lambda > \max(\sigma_e^2, \alpha^4)$. Then by (42) we may find $\varepsilon_0 > 0$ sufficiently small, and a constant $K > 0$ sufficiently large such that (12) holds with V defined in (38).

A modification of equations (39) and (40) may be used together with (37) to show that

$$\sup_{k \ge 0} \mathrm{E}_x[|y_k|^{2+\delta}] < \infty,$$

and applying Proposition 5.1 and Proposition 2.1 we obtain

Proposition 6.2. *The Markov chain* Φ *defined in (36) is geometrically ergodic, and for all initial conditions* $x \in \mathrm{X}$,

$$\lim_{k \to \infty} \mathrm{E}_x[y_k^2] = \int y^2 \, d\pi$$

at a geometric rate, and

$$\lim_{N \to \infty} \frac{1}{N} \sum_{k=1}^{N} y_k^2 = \int y^2 \, d\pi < \infty \qquad a.s.\ [\mathrm{P}_x]$$

□

References

Athreya, K.B. and P. Ney (1980) "Some aspects of ergodic theory and laws of large numbers for Harris recurrent Markov chains" Colloquia Mathematica Societatis János Bolyai **32** *Nonparametric statistical inference*, Budapest, Hungary, pp. 41-56.

Beneš, V.E. (1968) "Finite Regular Invariant Measures for Feller Processes", *J. Appl. Prob.*, *5*, p.203.

Bhatia, N.P. and G.P. Szegö (1970) *Stability theory of dynamical systems*, Springer-Verlag Berlin, Heidelberg, New York.

Billingsley, P. (1968). *Convergence of Probability Measures.* John Wiley & Sons, N.Y.

Brown, J.B. (1976) *Ergodic theory and topological dynamics*, Academic Press, N.Y.

Caines, P.E. (1988) *Linear stochastic systems*, John Wiley & Sons, N.Y.

Chan, K.S. (1986) *Topics in Nonlinear Time Series Analysis* Thesis, Dept. of Mathematics, Princeton University.

Cogburn, R. (1975) "A uniform theory for sums of Markov chain transition probabilities", *Ann. Probab.* **3** pp. 191-214.

Doob, J.L. (1953). *Stochastic Processes.* John Wiley & Sons, N.Y.

Foguel, S.R. (1969) "Positive operators on $C(X)$" Proceedings AMS **22** pp. 295-297.

Foster, F.G. (1953) "On the stochastic matrices associated with certain queueing processess" *Annals of Mathematical Statistics,* **24** pp. 355-360.

Goodwin, G.C., P.J. Ramadge and P.E. Caines, (1981). Discrete Time Stochastic Adaptive Control, *SIAM J. Contr. Optimiz., 19,* 829. Corrigendum, *20* No.6, 1982, p.893.

Guo, L. and S.P. Meyn (1988) "Short memory adaptive control" to appear in the *International J. of Adaptive Control and Signal Processing.*

Has'minskii, R.Z. (1980) *Stochastic stability of differential equations* sijthoff & Noordhoff Alphen an den Rijn, The Netherlands, Rockville, Maryland USA.

Jakubczyk, B. and E.D. Sontag (1988) "Controllability of nonlinear discrete time systems: A lie-algebraic approach" to appear *Siam J. Contr. Optimiz.*

Kalman, R.E. and J.E. Bertram (1960) "Control system analysis and design by the second method of Lyapunov" *Trans. ASME Ser. D: J. Basic Eng.* **82** No. 2 pp. 371-400.

Kushner, H.J. (1967). *Stochastic stability and control,* Academic Press, N.Y.

Meyn S.P. (1988a) "Ergodic theorems for discrete time stochastic systems using a generalized stochastic Lyapunov function" submitted to *Siam J. Control and Optimization.*

Meyn S.P. (1988b) "Ergodic theory and topological dynamics for discrete time controlled systems" in preperation.

Meyn, S.P. and P.E. Caines, (1987) "A new approach to stochastic adaptive control", *IEEE Transactions on Automatic Control, AC-32,* Number 3.

Meyn, S.P. and P.E. Caines (1988). Asymptotic Behavior of Stochastic Systems Possessing Markovian Realizations. submitted to *Siam J. Control and Optimization.*

Nummelin, E. (1984) *General Irreducible Markov Chains and Non-Negative Operators* Cambridge University Press, Cambridge.

Nummelin, E. and P. Tuominen. (1982) "Geometric ergodicity of Harris recurrent Markov chains with applications to renewal theory". *Stochastic Processes and their Applications,* **12** 187-202.

Orey, S. (1971) *Limit Theorems for Markov Chain Transition Probabilities,* Van Nostrand Reinhold Mathematical studies *34*, Van Nostrand, London.

Popov, V.M. (1973) *Hyperstability of Control Systems,* Springer-Verlag, Berlin, and Editura Academici Bucuresti.

Revuz, D. (1975) *Markov Chains,* North Holland Publishing Co., Amsterdam, 1975.

Rosenblatt, M. (1971) *Markov Processes: Structure and Asymptotic Behaviour.* Springer-Verlag, Berlin, Heidelberg, New York.

Safonov, M.G. (1980) *Stability and robustness of multivariable feedback systems* MIT Press, Cambridge, Mass.

Solo, V. (1978) *Time series recursions and stochastic approximation,* Ph.D. dissertation, The Australian National University, Sept. 1978.

Sontag, E.D. (1983) "A Lyapunov-like characterization of asymptotic controllability" *Siam J. Contr. Optimiz.* **21** No. 3 pp. 462-471.

Tuominen, P. and R.L. Tweedie (1979) "Markov chains with continuous components" *Proc. London Math. Society,* Series 3, Vol. 38.

Tweedie, R.L. (1983) "Criteria for rates of convergence of Markov chains, with application to queueing and storage theory", in *Probability, Statistics and Analysis,* London Math. Society Lecture Note Series **79** Ed. J.F.C. Kingman and G.E.H. Reuter, Cambridge University Press, Cambridge.

Zames, G. (1966) "On the Input-Output Stability of Time Varying Non-Linear Feedback Systems", Part I: *IEEE Transactions on Automatic Control, AC-11*, Number 2, pp. 228–238, April 1966. Part II: *IEEE Transactions on Automatic Control, AC-11*, Number 2, pp. 465–476, April 1966.

A SIMPLE STOCHASTIC GROWTH MODEL FOR FILAMENTARY CURRENT STRUCTURES IN SEMICONDUCTOR SYSTEMS

J. Parisi
Physikalisches Institut, Lehrstuhl Experimentalphysik II
Universität Tübingen, Morgenstelle 14, D-7400 Tübingen
Fed. Rep. Germany

Abstract

Low-temperature electric avalanche breakdown in extrinsic semiconductors is characterized by the spontaneous symmetry-breaking evocation of spatially inhomogeneous and temporally unstable current structures. Such kind of self-sustained complex discharge patterns can be well reproduced by computer simulations, employing the simplest nontrivial stochastic growth model developed previously for dielectric breakdown phenomena.

1. Introduction

Analogous to similar natural phenomena as the dielectric breakdown of gaseous, liquid, and solid insulators, low-temperature electric avalanche breakdown in extrinsic semiconductors frequently occurs by means of filamentary current flow channels that exhibit a strong tendency to branching into complicated stochastic patterns. The transport mechanism involved in the nondestructive avalanche breakdown phenomenon was found to be attributed to the autocatalytic impact ionization of impurities by mobile charge carriers heated through an applied electric field /1-5/.

In order to simulate numerically the self-generated development of ramified discharge patterns in semiconductors, we apply the simplest nontrivial stochastic model for dielectric breakdown introduced previously /6,7/. The essence of this stochastic model is that the growth probability depends upon the local electric field determined by the global discharge pattern via the well-known Laplace equation under appropriate boundary conditions. The model is based on the idea that the local electric field around a discharge pattern does not govern the growth directly, but through a stochastic process giving rise to the possibility of branching in a statistical sense. The origin of the stochastic nature is related to finite fluctuations of the electric field and the charge densities due to the incoherent propagation of avalanches in different points of space. Towards a more realistic description of the physical aspects of breakdown, we have generalized the above model – according as other groups /8,9/ – by introducing a critical electric field for growth and a characteristic internal field along the discharge paths, as it is observed experimentally.

2. The Model

Different electric and dielectric breakdown mechanisms have in common that they form stable tree-like, branched filamentary structures. The model which we will discuss in the following has no ambitions to explain the microscopic physics. It should be understood rather as a "minimum number of ingredients" model which is able to phenomenologically reproduce trees. A tree is at least at some level of approximation a fractal or self-similar object, in the sense that a branch of a tree again looks like a tree and so on /10/.

A fractal model for a tree growth has been introduced first by Niemeyer, Pietronero, and Wiesmann /6/. In their model the breakdown structure grows in steps on a lattice. The structure has zero internal resistance. The growth is controlled by the following rules: (1) The probability p that a new bond to an adjacent lattice point is added to the breakdown structure is a function of the local electric field ε, for instance a power law dependence $p \propto \varepsilon^{\eta}$ characterized by the exponent η. (2) After each growth step the discrete Laplace equation has to be solved for the new boundary conditions defined by the growing equipotential pattern. The growth algorithm produces branched fractal structures /7/, but needs to be generalized in order to reproduce physical aspects of breakdown /8/. The first generalization consists in the introduction of a threshold field ε_c for growth. Accordingly, $p(\varepsilon)$ is modified such that $p(\varepsilon) = 0$ for $\varepsilon < \varepsilon_c$ and $p(\varepsilon) \propto \varepsilon^{\eta}$ for $\varepsilon > \varepsilon_c$. This defines a breakdown condition in the sense that for a given electrode configuration no growth occurs below a critical voltage. The second generalization consists in the introduction of an internal field ε_s in the structure. The voltage at a given lattice point of the breakdown structure is now $\varepsilon_s \cdot s$, where s is the shortest path within the structure connecting the point to the electrode.

Computer simulations have shown that the topology of local breakdown is a function of the local electric field ε (specified by the exponent

η) and of both the critical and the internal electric field (defined by the finite values ε_c and ε_s, respectively). With the present stochastic growth model it becomes possible, therefore, to discuss complex breakdown patterns in terms of three simple parameters ($\eta,\varepsilon_c,\varepsilon_s$) which connect to microscopic mechanisms. We take note of the fact that there exists a close structural similarity within a large class of discharge types, the overall pattern of which closely resembles the tree-like, branched structure known from Lichtenberg figures.

In gas discharge systems the growth probability is often assumed to be proportional to the local electric field (η = 1). However, in other systems than gases (solids, liquids, and polymers) the microscopic relation between growth probability and local field may be more appropriately described by a nonlinear function (η different from 1). The corresponding fractal properties depend upon the parameter η /6,7/ and there is no universality in this respect. The existence of a finite critical field ε_c implies that a positive growth probability can only be defined if the local field is larger than the critical one. With increasing ε_c (and keeping $\varepsilon_s = 0$) the branching probability decreases, ending up with the situation that no branching is possible and growth occurs in the form of a single filament along the path of maximum instantaneous field gradient. Moreover, the fact of having $\varepsilon_c \neq 0$ introduces a length scale into the problem. Note that with $\varepsilon_c = 0$ we have two rather trivial length scales: one is the lattice spacing and another is the characteristic electrode spacing. Fractal properties are obtained only if the lattice spacing is very small compared with the electrode spacing. Now ε_c defines an additional nontrivial length scale which is the distance over which $\varepsilon/\varepsilon_c$ shows significant variation (see also /8,9/). Hence, the structure is only locally approximately fractal /10/. The effect of a finite internal field ε_s leads to a voltage drop along the breakdown paths. The idealized picture of an equipotential discharge pattern is thus improved by the more realistic description of conductive filaments – in accordance with experimental observations in distinct semiconductor sytems /11,12/. For the special case of equal critical and internal field ($\varepsilon_c = \varepsilon_s$) the breakdown structure is found to be completely space filling, i.e., the charge is distributed homogeneously between the electrodes. With decreasing ratio $\varepsilon_s/\varepsilon_c$ the symmetry of the charge distribution breaks up and tree-like, branched patterns are found.

Direct comparison between the simulation of a perspectival view of a three-dimensional tree and the experimental discharge pattern enables to extract information on the parameters η, ε_c, and ε_s. These macroscopic quantities, on the other hand, can eventually be interpreted in terms of the underlying microscopic processes. So far, the physical mechanisms of dielectric breakdown phenomena have been related to the present stochastic growth model /9/. We apply this model to electric avalanche breakdown occurring in extrinsic semiconductors at low temperatures. A typical spatial filament structure observed in p-germanium is shown in Fig. 1, the details of which are discussed in Sect. 3.

The probabilistic approach in the present growth model reproduces the stochastic nature of the ionization avalanche formation that starts the discharge progression. Keep in mind that the model is based on the idea that the local electric field around the discharge pattern does not govern the growth directly, but through a stochastic process. This means that the pattern does not just grow at the point of maximum local field, but that at this point the probability of growing is the highest. This changes the process from deterministic into stochastic and gives rise to the possibility of branching in a statistical sense. The physical reason for this probabilistic description of the phenomenon lies in the following fact: the electric field as it is calculated in the model is determined by the global structure of the discharge pattern. It reflects the quasistatic influence of the pattern on the local conditions for its growth. But it does not contain the fluctuations of the field and charge densities as they occur in the real dynamic process, especially at the tips of the filaments. These fluctuations are the origin of the stochastic nature of the process and make branching possible. The particular relation between probability and field reflects, therefore, the fact that the microscopic mechanism of incoherent propagation of avalanches in different points of space is modulated by the global structure of the discharge. At this point the model clearly illustrates the subtle interplay between competing local stochastic and global deterministic aspects of electric instabilities.

The physical interpretation of the three model parameters η, ε_c, and ε_s may thus be the following. The characteristic nonlinearity exponent η of the probability law now describes the intricate relationship between avalanche velocity and local electric field. The critical and internal electric field ε_c and ε_s, respectively, have their interpretation in terms of an S-shaped negative differential conductivity characteristic. The current-density versus electric-field relation shows a threshold character at the critical field ε_c, beyond that a current-controlled negative differential resistivity region is reached. It is well known that a negative differential resistivity has a qualitative influence on the current distribution /13,14/. The smooth and continuous current distribution decays into filaments. Hence, ε_c is the critical field required to reach the nonlinear regime of negative differential resistivity and ε_s is the "hold" field, i.e., the minimum field to sustain the filamentary state ($\varepsilon_c > \varepsilon_s$ assumed).

Finally, we add some general remarks on the physical significance of the probability law applied, which is at least qualitatively evident. For $\eta = 1$, i.e., $p(\varepsilon) \propto \varepsilon$, and $\varepsilon_c = \varepsilon_s = 0$ the present model is formally equivalent /7/ to the nonequilibrium diffusion-limited aggregation model introduced by Witten and Sander /15/. In this model randomly walking particles launched from distant points stick to the surface of the growing aggregate whenever they hit it. There is an analogy between such a particle and the seed electron for forming a microavalanche /7/. $\eta = 1$ implies that the probability that a given high field site is reached by a seed electron is proportional to the local field. A power-law-type probability law seems to be a reasonable generalization. Preliminary numerical simulations indicate that to obtain tree-like structures static disorder, for instance in the form of a statistically distributed critical field, is not sufficient (see also /16/). The probabilistic element required to generate a probabilistic breakdown pattern has to be introduced explicitly in the growth law.

3. The System

Complex nonlinear dynamics comprising the self-generated formation of both spatial and temporal dissipative structures during nondestructive avalanche breakdown in extrinsic semiconductors have been extensively studied in recent years under a variety of experimental conditions (for an overview, see /1,17/). In the following, we concentrate on an exemplary semiconductor system consisting of homogeneously doped p-type germanium single crystals, electrically driven into the low-temperature breakdown regime via impurity impact ionization (for details, see /2-5/ with references therein).

Similar to the corresponding processes of structure formation in gaseous plasma discharges, impact ionization of the shallow impurity acceptors can be achieved in the bulk of the homogeneously doped semiconductor at low temperatures. In the temperature regime of liquid helium most of the charge carriers are frozen out at the impurities. Since the ionization energy is only a few meV and electron-phonon scattering is strongly reduced, avalanche breakdown already takes place at electric fields of a few V/cm and persists until nearly all impurities are ionized /18/. The underlying nonequilibrium phase transition from a low conducting state to a high conducting state is directly reflected in strongly nonlinear regions of negative differential resistivity in the microscopic current-density versus electric-field characteristic /1/. Accordingly, the autocatalytic process of impurity impact ionization also leads to a strongly nonlinear curvature of the macroscopic (measured) current-voltage characteristic (with sometimes S-shaped negative differential resistance /14/), the nonlinearity occurring just beyond the voltage corresponding to the critical electric field where the current increases by many orders of magnitude (typically, from a few nA in the pre-breakdown up to a few mA in the post-breakdown region).

Under slight variation of distinct control parameters (electric field, magnetic field, temperature, electromagnetic irradiation, and/or electron-beam irradiation) the resulting electric current flow displays a wide variety of spatio-temporal nonlinear transport behavior /2-5/. In particular, we have observed the spontaneous symmetry-breaking emergence of current filaments and current oscillations in the highly nonlinear post-breakdown region of the current-voltage characteristic. The oscillatory behavior of distinct state variables was found to be closely linked to the filamentary flow pattern developing during avalanche breakdown. Apparently, the break-up of spatial order during current filamentation is correlated with the onset of low-dimensional temporal chaos in the current oscillations.

The complex spatial behavior of our semiconductor system can be globally visualized by means of low-temperature scanning electron microscopy /12/. This imaging method combines a liquid-helium cryostage with a commercial scanning electron microscope. Two-dimensional images of current filament patterns are obtained by scanning the sample surface with an electron beam and by recording the beam-induced changes of the sample conductance as a function of the coordinates (x,y) of the beam focus. In voltage-biased samples the beam-induced current changes $\Delta I(x,y)$ are recorded. Figure 1 clearly indicates a typical filament structure developing in the S-shaped negative differential resistance region of the current-voltage characteristic. As reported elsewhere /4,11/ in detail, the multifilamentary current flow becomes more and more homogeneous if the semiconductor system is driven further into its linear post-breakdown region at higher electric fields. Nucleation of additional filaments is often accompanied by abrupt changes between different stable filament configurations via noisy current instabilities.

Two-dimensional images of stable tree-like, branched filamentary structures similar to those presented in Fig. 1 have been reproduced

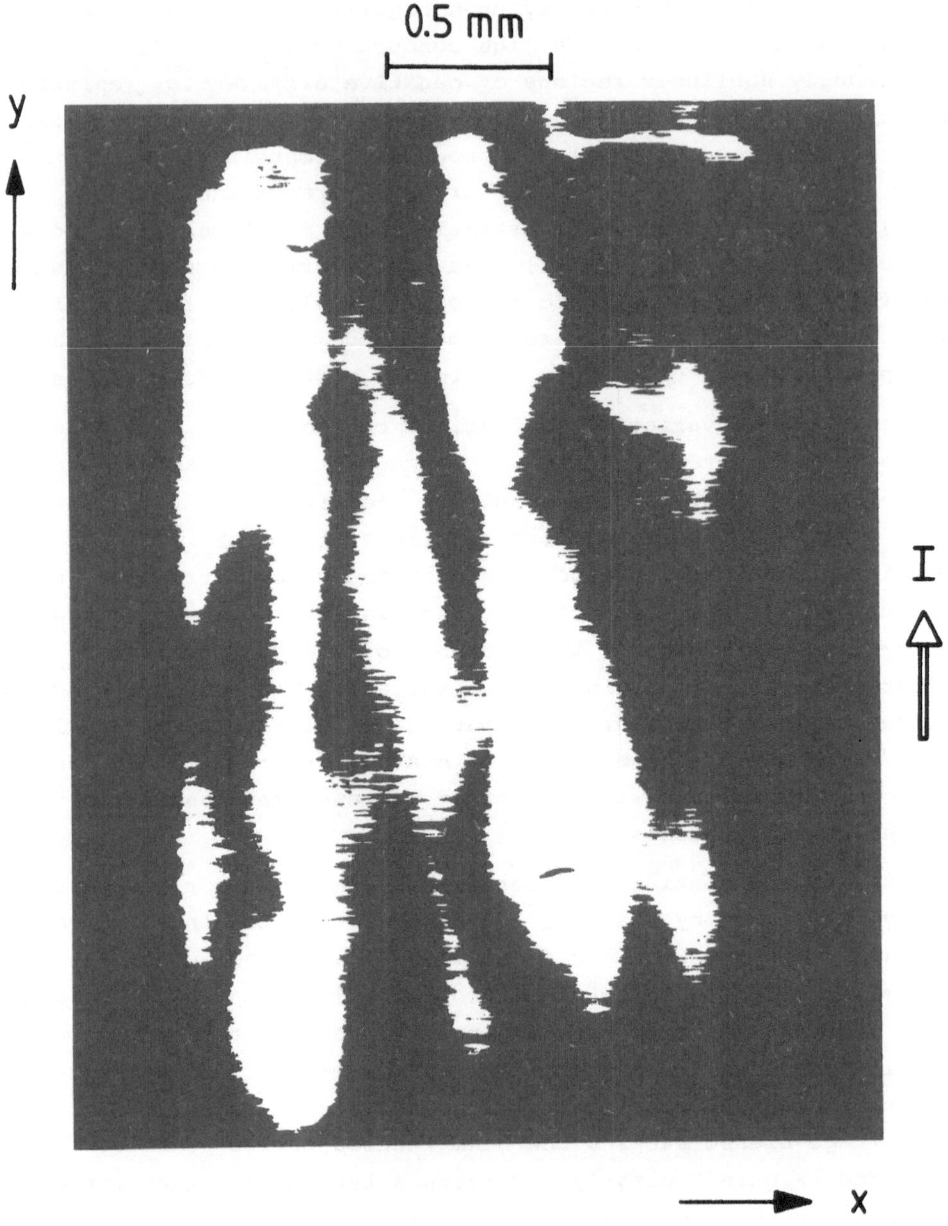

Fig. 1. Brightness-modulated image of the filamentary current flow in a homogeneously doped semiconductor during avalanche breakdown obtained by low-temperature scanning electron microscopy. The dark regions correspond to the filament channels extending along the y-direction.

by first numerical simulations. In particular, the present stochastic growth model is capable of eliciting the overall evolution of such complex filament patterns as a function of the applied electric field, ranging from the initial abrupt nucleation of smallest possible single filaments to the subsequent nonlinear growth and branching behavior of distinct multifilament configurations. According to the different experimental situations analyzed, the model parameters have been varied in the ranges $2 < \eta < 3$ and $0.7 < \varepsilon_s/\varepsilon_c < 1.0$. We emphasize that the values of the critical and internal electric field could be directly extracted from the corresponding experimental current-voltage characteristics, whereas the values of the nonlinearity exponent of the probability law were adapted numerically thereafter.

References

/1/ E. Schöll, Nonequilibrium Phase Transitions in Semiconductors (Springer, Berlin, 1987).

/2/ J. Parisi, J. Peinke, B. Röhricht, and K.M. Mayer, Z. Naturforsch. 42 a, 329 (1987).

/3/ R.P. Huebener, K.M. Mayer, J. Parisi, J. Peinke, and B. Röhricht, Nuclear Physics B (Proc. Suppl.) 2, 3 (1987).

/4/ K.M. Mayer, J. Peinke, B. Röhricht, J. Parisi, and R.P. Huebener, Physica Scripta T 19, 505 (1987).

/5/ J. Peinke, J. Parisi, B. Röhricht, K.M. Mayer, U. Rau, and R.P. Huebener, Solid State Electronics 31, 817 (1988).

/6/ L. Niemeyer, L. Pietronero, and H.J. Wiesmann, Phys. Rev. Lett. 52, 1033 (1984).

/7/ L. Pietronero and H.J. Wiesmann, J. Stat. Phys. 36, 909 (1984).

/8/ H.J. Wiesmann and H.R. Zeller, J. Appl. Phys. 60, 1770 (1986).

/9/ L. Pietronero and H.J. Wiesmann, Z. Phys. B - Condensed Matter 70, 87 (1988).

/10/ B.B. Mandelbrot, The Fractal Geometry of Nature (Freeman, San Francisco, 1982).

/11/ K.M. Mayer, R. Gross, J. Parisi, J. Peinke, and R.P. Huebener, Solid State Commun. 63, 55 (1987).

/12/ K.M. Mayer, J. Parisi, and R.P. Huebener, Z. Phys. B - Condensed Matter 71, 171 (1988).

/13/ B.K. Ridley, Proc. Phys. Soc. 82, 954 (1963).

/14/ J. Peinke, D.B. Schmid, B. Röhricht, and J. Parisi, Z. Phys. B - Condensed Matter 66, 65 (1987).

/15/ T.A. Witten and L.M. Sander, Phys. Rev. Lett. 47, 1400 (1981).

/16/ H.R. Zeller, preprint (to be published).

/17/ R.P. Huebener, J. Peinke, and J. Parisi, Appl. Phys. A (to be published).

/18/ K. Seeger, Semiconductor Physics (Springer, Berlin, 1985).

THE RATE OF CONVERGENCE AND THE ASYMPTOTIC NORMALITY OF AN ESTIMATOR IN A CONTROLLED INVESTMENT MODEL WITH TIME-VARYING PARAMETERS*

Bożenna Pasik-Duncan
Department of Mathematics, University of Kansas
Lawrence, Kansas 66045, U.S.A.

The sample path rate of convergence is obtained for a strongly consistent, recursive estimator of a parameter in a bilinear stochastic differential equation. The bilinear stochastic differential equation arises in a model of portfolio selection and consumption. The parameters of this equation change with time and converge to limits. It is assumed that one of these parameters is unknown. In this case it is necessary simultaneously to estimate the unknown parameter and to control the state equation so there is the problem of adaptive control. Asymptotic normality for a parameter estimator is also given.

1. Introduction.

While there has been a sizable amount of research in stochastic identification and adaptive control to show the strong consistency of parameter estimators there are relatively few results on the sample path rate of convergence of these estimators (e.g., [6-7]). In the paper [4] the sample path rate of convergence is obtained for a strongly consistent, recursive estimator of a parameter in a bilinear stochastic differential equation. The bilinear stochastic differential equation arises in a model of Merton [9] of portfolio selection and consumption. The solution of this stochastic differential equation is the wealth of an individual investor. In the case of only two assets, one riskless and one risky, where the parameters of the stochastic differential equation are constants, Merton [9] obtained explicitly the rules or controls for the distribution of the wealth in the two assets and the consumption rate for a class of utility functions. Generalizations of this model are considered in [5] and some generalizations very related to this work are considered in [3] where the parameters of the stochastic

* This work was supported by NSF Grant ECS-8718026.

differential equation for the wealth are functions of time. Allowing the time dependence of the parameters produces models that are better descriptions of the economic reality. T. Duncan and B. Pasik-Duncan [2] obtained strong consistency of a recursive estimator of the average return rate of the risky asset.

In [2] it is assumed that the average return rate of the risky asset in Merton's model is unknown. In this case it is necessary simultaneously to estimate the unknown average return rate and to control the wealth equation by determining the distribution of wealth in two assets and determining the consumption rate so there is the problem of adaptive control. Since the expected discounted total utility is maximized it is not meaningful to consider the optimality of the average cost as is often done in adaptive control problems.

The recursive estimators are formed [2-3] based on the observations of the wealth. It seems that these are the most natural observations because one must perform computations with the wealth to form the control and the wealth is the natural state for these economic models. If the price of the risky asset is observed then a recursive family of estimators can be formed based on the price in analogy with the estimator based on the wealth. However there are economic situations where the price per share of the risky asset is not directly observable. For example, suppose that the fund for the risky asset is not publicly traded and one is still allowed to determine how much to invest but one only knows the value of the investment at each time and not the price per share. In this case it is appropriate and even necessary to form the family of estimators from the wealth.

The unknown parameter that was estimated is the average return rate of the risky asset and it changes with time and converges to a limit. P. Mandl and G. Hübner [8] consider quite different systems described by Markov chains but they investigate the same situation when the parameters of the system change with time and converge to limits.

The rate of convergence of the estimator is described in a sample path sense from the law of the iterated logarithm for Brownian motion and the form of the estimator. It is clear from the proof of this result that the rate is optimal in a natural sense.

Asymptotic normality also will be given for a strongly consistent recursive estimator of the unknown parameter.

2. Preliminaries.

To fix the notation and some ideas for the rate of convergence of a recursive estimator of the unknown time-varying parameter a formulation of Merton [9] for the wealth in a portfolio and consumption model is reviewed as well as our results [2-3] for the strongly consistent recursive estimator of the average return rate of the risky asset based on the observations of the wealth.

It is supposed that the individual invests his money in two types of assets. He can put his money in the bank where it earns a constant interest rate r. This is the safe or riskless asset. He can also invest in a single stock. This is the risky asset. The value P(t) of the stock is supposed to grow at a rate $\alpha > r$ and it changes according to the stochastic differential equation

$$dP(t) = \alpha P(t)dt + \sigma P(t)dB(t) \tag{1}$$

where $(B(t), t \geq 0)$ is a real-valued standard Brownian motion process. If the initial investment in the stock is P then in the absence of any trading the value $S_1(t)$ of this holding evolves according to

$$\begin{aligned} dS_1(t) &= \alpha S_1(t)dt + \sigma S_1(t)dB(t) \\ S_1(0) &= P \end{aligned} \tag{2}$$

Meanwhile the holding $S_0(t)$ in the bank account satisfies

$$dS_0(t) = (rS_0(t) - U_2(t))dt \tag{3}$$

where $U_2(t)$ is the consumption rate.

We now need to consider how funds are transferred from bank to stock and vice versa. Let

$$W(t) := S_0(t) + S_1(t) \tag{4}$$

$$U_1(t) := S_1(t)/W(t) \tag{5}$$

be the total wealth at time $t \geq 0$ of the investor and $U_1(t)$ the fraction of total wealth held in stock.

Adding (2) and (3) we get the stochastic differential equation for the wealth W(t) at time t and we assume that the parameters change in time

$$\begin{aligned} dW(t) = {} & (r(t)W(t) + (\alpha(t) - r(t))U_1(t)W(t) - U_2(t))dt \\ & + U_1(t)W(t)\sigma(t)dB(t) \end{aligned} \tag{6}$$

where $W(0) = y > 0$.

We can now regard both $(U_1(t))$ and $(U_2(t))$ as decision variables, it means that if $(\Omega,\mathcal{F},(\mathcal{F}_t),P)$ is a probability space with the Brownian motion process (B(t), t

≥ 0) the portfolio selection-consumption problem is to choose $\mathcal{F}_t$-adapted processes $(U_1(t), U_2(t))$ so as to maximize the expected discounted total utility

$$\mathcal{J}(U_1,U_2) := E_y \int_0^T e^{-\int^t \rho} F(U_2(t))dt \tag{7}$$

where $F(u) = u^\gamma$ with $0 < \gamma < 1$ is the utility function, $\rho > 0$ is the discount factor and y is the initial endowment.

The constraints are

(i) $W(t) > 0 \quad \forall t$

(ii) $U_2(t) \geq 0 \quad \forall t$

(iii) $0 \leq U_1(t) \leq 1 \quad \forall t$ (We do not need to impose the condition $U_1(t) \in [0,1]$. $U_1(t) < 0$ implies the risky asset is shortsold and $U_1(t) > 1$ implies borrowing. For more details see [1].)

(A1) We assume that r, α, σ and ρ are piecewise continuous functions of time such that $\alpha(t) > r(t)$ and $\sigma(t) > 0$, $\rho(t) > 0$ for all $t \geq 0$.

(A2) Furthermore it is assumed that δ and ρ are uniformly bounded and

$$\lim_{t\to\infty} \sigma(t) = \sigma_0$$

$$\lim_{t\to\infty} \rho(t) = \rho_0$$

It turns out that maximizing (7) over policies (U_1,U_2) for the wealth equation (6) is one of the few nonlinear control problems which can be explicitly solved.

This stochastic optimal control problem can be solved by finding a smooth solution to the Hamilton-Jacobi equation for this control problem and using the verification theorem (p. 153 [10]). This solution is described in the following result [3].

<u>Lemma</u>. *For the stochastic control problem described by (6-7) if*

$$0 < \alpha(t) - r(t) \leq \sigma^2(t)(1 - \gamma) \tag{8}$$

for all $t \geq 0$, $T < \infty$ *and* $\gamma \in (0,1)$ *then the optimal controls are*

$$U_1^*(t,y) = \frac{\alpha(t) - r(t)}{(1 - \gamma)\sigma^2(t)} \tag{9}$$

$$U_2^*(t,y) = \frac{y}{\int_t^T e^{\int_t^s \mu} ds} \qquad (10)$$

where

$$\mu(t) = \frac{1}{1-\gamma}\left[\rho(t) - \frac{\gamma(\alpha(t) - r(t))^2}{2\sigma^2(t)(1-\gamma)} - \gamma r(t)\right] \qquad (11)$$

Corollary. *Let* $T = +\infty$ *in (7). If (8) is satisfied and* μ *given by (11) satisfies* $\mu(t) \geq c \geq 0$ *for all* $t \geq 0$ *where* c *is a positive constant, then the optimal controls for the control problem (7-8) are*

$$U_1^*(t,y) = \frac{\alpha(t) - r(t)}{\sigma^2(t)(1-\gamma)} \qquad (12)$$

$$U_2^*(t,y) = \frac{y}{\int_t^\infty e^{-\int_t^s \mu} ds} \qquad (13)$$

For the adaptive control problem it is assumed that the average return rate function $\alpha(t)$ is unknown such that

(i) if $\alpha \in \mathcal{A}$ = (family of average return rate function) then

$$\lim_{t\to\infty} \alpha(t) = \alpha_0 \qquad (14)$$

(ii) $\exists c_i: [0,\infty) \to (0,+\infty)$ for $i = 1,2$ $\forall \alpha \in \mathcal{A}$ $\forall t \geq 0$

$$c_1(t) \leq \alpha(t) - r(t) \leq c_2(t) \qquad (15)$$

and

$$\lim_{t\to\infty} c_i(t) = c_i^0 > 0, \quad i = 1,2 \qquad (16)$$

$$\lim_{t\to\infty} r(t) = r_0 > 0 \qquad (17)$$

$$c_2^0 \leq \sigma_0^2 (1-\gamma) \qquad (18)$$

$$\exists k > 0 \qquad \rho_0 - \gamma\left[\frac{(c_2^0)^2}{2\sigma_0^2(1-\gamma)} + r_0\right] \geq k. \qquad (19)$$

If $\alpha \in \mathcal{A}$ is the true average return rate function then an estimate $\hat{\alpha}(t)$ of $\alpha(t)$ is constructed that is a measurable function of the past of the wealth. This estimate $\hat{\alpha}(t)$ is used for the unknown $\alpha(t)$ in the optimal controls (12-13) and these resulting controls are substituted into (6) to obtain

$$dW(t) = \frac{(\alpha(t) - r(t))(\hat{\alpha}(t) - r(t))}{\sigma^2(t)(1-\gamma)} W(t)dt + \left[r(t) - \frac{1}{\int_t^\infty e^{-\int_t^s \hat{\mu}_t} ds} \right] W(t)dt + \frac{\sigma(t)(\hat{\alpha}(t) - r(t))}{\sigma^2(t)(1-\gamma)} W(t)dB(t) \tag{20}$$

where

$$\hat{\mu}_t(s) = \frac{1}{1-\gamma}\left[\rho(t) - \frac{\gamma(\hat{\alpha}(t) - r(t))^2}{2\sigma^2(t)(1-\gamma)} - \gamma r(t)\right] \tag{21}$$

for all $s \geq t$.

Let $\beta(t)$ and $\hat{\beta}(t)$ be defined as

$$\beta(t) = \alpha(t) - r(t) \tag{22}$$

$$\hat{\beta}(t) = \hat{\alpha}(t) - r(t) \tag{23}$$

A recursive family of estimators are defined for $\beta^2(t)$. Fix $\delta > 0$. Define $\tilde{\beta}^2(t)$, an estimate of $\beta^2(t)$, that is a measurable function of the past of the wealth as

$$d\tilde{\beta}^2(t) = \frac{-\tilde{\beta}^2(t)}{t} dt + \frac{\sigma^2(t-\delta)}{tW(t-\delta)}\left[dW(t-\delta) - \left(r(t-\delta) - \frac{1}{\int_{t-\delta}^\infty e^{-\int_{t-\delta}^s \hat{\mu}_{t-\delta}} ds} \right) W(t-\delta)dt \right] \tag{24}$$

where $t \geq \delta > 0$, $\tilde{\beta}^2(\delta) = 0$ and $\hat{\beta}(t)$ is defined by

$$\hat{\beta}(t) = \begin{cases} c_1(t) & \tilde{\beta}^2(t) < c_1^2(t) \quad \text{or} \quad t \le \delta \\ \sqrt{\tilde{\beta}^2(t)} & c_1^2(t) \le \tilde{\beta}^2(t) \le c_2^2(t) \\ c_2(t) & \tilde{\beta}^2(t) > c_2^2(t) \end{cases} \tag{25}$$

Since the stochastic differential equation (24) can be written as

$$d\tilde{\beta}^2(t) = -\frac{\tilde{\beta}^2(t)}{t}dt + \frac{1}{t}[\beta(t)\hat{\beta}(t-\delta)dt + \sigma(t)\hat{\beta}(t-\delta)dB(t-\delta)]$$

it is elementary to verify that the unique solution of (24) is

$$\tilde{\beta}^2(t) = \frac{1}{t}\int_{\delta}^{t-\delta} \beta(s)\hat{\beta}(s)ds + \frac{1}{t}\int_{\delta}^{t-\delta} \sigma(s)\hat{\beta}(s)dB(s) \tag{26}$$

The introduction of $\delta > 0$ ensures that the solution of the equation (6) for the wealth exists and is unique.

It was shown [3] that the family of estimates $(\hat{\beta}(t), t \ge 0)$ converge almost surely to $\beta_0 = \alpha_0 - r_0$.

<u>Theorem</u>. *Let* $r(\cdot)$ *be the return rate function for the safe asset in (6) given in (17). If* $\alpha \in \mathcal{A}$ *is the unknown average return rate function for the risky asset in (6), then the recursive family of estimators given by (26) converge almost surely to* $\beta_0 = \alpha_0 - r_0$*, that is,*

$$\lim_{t\to\infty} \hat{\beta}(t) = \lim_{t\to\infty} \tilde{\beta}(t) = \beta_0 \qquad \text{a.s.} \tag{27}$$

3. Main results.

The rate of convergence and the asymptotic normality of the estimator.

The following result provides a sample path rate of convergence of $(\tilde{\beta}^2(t), t \ge \delta)$ to the true value β^2. Since a stochastic integral appears in the expression (26) for $\tilde{\beta}^2(t)$ it easily follows from the law of the iterated logarithm for Brownian motion (e.g., [11]) that the rate of convergence is optimal in an obvious sense.

Theorem. *Let* $\forall\varepsilon > 0$

$$\limsup_{t\to\infty} \frac{|\beta(t) - \beta|}{t^{-1/2+\varepsilon}} = 0 \tag{28}$$

Then $\forall\varepsilon > 0$ *the family of estimators given by (26) satisfy*

$$P\left(\limsup_{t\to\infty} \frac{|\tilde{\beta}^2(t) - \beta^2|}{t^{-1/2+\varepsilon}} \leq K\right) = 1$$

where K *is a constant.*

Proof. If $(B(t), t \geq 0)$ is a real-valued standard Brownian motion process then the law of the iterated logarithm (e.g., [11]) implies that

$$P\left(\limsup_{t\to\infty} \frac{|B(t)|}{(2t \log\log t)^{1/2}} = 1\right) = 1 \tag{29}$$

Let $\tau(t)$ be defined as

$$\tau(t) = \int_0^t \hat{\beta}^2(s)ds \tag{30}$$

so that

$$\int_0^t \hat{\beta}(s)dB(s) = \int_0^{\tau(t)} dB(s) \tag{31}$$

Let Λ be a set of probability zero such that $\tilde{\beta}^2(t)$ converges to β^2 as $t \to \infty$ on $\Omega\setminus\Lambda$, the event in (29) contains $\Omega\setminus\Lambda$ and $\frac{1}{t}\int_0^\delta \hat{\beta}(s)dB(s)$ converges to 0 as $t \to \infty$ on $\Omega\setminus\Lambda$.

Choose $\varepsilon > 0$ and fix it. Let $\varepsilon_1 \in (0,1)$ have the property that

$$\left(\frac{1}{2 - \varepsilon_1}\right)\left(\frac{2}{1 + 2\varepsilon}\right) < 1 \tag{32}$$

and let M satisfy

$$M = (2 - \varepsilon_1)\beta \tag{33}$$

For $\omega \in \Omega \backslash \Lambda$ it follows that there is a $T(\omega) \in [1,+\infty)$ such that if $t \geq T(\omega)$ then the following five inequalities are satisfied

$$\tilde{\beta}(t,\omega) > (1 - \varepsilon_1)\beta \tag{34}$$

$$|\tilde{\beta}^2(t,\omega) - \beta^2| \leq 1 \tag{35}$$

$$\frac{2\delta\beta^2}{t} + \frac{2\sigma_0}{t}\left[2(a_2 - r)^2 t \log\log(t(a_2 - r)^2)\right]^{1/2} \leq \frac{1}{t^{1/2-\varepsilon}}$$

$$\frac{\left|\int_0^t \sigma(s,\omega)\hat{\beta}(s,\omega)dB(s,\omega)\right|}{[2\tau(t,\omega)\log\log\tau(t,\omega)]^{1/2}} \leq 1 + \varepsilon_1 \tag{37}$$

$$\frac{1}{t}\left|\int_0^\delta \sigma(s,w)\hat{\beta}(s,\omega)dB(s,\omega)\right| \leq \frac{1}{t^{1/2-\varepsilon}} \tag{38}$$

Since it is clear that

$$|\hat{\beta}^2(s,\omega) - \beta^2| \leq |\tilde{\beta}^2(s,\omega) - \beta^2| \tag{39}$$

it easily follows from (26) and the assumption of the theorem that

$$\left|\tilde{\beta}^2(t,\omega) - \beta^2\right| \leq \frac{1}{t}\int_\delta^{t-\delta}\left|\beta(s,\omega)\hat{\beta}(s,\omega) - \beta^2\right| ds + \frac{2\delta\beta^2}{t} +$$

$$+ \frac{1}{t}\left|\int_\delta^{t-\delta}\sigma(s,\omega)\hat{\beta}(s,\omega)dB(s,\omega)\right| \leq \tag{40}$$

since

$$|\beta(s,\omega)\hat{\beta}(s,\omega) - \beta^2| = |\beta(s,\omega)\hat{\beta}(s,\omega) - \beta\hat{\beta}(s,\omega) + \beta\hat{\beta}(s,\omega) - \beta^2|$$

$$= |\beta(\hat{\beta}(s,\omega) - \beta) + \hat{\beta}(s,\omega)(\beta(s,\omega) - \beta)|$$

$$\leq \frac{\beta}{t}\int_\delta^{t-\delta}|\hat{\beta}(s,\omega) - \beta|ds + \frac{1}{t}\int_\delta^{t-\delta}|\hat{\beta}(s,\omega)||\beta(s,\omega) - \beta|ds$$

$$+ \frac{2\delta\beta^2}{t} + \frac{1}{t}\left|\int_\delta^{t-\delta}\sigma(s,\omega)\hat{\beta}(s,\omega)dB(s,\omega)\right|$$

$$\leq \frac{\beta}{t}\int_{\delta}^{T(\omega)} |\hat{\beta}(s,\omega) - \beta| ds + \frac{\beta}{t}\int_{T(\omega)}^{t-\delta} \frac{|\tilde{\beta}^2(s,\omega) - \beta^2|}{|\tilde{\beta}(s,\omega) + \beta|} ds$$

$$+ \frac{1}{t}\int_{\delta}^{t-\delta} |\hat{\beta}(s,\omega)||\beta(s,\omega) - \beta| ds + \frac{2\delta\beta^2}{t} +$$

$$+ \frac{1}{t}\left|\int_{0}^{t-\delta} \sigma(s,\omega)\hat{\beta}(s,\omega)dB(s,\omega)\right| + \frac{1}{t}\left|\int_{0}^{\delta} \sigma(s,\omega)\hat{\beta}(s,\omega)dB(s,\omega)\right|$$

$$\leq \frac{\beta}{t}\int_{\delta}^{t-\delta} |\hat{\beta}(s,\omega) - \beta| ds + \frac{\beta}{tM}\int_{T(\omega)}^{t-\delta} |\tilde{\beta}^2(s,\omega) - \beta^2| ds$$

$$+ \frac{2\delta\beta^2}{t} + \frac{1}{t}\left|\int_{0}^{t-\delta} \sigma(s,\omega)\hat{\beta}(s,\omega)dB(s,\omega)\right|$$

$$+ \frac{1}{t}\left|\int_{0}^{\delta} \sigma(s,\omega)\hat{\beta}(s,\omega)dB(s,\omega)\right|$$

Let $a(t) := |\tilde{\beta}^2(t) - \beta^2|$. The above inequality can be simplified using (34-38) and the definition of $\hat{\beta}$ as

$$a(t,\omega) \leq \frac{\beta}{t}(a_2 - a_1)T(\omega) + \frac{\beta}{tM}\int_{T(\omega)}^{t-\delta} a(s,\omega)ds$$
$$+ \frac{1}{t^{1/2-\varepsilon}} + \frac{1}{t^{1/2-\varepsilon}} \leq \frac{L(\omega)}{t^{1/2-\varepsilon}} + \frac{\beta}{tM}\int_{T(\omega)}^{t} a(s,\omega)ds \tag{41}$$

where $L(\omega) = 2 + \beta(a_2 - a_1)T(\omega)$.

Assume by induction that for $t \geq T(\omega)$

$$a(t,\omega) \leq \left(\frac{\beta}{M}\right)^n + L(\omega)t^{-1/2+\varepsilon}\sum_{j=0}^{n-1} \alpha^j \tag{42}$$

where

$$\alpha = \frac{\beta}{M}\left(\frac{2}{1 + 2\varepsilon}\right)$$

To verify the induction hypothesis which is satisfied for n = 1 by (35, 41) we substitute (42) in the right hand side of (41) to obtain

$$a(t,\omega) \le \frac{\beta}{tM} \int_{T(\omega)}^{t} \left(\frac{\beta}{M}\right)^{n} + L(\omega)s^{-1/2+\varepsilon} \sum_{j=0}^{n-1} \alpha^{j} ds + \frac{L(\omega)}{t^{1/2-\varepsilon}} \tag{43}$$

so that

$$\begin{aligned} a(t,\omega) &\le \left(\frac{\beta}{M}\right)^{n+1} + L(\omega)\frac{\beta}{M}\left(\frac{1}{\frac{1}{2}+\varepsilon}\right)t^{-1/2+\varepsilon}\sum_{j=0}^{n-1}\alpha^{j} + \frac{L(\omega)}{t^{1/2-\varepsilon}} \\ &= \left(\frac{\beta}{M}\right)^{n+1} + L(\omega)t^{-1/2+\varepsilon}\sum_{j=0}^{n}\alpha^{j} \end{aligned} \tag{44}$$

Since $\alpha \in (0,1)$ we can let $n \to \infty$ in (44) to obtain

$$a(t,\omega) \le \frac{L(\omega)}{1-\alpha} t^{-1/2+\varepsilon} \tag{45}$$

By choosing ε less than the epsilon in the statement of the theorem the proof is complete. □

This theorem provides the sample path rate of convergence for the estimator (26) that is a strongly consistent, recursive estimator of $\beta^2(t) = (\alpha(t) - r(t))^2$ and thereby a recursive estimator of the unknown average return rate function of the risky asset in (6).

Remark. Consider a > 0 such that

$$\limsup_{t\to\infty} (\beta^2(t) - \beta^2|t^a = 0.$$

If $\sup a < \frac{1}{2}$ then $(\beta^2(t))$ converges slowly to β^2 and the rate of convergence of the estimators $(\tilde{\beta}^2(t))$ to the true value β^2 is the same as the rate of convergence of $(\beta^2(t))$ to β^2.

In the theorem $\sup a \ge \frac{1}{2}$ in which case $(\beta^2(t))$ converges quickly to β^2.

Let us now consider the question of the asymptotic normality of the estimator (26).

Proposition 4.1. *Let* $(\tilde{\beta}^2(t))$ *be the family of estimators of* β^2. *If* $(\tilde{\beta}^2(t))$ *is of the form (26) then*

$$\sqrt{t}\left(\tilde{\beta}^2(t) - \frac{1}{t}\int_{\delta}^{t-\delta} \beta(s)\hat{\beta}(s)ds\right) \sim N(0,\sigma_0^2\beta^2) \tag{46}$$

Proof: To prove (46) it suffices to show that

$$\frac{1}{t}\int_0^t \sigma^2(s)\hat{\beta}^2(s)ds \underset{t\to\infty}{\to} \sigma_0^2\beta^2 \quad \text{a.s.} \tag{47}$$

Since

$$c_1(t) \le \alpha(t) - r(t) \le c_2(t)$$

and

$$\lim_{t\to\infty} c_i(t) = c_i^0 > 0$$

so

$$(c_1^0)^2 \le \beta^2 \le (c_2^0)^2 \tag{48}$$

Note that by (A2)

$$\limsup_{t\to\infty} \frac{1}{t}\int_0^t \sigma^2(s)\hat{\beta}^2(s)ds \ge \sigma_0^2(c_1^0)^2$$

and

$$\limsup_{t\to\infty} \frac{1}{t}\int_0^t \sigma^2(s)\hat{\beta}^2(s)ds \le \sigma_0^2(c_1^0)^2$$

it easily follows directly from above and (48) that

$$\lim_{t\to\infty} \frac{1}{t}\int_0^t \sigma^2(s)\hat{\beta}^2(s) = \sigma_0^2\beta^2 \quad \text{a.s.}$$

This completes the proof of the Proposition.

References

[1] Davis, M. H. A., Local Time on the Stock Exchange, Stochastic Calculus in Application, ed. J. R. Norris, Pitman Notes in Mathematics, Longman, London 1988, to appear.

[2] Duncan, T. E. and Pasik-Duncan, B., Adaptive Control of Continuous Time Portfolio and Consumption Model, Journal of Optimization Theory and Applications, Vol. 61, April, 1989.

[3] Duncan, T. E. and Pasik-Duncan, B., Adaptive Control of Three Continuous Time Portfolio and Consumption Models, Journal of Optimization Theory and Applications, Vol. 61, June 1989.

[4] Duncan, T. E. and Pasik-Duncan, B., The Rate of Convergence for an Estimator in a Portfolio and Consumption Model, Journal of Optimization Theory and Applications, Vol. 61, May 1989.

[5] Karatzas, I., Lahoczky, J. P. and Shreve, S. E., Optimal portfolio and consumption decision for a "small investor" on a finite horizon, SIAM J. Control & Optimization, 25, 1557-1586, 1987.

[6] Kumar, P. R., Adaptive Control with a Compact Parameter Set, SIAM Journal on Control and Optimization, Vol. 20, p. 9-13, 1982.

[7] Kushner, H. and Kumar, R., Convergence and Rate of Convergence of a Recursive Identification and Adaptive Control Model Which Uses Truncated Estimates, IEEE Transactions on Automatic Control, Vol. AC-27, pp. 775-782, 1982.

[8] Mandl. P. and Hübner, G., Transient Phenomena and Self-Optimizing Control of Markov Chains, Acta Universitatis Carolinae-Mathematica et Physica, Vol. 26, No. 1, pp. 35-51, 1985.

[9] Merton, R. C., Optimum Consumption and Portfolio Rules in Continuous-Time Model, Journal of Economic Theory, Vol. 3, pp. 373-413, 1971.

[10] Fleming, W. H. and Rishel, R. W., Deterministic and Stochastic Optimal Control, Springer, Berlin, Germany 1975.

[11] McKean, H. P., Jr., Stochastic Integrals, Academic Press, New York, 1960.

On Invariant Measures of Filtering Processes

Lukasz Stettner
Institute of Mathematics Polish Academy of Sciences
Śniadeckich 8, 00-950 Warsaw, Poland

Abstract. In the paper invariant measures of both discrete and continuous time filtering processes are studied. A generalization of the famous Kunita's result [6] to the case of locally compact signal state space is obtained. Moreover some approximations of invariant measures of filtering processes are considered. Examples which explain the assumptions imposed in the paper are also given.

1. Introduction. Filtering processes.

Let $(E,\mathcal{E})$ be a locally compact separable, endowed with a Borel σ-field state space E. Denote by

- $\mathcal{P}(E)$, $\mathcal{P}(\mathcal{P}(E))$- the spaces of all probability measures on E, $\mathcal{P}(E)$, endowed with weak topology respectively,
- $C(E)$, $C(\mathcal{P}(E))$- the spaces of continuous bounded functions on E, $\mathcal{P}(E)$ resp.,
- $C_o(E)$ - the subspace of $C(E)$ consisting of functions vanishing at infinity,
- $C_c(\mathcal{P}(E))$ - the subset of $C(\mathcal{P}(E))$ consisting of convex functions,
- $\mathcal{B}(\mathcal{P}(E))$ - σ-field of Borel subsets of $\mathcal{P}(E)$.

Let $X=(\Omega_1,F_t^1,x_t,\Theta_t^1,\bar{P}_x)$ be a right continuous, homogeneous Markov signal process on E, with transition semigroup (T_t), satisfying Feller property i.e. for $t\geq 0$, $T_tC(E)\subset C(E)$. Assume $W=(\Omega_2,F_t^2,w_0=0,w_t,\Theta_t^2,P')$ is an independent of X Brownian motion. Define a probability space $(\Omega=\Omega_1\times\Omega_2,F_t^1\times F_t^2,P_x=\bar{P}_x\otimes P')$, and a R^d- valued observation process (y_t),

$$y_t \stackrel{\text{def}}{=} \int_0^t h(x_s)ds + w_t, \text{ where } h\in C(E) \tag{1}$$

Let $G_t=\sigma\{\, y_s,\ s\leq t\,\}$. Then

Proposition 1. For every $\nu\in\mathcal{P}(E)$ there exists a continuous $\mathcal{P}(E)$- valued *filtering process* $(\pi_t^{(\nu)})$ such that for $f\in C(E)$, $E_\nu\{f(x_t)|G_t\} \stackrel{\text{def}}{=} \pi_t^{(\nu)}(f)$, P_ν a.s.. Moreover $(\pi_t^{(\nu)})$ is unique up indistinguishable processes and is Feller Markov with transition semigroup

$$\Pi_t(\nu,\Lambda)=E_\nu\{\pi_t^{(\nu)}\in\Lambda\}, \text{ for } \Lambda\in\mathcal{B}(\mathcal{P}(E)) \tag{2}$$

Proof. All results except Feller property of (Π_t) follow from [6]. The fact that $\Pi_t C(\mathcal{P}(E))\subset C(\mathcal{P}(E))$ is proved in [10, Thm. 3.1].

Together with continuous time we shall study also discrete time model, which we define throughout the reference probability method.

Let $X=(\Omega_1=E^N, F_n^1, X_n, \theta_n, \bar{P}_x)$ be a homogeneous discrete time Markov process with transition operator $P(x,.)$, and $Y=(\Omega_2=(R^d)^N, F_n^2, Y_n, n=1,2,..,P')$ be a sequence of independent of X i.i.d. random variables with positive density $g(x)$ with respect to d-dimensional Lebesgue measure λ^d. Define a product probability space $(\Omega=\Omega_1\times\Omega_2, F_n=F_n^1\times F_n^2, P_x^o=\bar{P}_x\otimes P')$. Assume that for a fixed x, the function $w(x,.)$ is a diffeomorphism of R^d i.e. 1-1, C^1 transformation of R^d with Jacobian $|\Delta(x,y_1,...,y_d)|\neq 0$.

Put

$$L_n \overset{def}{=} \prod_{i=1}^{n} \frac{g(w(X_n,Y_n))\ |\Delta(X_n,Y_n)|}{g(Y_n)} \tag{3}$$

Then using the integration by substitution in R^d we obtain

Lemma 1. L_n is P_x^o, (F_n) martingale.

Thus we can define a new measure P_x on Ω, such that its restriction P_x^n to F_n satisfies $P_x^n=L_nP_x^o$, and a standard verification shows

Lemma 2. Under P, X is again Markov process with transition operator $P(x,.)$. Moreover $W_n=w(X_n,Y_n)$ are for $n=1,2,...$, i.i.d. with density g, independent of X.

Let $h(x,.)$ be inverse function to $w(x,.)$. Then

$$Y_n=h(X_n,W_n) \tag{4}$$

and we can define the *filtering process* $(\pi_n^{(\nu)})$

$$\pi_n^{(\nu)}(f) \overset{def}{=} E_\nu\{f(X_n)|Y_1,...,Y_n\}, \ P_\nu \text{ a.s. for } f\in C(E) \tag{5}$$

From the Kallianpur Striebel formula we obtain for $n=1,2,...,$

$$\pi_{n+1}^{(\nu)}(f)=\int_E\int_E f(z)g(w(z,Y_{n+1}))|\Delta(z,Y_{n+1})|P(x,dz)\ \pi_n^{(\nu)}(dx)$$
$$(\int_E\int_E g(w(z,Y_{n+1}))|\Delta(z,Y_{n+1})|P(x,dz)\ \pi_n^{(\nu)}(dx))^{-1} = \tag{6}$$
$$\overset{def}{=} S(f,Y_{n+1},\pi_n^{(\nu)})$$

Lemma 3. $(\pi_n^{(\nu)})$ is $G_n=\sigma\{Y_1,Y_2,...,Y_n\}$ Markov process on $\mathcal{P}(E)$ with transition operator

$$\Pi(\nu,\Lambda)=\int_{R^d}\int_E(\int_E g(w(z,u))|\Delta(z,u)|P(x,dz))\nu(dx)\ 1_\Lambda(S(.,u,\nu))du \tag{7}$$

Proof. The proof is based on the following identity

$$E_\nu\{\pi_{n+1}^{(\nu)}\in\Lambda|Y_1,...,Y_n\} = E^o\{L_{n+1}\ 1_\Lambda(\pi_{n+1}^{(\nu)})|Y_1,...,Y_n\} \tag{8}$$
$$(E^o\{L_n|Y_1,...,Y_n\})^{-1}$$

The remaining computations are straightforward and therefore left to the reader.

Because of further applications to ergodic theory we shall need the Feller property of the transition operator Π.

Proposition 1. Assume the transition operator $P(x,.)$ is Feller i.e. for $f\in C(E)$, $Pf(x)= \int f(y)P(x,dy)\in C(E)$. If for fixed $y\in R^d$, $w(.,y)$ and $|\Delta(.,y)|\in C(E)$, and $g\in C(R^d)$, then $(\pi_n^{(\nu)})$ is Feller Markov with transition operator Π.

Proof. In view of Lemma 3 it remains only to show Feller property of Π. Let (f_i) be a countable dense sequence in $C_o(E)$. Then one can easily see that weak convergence of measures is equivalent to the convergence of the values of their integrals of f_i for $i=1,2,\ldots$ If $\nu_n\Rightarrow\nu$, then we have $\pi_1^{(\nu_n)}(f_i)=S(f_i,Y_1,\nu_n)\longrightarrow S(f_i,Y_1,\nu)=\pi_1^{(\nu)}(f_i)$, P a.s., for each i, as $n\rightarrow\infty$. Thus $\pi_1^{(\nu_n)}\Rightarrow\pi_1^{(\nu)}$, P a.s., and for $F\in C(\mathcal{P}(E))$ $E\{F(\pi_1^{(\nu_n)})\} = \Pi F(\nu_n)\longrightarrow E\{F(\pi_1^{(\nu)})\} = \Pi F(\nu)$, which is just the Feller property of Π.

Thus we have obtained Feller Markov properties of filtering processes in discrete and continuous time cases. It seems to be natural to ask about ergodic properties of filtering processes. There are several reasons to study invariant measures of filtering processes. One of them is to characterize so called Cesaro square mean filtering error i.e. $t^{-1}\int_0^t E_\nu\{(f(x_s)-\pi_s^{(\nu)}(f))^2\}\,ds$, or in discrete time case $n^{-1}\sum_{i=1}^n E_\nu\{(f(x_i)-\pi_i^{(\nu)}(f))^2\}$, as t or $n\rightarrow\infty$. Another is associated with identification problems i.e. when some parameters corresponding to signal process or noise are unknown and have to be identified.

In the case when the signal state space E is compact, a complete characterization of invariant measures of filtering processes and filtering errors was obtained by Kunita [6]. Some approximation and limit results for wide bandwidth observation noise were studied in [8]. Below, in section 2 we generalize Kunita's result to the case of locally compact state space E. Then in section 3 we provide examples and explain the meaning of assumptions which we imposed in section 2. Finally (section 4) we study approximations of invariant measures of filtering processes. Moreover in Appendix we sketch the proof of Stone Weierstrass type theorem applied in section 2.

It should be pointed out here, that in the discussion after the presentation of the above result prof. H. Kushner remarked that he had also obtained a similar generalization of Kunita's paper as in section 2. However this generalization has not been published so far.

2. Invariant measures.

For simplicity of notation throughout this section we shall write continuous time filtering processes notation only. Nevertheless under the assumptions of section 1, all results proved below can be easily rewritten to discrete time case, replacing continuous parameter t by discrete time n, the semigroups T_t, Π_t by the iterations of operators $P(x,.)$ and $\Pi(\nu,.)$ respectively.

We start with the definition of a pair of measures on $\mathcal{P}(E)$

Definition 1. For $\nu\in\mathcal{P}(E)$ let

$$m_t^\nu(\Lambda) \stackrel{\text{def}}{=} \Pi_t(\nu,\Lambda) = P_\nu\{ \pi_t^{(\nu)}\in \Lambda \} \tag{9}$$

$$M_t^\nu(\Lambda) \stackrel{\text{def}}{=} \int_E \Pi_t(\delta_x,\Lambda)\ \nu(dx), \tag{10}$$

for $\Lambda\in\mathcal{B}(\mathcal{P}(E))$, $t\ge0$, where δ_x is Dirac measure at point x.

Remark. If we define another filtering process $\tilde{\pi}_t^{(\nu)}(f)=E_\nu\{f(x_t)|G_t\vee\sigma(x_0)\}$ for $f\in C(E)$, then $M_t^\nu(\Lambda)=P_\nu\{ \tilde{\pi}_t^{(\nu)}\in\Lambda \}$.

Definition 2. A measure $\nu\in\mathcal{P}(E)$ is a *barycenter of measure* $\Phi\in\mathcal{P}(\mathcal{P}(E))$ (we denote $b(\Phi)=\nu$) if and only if for every $\phi\in C(E)$ we have

$$\nu(\phi)=\int_{\mathcal{P}(E)} \nu'(\phi)\ \Phi(d\nu')$$

A trivial verification shows that the following two Corollaries hold

Corollary 1. $b(m_t^\nu) = b(M_t^\nu) = \nu T_t$, for $t\ge0$.

Corollary 2. If Φ is (Π_t) invariant measure, then its barycenter $b(\Phi)$ is (T_t) invariant measure.

The next Proposition and its proof explain the main technics applied in the paper.

Proposition 3. Suppose μ is (T_t) invariant. Then for every $\varepsilon>0$ there exists a compact set $\Gamma(\varepsilon)\subset\mathcal{P}(E)$ such that for $t\ge0$

$$m_t^\nu(\Gamma(\varepsilon))\ge1-\varepsilon \qquad M_t^\nu(\Gamma(\varepsilon))\ge1-\varepsilon \tag{11}$$

Moreover, there exists a (Π_t) invariant measure Φ with barycenter μ.

Proof. Choose an increasing sequence of compact sets $L_n\subset E$, $L_n\subset L_{n+1}$, such that $\mu(L_n)\ge1-2^{-2n}\varepsilon$. Then

$$E_\mu\{\pi_t^{(\mu)}(L_n^c)\ge2^{-n}\} \le 2^n\ E_\mu\{\pi_t^{(\mu)}(L_n^c)\} = 2^n\ \mu(L_n^c) \le 2^{-n}\ \varepsilon$$

Let $\Gamma(\varepsilon)=\{ \nu\in\mathcal{P}(E):\ \nu(L_n)\ge1-2^{-n}$ for $n=1,2,\dots\}$. Clearly (see Lemma 3-3 [10]), $\Gamma(\varepsilon)$ is compact set in $\mathcal{P}(E)$. Moreover

$$m_t^\mu(\Gamma(\varepsilon)) = P_\mu\{\pi_t^{(\mu)}(L_n^c)\in\Gamma(\varepsilon)\} = P_\mu\{ \bigcap_{n=1}^{\infty} (\pi_t^{(\mu)}(L_n^c)\le2^{-n})\} =$$

$$= P_\mu\{\Omega\setminus \bigcup_{n=1}^{\infty} (\pi_t^{(\mu)}(L_n^c)>2^{-n})\} \ge 1- \varepsilon \sum_{n=1}^{\infty} 2^{-n} = 1-\varepsilon$$

In much the same way we obtain also $M_t^{(\mu)}(\Gamma(\varepsilon))\ge 1-\varepsilon$. Consider now a sequence of measures $\Phi_n(\Lambda)=n^{-1}\int_0^n m_t^{(\mu)}(\Lambda)dt$. By (11), (Φ_n) is tight, so from Prohorov theorem [Thm. 6.1, 2], there exists a subsequence (n_k) and $\Phi\in\mathcal{P}(E)$ such that $\Phi_{n_k} \Rightarrow \Phi$, as $k\to\infty$. Since (Π_t) is Feller, then Φ is (Π_t) invariant measure.

We have to recall now some facts from convex analysis.

Definition 3. A function $\psi:\mathcal{P}(E)\to R$ is *affine* if and only if there exists a real constant c and a function $\phi\in C(E)$, such that for every $\nu\in\mathcal{P}(E)$, $\psi(\nu)=c+\nu(\phi)$.

Proposition 4. If $F\in C_c(\mathcal{P}(E))$, then

$$\forall_{\nu\in\mathcal{P}(E)} \quad F(\nu) = \sup_{G\in\mathcal{A}_F} G(\nu),$$ where $\mathcal{A}_F$ is the set of all affine

functions ψ majorized by F i.e. such that $\psi(\nu)\le F(\nu)$ for every $\nu\in\mathcal{P}(E)$.

Proof. We apply Proposition I.3.1 of [4].

In the same way exactly as in [6, Lemma 3.2] we can prove

Lemma 4. If $F\in C_c(\mathcal{P}(E))$, then $\Pi_t F\in C_c(\mathcal{P}(E))$.

From Lemma 4, Proposition 4, using a version of Jensen's Lemma we obtain

Corollary 3. For $F\in C_c(\mathcal{P}(E))$, $\nu\in\mathcal{P}(E)$, $m_t^{\nu}(F) \le M_t^{\nu}(F)$.

Recall now Lemma 3.3 [6]

Lemma 5. Assume μ is (T_t) invariant and $F\in C_c(\mathcal{P}(E))$. Then for $h\ge 0$

$$m_t^{\mu}(F) \le m_{t+h}^{\mu}(F) \le M_{t+h}^{\mu}(F) \le M_t^{\mu}(F) \tag{12}$$

We are in position now to prove the main result of the section

Theorem 1. Assume μ is (T_t) invariant. Then

$$m_t^{\mu} \Rightarrow m^{\mu}, \text{ and } M_t^{\mu} \Rightarrow M^{\mu} \text{ as } t\to\infty, \tag{13}$$

and m^{μ}, M^{μ} are (Π_t) invariant with barycenter μ. Moreover, if $\Phi\in\mathcal{P}(\mathcal{P}(E))$ is (Π_t) invariant and $b(\Phi)=\mu$, then for $F\in C_c(\mathcal{P}(E))$

$$m^{\mu}(F) \le \Phi(F) \le M^{\mu}(F) \tag{14}$$

Proof. By Proposition 3, $\{m_t^{\mu}, t\ge 0\}$, $\{M_t^{\mu}, t\ge 0\}$ are tight. Thus from any sequence (t_n), $t_n\to\infty$ one can choose a subsequence (t_{n_k}), such that $m_{t_{n_k}}^{\mu} \Rightarrow m^{\mu}((t_{n_k}))$ weakly. So far we haven't proved that the limit measure m^{μ} has not depended on chosen subsequence. But for $F\in C_c(\mathcal{P}(E))$, by virtue of Lemma 5, for any two subsequences (t_{n_k}) and (t_{n_k}') we have $m^{\mu}((t_{n_k}))(F)=m^{\mu}((t_{n_k}'))(F)$. Thus from Proposition A1, proved in the Appendix, $m^{\mu}((t_{n_k}))=m^{\mu}((t_{n_k}'))$, and applying Theorem 2.3 [2] we

obtain the first convergence in (13). Similarly we show that $M_t^{\mu} \Rightarrow M^{\mu}$.
The inequalities (14) follow from Lemma 1, Jensen inequality and the definition of barycenter.

Analysis similar to the proof of Theorem 3.3 [6] shows that the following result characterizes uniqueness of invariant measures of filtering processes

Theorem 2. Let μ be (T_t) invariant. Then (Π_t) invariant measure with barycenter μ is unique if and only if

$$\forall_{f\in C(E)} \quad \lim_{t\to\infty} \sup \int_E |T_t f(x)-\mu(f)|\, \mu(dx) = 0 \tag{ER}$$

To approximate the Cesaro mean square error of filtration we need the following Proposition

Proposition 5. Suppose for some $\nu\in\mathcal{P}(E)$

$$t^{-1} \int_0^t E_\nu\{x_s\in .\}ds \Rightarrow \mu(.), \text{ as } t\longrightarrow \infty \tag{F1}$$

Then for every $\varepsilon>0$, there exists a compact set $\Gamma(\varepsilon)\subset\mathcal{P}(E)$ such that

$$t^{-1} \int_0^t m_s^{\nu}(\Gamma(\varepsilon))ds\geq 1-\varepsilon, \qquad t^{-1} \int_0^t M_s^{\nu}(\Gamma(\varepsilon))ds\geq 1-\varepsilon, \text{ for } t\geq 1 \tag{15}$$

Proof. By Feller property of (T_t) and (F1) the family $\{t^{-1}\int_0^t E_\nu\{x_s\in .\}ds,\ t\geq 1\}$ is relatively compact. Thus from the inverse Prohorov theorem [2. Thm. 6.2] is tight. Choose an increasing sequence of compact sets $L_n\subset L_{n+1}\subset E$, such that $t^{-1}\int_0^t P_\nu\{x_s\notin L_n\}$ ds $\leq 2^{-2n}\varepsilon$, for $t\geq 1$. Then clearly $t^{-1}\int_0^t P_\nu\{\pi_s^{(\nu)}(L_n^c)\geq 2^{-n}\}ds\leq \varepsilon 2^{-n}$, and for a compact set $\Gamma(\varepsilon)=\{\nu'\in\mathcal{P}(E),\ \nu'(L_n^c)\leq 2^{-n},\ n=1,2,..\}$ we obtain (15) in much the same way as in the proof of Proposition 3.

Now, by an easy computation based on Feller property of filtering process and inequalities (15) we can obtain filtering errors

Corollary 4. Assume μ is a unique (T_t) invariant measure and (F1) is satisfied for some $\nu\in\mathcal{P}(E)$. Then for $f\in C(E)$

$$\mu(f^2)-\int_{\mathcal{P}(E)} [\nu'(f)]^2\, m^{\mu}(d\nu')\geq\lim_{t\longrightarrow\infty}\sup\, t^{-1}\int_0^t E_\nu\{(f(x_s)-\pi_s^{(\nu)}(f))^2\}ds \tag{16}$$
$$\geq\lim_{t\longrightarrow\infty}\inf\, t^{-1}\int_0^t E_\nu\{(f(x_s)-\pi_s^{(\nu)}(f))^2\}ds\geq\mu(f^2)-\int_{\mathcal{P}(E)} [\nu'(f)]^2\, M^{\mu}(d\nu')$$

Replacing the assumption (F1) by

$$\nu T_t \Rightarrow \mu, \text{ as } t\longrightarrow \infty \tag{F2}$$

we obtain a stronger result

Proposition 6. Suppose (F2) holds for some $\nu\in\mathcal{P}(E)$. Then the family of measures $\{m_t^{\nu},\ M_t^{\nu},\ t\geq 0\}$ is tight.

Proof. We apply the same consideration as in Proposition 5. Namely, because of (F2), the family $\{\nu T_t,\ t\geq 0\}$ is relatively compact, hence tight, and there exists an increasing sequence of compact sets L_n such that for each $t\geq 0$, $\nu T_t(L_n^c)\leq 2^{-2n}\varepsilon$. Put $\Gamma(\varepsilon)=\{\nu'\in\mathcal{P}(E)$, for each n, $\nu'(L_n^c)\leq 2^{-n}\}$. Then we obtain $m_t^\nu(\Gamma(\varepsilon))\geq 1-\varepsilon$ and $M_t^\nu(\Gamma(\varepsilon))\geq 1-\varepsilon$.

In (13) we proved the weak convergence of the transition measures $\Pi_t(\mu,.)$, starting from invariant measure μ, to (Π_t) invariant measure m^μ. Under certain assumptions the same holds for any initial measure ν.

Theorem 3. Let μ be (T_t) invariant. Assume (ER) holds. Then if $\nu\in\mathcal{P}(E)$ satisfies (F2), we have

$$m_t^\nu \Rightarrow m^\mu=M^\mu \text{ and } M_t^\nu \Rightarrow M^\mu \text{ as } t\longrightarrow\infty \tag{17}$$

Proof. From the proof of Lemma 3.3 [6] and Corollary 3 we obtain for $f\in C_c(\mathcal{P}(E))$

$$m_t^{\nu T_s}(F) \leq m_{t+s}^\nu(F) \leq M_{t+s}^\nu(F) \leq M^{\nu T_s}(F) \tag{18}$$

By the Feller property $M_t^{\nu T_s}(F)\longrightarrow M_t^\mu(F)$ and $m_t^{\nu T_s}(F)\longrightarrow m_t^\mu(F)$, as $s\longrightarrow\infty$. Letting in (18) $s\longrightarrow\infty$, then $t\longrightarrow\infty$, from (13) and (ER) we obtain $\lim_{t\to\infty} m_t^\nu(F)=\lim_{t\to\infty} M_t^\nu(F)=m^\mu(F)$, for $F\in C_c(\mathcal{P}(E))$. Taking into account the compactness of the family $\{m_t^\nu,\ M_t^\nu,\ t\geq 0\}$ (Proposition 6) and Proposition A1, we can repeat the similar consideration as in the proof of Theorem 1, to obtain (17).

3. Remarks on assumptions and examples.

An analysis of the proofs of section 2 shows that many of results can be extended to continuous time model with more general observation than (1), i.e. h-unbounded, noise (w_t) nonnecessary additive. To obtain Theorem 1, Proposition 5, Corollary 4 we had to know that filtering process was Feller Markov. In the case additive noise, but h continuous unbounded, Markov property of filter corresponds to the uniqueness of the nonlinear filtering equation. If it holds (for conditions see [7]), then also Theorem 2 and 3 are satisfied, provided we can show the Feller property of filtering process.

For better understanding of the ergodic assumption (ER) consider so called *Harris processes*. If there exists a measure η such that

$$\forall_{A\in\mathcal{E}}\ \eta(A)>0 \Rightarrow \forall_{x\in E}\ P_x\{\int_0^\infty 1_A(x_s)ds=\infty\}=1 \tag{H}$$

then continuous time Markov process X is called Harris.

It is known ([1]) that under (H) there exists a unique σ-finite (T_t) invariant measure μ. Consider now so called Lebesgue decomposition of transition kernel i.e., for $A\in\mathcal{E}$, $T_t 1_A(x)= \int_A p_t(x,y)\mu(dy)+P_t^s(x,A)$, where measure $P^s_t(x,.)$ is singular with respect to μ.

Proposition 7. Assume (H) and (T_t) invariant measure $\mu\in\mathcal{P}(E)$. Then

$$(ER) \Leftrightarrow \forall_{x\in E} \lim_{t\to\infty} P^s_t(x,E) = 0 \qquad (19)$$

Proof. $\Leftarrow$ If the right hand side of (19) is satisfied, then by virtue of Proposition I.1 of [3], X is Harris regular and from Theorem II.4 of [3], (ER) is obviously satisfied.

$\Rightarrow$ Assume the right hand side of (19) does not hold. Then again from Proposition I.1 of [3] we obtain for for every $t>0$, $P^s_t(x,E)=1$, μ a.e.. Since (ER) is equivalent to $\limsup_{t\longrightarrow\infty} \int_E |T_tf(x)-\mu(f)|\mu(dx)=0$ for every bounded measurable with respect to μ-completed σ-field $\mathcal{E}^\mu$ function f, then we obtain this way that (ER) is not satisfied. Thus implication $\Rightarrow$ is proved.

Below we show 4 examples of Feller Markov processes. First three concern state space $[0,1[$, endowed with the topology generated by open subintervals of $[0,1[$ and sets $[0,\alpha[\ \cup\ ]1-\beta,1[$, $0<\alpha,\beta<1$. The sign $\oplus$ means the summation modulo 1.

Example 1. Deterministic motion. Let $x_t=x\oplus t$, and $E=[0,1[$. Then X is Harris. Lebesgue measure λ is a unique invariant measure. Nevertheless

$$m^\lambda(\Lambda)=m^\lambda_t(\Lambda)=1_\Lambda(\lambda) \neq M^\lambda(\Lambda)=M^\lambda_t(\Lambda)=\int_E 1_\Lambda(\delta_x)\lambda(dx)$$

Of course the right hand side of (19) is not satisfied.

Example 2. Brownian motion. Let $x_t=x\oplus w_t$, where w_t is a Brownian motion starting from 0, and $E=[0,1[$. Then X is again Harris with unique invariant measure λ. By Proposition 7 the condition (ER) is satisfied, and we have also unique invariant measure corresponding to filtering process.

Example 3. $E=[0,1[$ and Markov process X has a semigroup

$$T_tf(x)=\sum_{k=0}^{\infty} e^{-t}\frac{t^n}{n!}P^nf(x), \text{ where } Pf(x)=\int_0^1 f(x\oplus y)F\{dy\}$$

and $F=1/2\ \delta_{\{y_1\}} + 1/2\ \delta_{\{y_2\}}$, y_1 is rational, and y_2 is irrational number from $[0,1[$. Then one can show that for each $x\in E$, $P^n(x,.)\Rightarrow\lambda$ -Lebesgue measure as $n\to\infty$. Thus λ is a unique (T_t) invariant measure and (ER) is satisfied. Since if Markov process is Harris, then (see [1]) (H) also holds for measure $\eta=\mu$, where μ is invariant, and (H) in the above example does not hold for $\eta=\lambda$, therefore X is not Harris.

Example 4. Assume a discrete time R^m-valued Feller Markov process is given by the formula

$$X_{n+1} = a(X_n) + v_n \qquad (20)$$

where $a:R^m\longrightarrow R^m$ is continuous bounded, v_n are i.i.d. with a continuous, strictly positive density $d(x)$ with respect to m-dimensionnal Lebesgue measure λ^m. Then one can show that X is uniformly ergodic i.e. there exists a unique invariant measure μ and constants $K,\gamma>0$, $\gamma<1$, such that for every bounded Borel function f we have $\sup_y |P^n f(y)-\mu(f)|\le K\gamma^n \|f\|$. Obviously, (ER) is also satisfied.

4. Approximations of invariant measures.

We finish the paper with some approximation results. Namely, given continuous time (y_t), or discrete time observation (Y_n), we approximate signal process X by another Feller Markov process $X^{(k)}=(x^k_t$, or $X^k_n)$, such that $y_t=\int_0^t h_k(x^k_s)ds+\bar{w}^k_t$, or $Y_n=h_k(X^k_n,\bar{W}^k_n)$ $((\bar{w}^k_t)$, $(\bar{W}^k_n)$ independent of $X^{(k)}$, h_k satisfies the same assumptions as h), and construct an approximated filtering process $\pi^{k(\nu)}_t(f)=E_\nu\{f(x^k_t)|G_t\}$, or $\pi^{k(\nu)}_n(f)=E_\nu\{f(X^k_n)|Y_1,\dots,Y_n\}$, in continuous or discrete time respectively, for $f\in C(E)$. Let $\Pi^{(k)}_t(\nu,\Lambda)=E_\nu\{\pi^{k(\nu)}_t\in\Lambda\}=m^{k,\nu}_t(\Lambda)$, and $\Pi^{(k)}(\nu,\Lambda)=E_\nu\{\pi^{k(\nu)}_1\in\Lambda\}=m^{k,\nu}_1(\Lambda)$, for $\Lambda\in\mathcal{B}(\mathcal{P}(E))$.

Theorem 4. Assume the processes $X^{(k)}$, X have unique invariant measures μ_k, μ respectively, X satisfies (ER) and $\mu_k\Rightarrow\mu$, as $k\to\infty$. Moreover suppose for every $F\in C(\mathcal{P}(E))$

$$\Pi^{(k)}_t F(\nu)\to \Pi_t F(\nu), \tag{21}$$

(or $\Pi^{(k)n}F(\nu)\to\Pi^n F(\nu)$ in discrete time case resp.), uniformly on compact sets from $\mathcal{P}(E)$, as $k\to\infty$.

Then $m^{k,\mu_k}\Rightarrow m^\mu$, as $k\to\infty$, where m^{k,μ_k} and m^μ are defined by (13) for processes $X^{(k)}$, X resp., and m^μ is a unique (Π_t), $((\Pi))$ invariant measure for filtering process (π_t) $((\pi_n))$.

Proof. Step 1. The family $\{m^{k,\mu_k}$, $k=1,2,\dots\}$ is tight. In fact, since $\{\mu_k$, $k=1,2,\dots\}$ is tight, then for any $\varepsilon>0$ there exists an increasing sequence of compact sets L_i, such that for each k, $\mu_k(L_i)\ge 1-2^{-2n}\varepsilon$. In much the same way as in the proof of Proposition 3, we obtain that the family $\{m^{k,\mu_k}_t$, $k=1,2,\dots$, $t\ge 0\}$ is tight. By Theorem 1 also $\{m^{k,\mu_k}$, $k=1,2,\dots\}$ is tight.

Step 2. From Prohorov theorem [2, Thm. 6.1] from any sequence (k_r) one can choose a subsequence, for simplicity again denoted by (k_r), such that $m^{k_r,\mu_{k_r}}\Rightarrow m\in\mathcal{P}(E)$, as $r\to\infty$. Because (π_t) has a unique invariant measure (Thm. 2), it remains to show that m is (Π_t) invariant. By step 1, for any $\varepsilon>0$, there exists a compact set $\Gamma(\varepsilon)\subset\mathcal{P}(E)$ such that $m^{k,\mu_k}(\Gamma(\varepsilon)\ge 1-\varepsilon$. Therefore for $F\in C(\mathcal{P}(E))$ we have

$$|m(\Pi_t F)-m(F)| \le |m(\Pi_t F)-m^{k_r,\mu_{k_r}}(\Pi_t F)| + 2\|F\|\varepsilon +$$

$$+ \sup_{\nu\in\Gamma(\varepsilon)} |\Pi_t F(\nu)-\Pi_t^{(k_r)} F(\nu)| + |m^{k_r,\mu_{k_r}}(F)-m(F)| \longrightarrow 2\|F\|\varepsilon, \text{ as } r\longrightarrow \infty$$

Since ε was arbitrary , the m is (Π_t) invariant.

The condition (21) looks at first glance to be difficult for verification. The remaining part of the paper is devoted to the proof of sufficient conditions for (21), first in continuous time then discrete time case.

Theorem 5. Suppose $h_k, h\in C(E)$, $\|h_k-h\|\longrightarrow 0$, the semigroups $T_t^{(k)}$, T_t corresponding to $X^{(k)}$, X satisfies $T_t^{(k)}C_o(E)\subset C_o(E)$, $T_tC_o(E)\subset C_o(E)$, for $t\ge 0$ and for $f\in C_o(E)$, $\|T_t^{(k)}f-T_tf\|\longrightarrow 0$, as $k\longrightarrow\infty$. Then (21) holds.

Proof. We sketch the main steps of the proof.

Step 1. For every compact set $\Gamma\subset\mathcal{P}(E)$, $\varepsilon>0$, $t\ge 0$, there exists a compact set $\Gamma_t(\varepsilon)$, such that for $\nu\in\Gamma$, $k=1,2,\ldots$

$$P_\nu\{\pi_t^{k(\nu)}\in \Gamma_t(\varepsilon)\} \ge 1-\varepsilon, \quad P_\nu\{\pi_t^{(\nu)}\in \Gamma_t(\varepsilon)\} \ge 1-\varepsilon \tag{22}$$

In fact, the family of measures $\{\nu T_t^{(k)}, \nu T_t, \nu\in\Gamma, k=1,2,\ldots\}$ is tight. Thus $\Gamma_t(\varepsilon)$ can be constructed using the same technics as in the proof of Proposition 3.

Step 2. For $\alpha>0$, $f\in C_o(E)$

$$\sup_{\nu\in\mathcal{P}(E)} P(|\pi_t^{k(\nu)}(f)-\pi_t^{(\nu)}(f)|>\alpha)\longrightarrow 0, \text{ as } k\longrightarrow\infty \tag{23}$$

Let $dP^t=\exp\{\int_0^t h(x_s)\,dy_s - 1/2\int_0^t h^2(x_s)ds\}\,dP^o$, where P^t stands for restriction of measure P to F_t. Clearly under P^o, (y_s) is Brownian motion independent of X. Then

$$P(|\pi_t^{k(\nu)}(f)-\pi_t^{(\nu)}(f)|>\alpha)\le e^{1/2\,\|h\|^2 t}\, P^o\{|\pi_t^{k(\nu)}(f)-\pi_t^{(\nu)}(f)|>\alpha\}^{1/2} \tag{24}$$

Denote by $\sigma_t^{k(\nu)}(.)$, $\sigma_t^{(\nu)}(.)$ the unnormalized conditional probabilities corresponding to filter $\pi_t^{k(\nu)}$, $\pi_t^{(\nu)}$ resp..

The following estimation can be obtained

$$P^o\{|\pi_t^{k(\nu)}(f)-\pi_t^{(\nu)}(f)|>\alpha\} \le \alpha^{-1}(e^{3/2t\|h_k\|^2}((E^o(\sigma_t^{k(\nu)}(f) - \sigma_t^{(\nu)}(f))^2)^{1/2} + \|f\|\,(E^o(\sigma_t^{(\nu)}(1)-\sigma_t^{k(\nu)}(1))^2)^{1/2})) \tag{25}$$

Since by an easy computation $E^o(\sigma_t^{k(\nu)}(1)-\sigma_t^{(\nu)}(1))^2\longrightarrow 0$, then it remains to estimate $E^o(\sigma_t^{k(\nu)}(f)-\sigma_t^{(\nu)}(f))^2$, as $k\longrightarrow\infty$. But both $\sigma_t^{k(\nu)}$ and $\sigma_t^{(\nu)}$ are solutions to the following equations (see [12, Thm. IV1, IV2])

$$\sigma_t^{k(\nu)}(f)=\nu(T_t^{(k)}f) + \int_0^t \sigma_s^{k(\nu)}(h_k T_{t-s}^{(k)}f)\, dy_s,$$

$$\sigma_t^{(\nu)}(f)=\nu(T_t f) + \int_0^t \sigma_s^{(\nu)}(hT_{t-s}f)\, dy_s$$

Thus one can show that

$$\rho_t^{k,\nu}(f) \overset{\text{def}}{=\!=} E^o|\sigma_t^{k(\nu)}(f)-\sigma_t^{(\nu)}(f)|^2 \le c_k^t(f) + \int_0^t \rho_s^{k,\nu}(hT_{t-s}f)ds \qquad (26)$$

where $c_k^t(f) = 4\|T_t^{(k)}f-T_tf\|^2 + (4\|f\|^2\|h_k-h\|^2 +$

$$+4\|h\|^2 \sup_{0\le s\le t} \|T_s^{(k)}f-T_sf\|^2) \int_0^t e^{s\|h\|^2} ds)$$

Now by r-th iteration of (26), similar as in [6]-(2.18)-(2.20), (see also the proof of Lemma 5 in [11]), we obtain on the right hand side of (26) a series, the last, r+1 term of which converges uniformly in k to zero, as $r\longrightarrow\infty$, and the first terms are small since for $\phi\in C_o(E)$, $c_k^t(\phi)\longrightarrow 0$. But this means $\rho_t^{k,\nu}(f)\longrightarrow 0$ for $f\in C_o(E)$, as $k\longrightarrow\infty$.

Step 3. Fix a compact set $\Gamma\subset\mathcal{P}(E)$. From step 1, for every $\varepsilon>0$, there exists a compact set $\Gamma_t(\varepsilon)$ such that (22) holds, for $\nu\in\Gamma$. Since $F\in C(\mathcal{P}(E))$ is uniformly continuous on $\Gamma_t(\varepsilon)$, then there exists $\eta>0$ such that for ν, $\bar\nu\in\Gamma_t(\varepsilon)$, $\rho(\nu,\bar\nu)\le\eta \Rightarrow |F(\nu)-F(\bar\nu)|<\varepsilon$, where

$\rho(\nu,\bar\nu) \overset{\text{def}}{=\!=} \sum_{i=1}^{\infty} 2^{-(i+1)}(\|\phi_i\|\vee 1)^{-1}|\nu(\phi_i)-\bar\nu(\phi_i)|$ is a metric on $\mathcal{P}(E)$, (ϕ_i) is a dense sequence in $C_o(E)$. Let r be a positive integer such that $2^{-r}\le\eta/2$. By step 2 for $k\ge N$, $i=1,2,\dots r$,

$$\sup_{\nu\in\mathcal{P}(E)} P\{ |\pi_t^{k(\nu)}(\phi_i)-\pi_t^{(\nu)}(\phi_i)| \ge \eta/2 \} \le \varepsilon/r$$

Put $G_{k,i}^{(\nu)}(t)=\{ |\pi_t^{k(\nu)}(\phi_i)-\pi_t^{(\nu)}(\phi_i)| \ge \eta/2 \}$ and $G_k^{(\nu)}(t)=\bigcup_{i=1}^{r} G_{k,i}^{(\nu)}(t)$.

Then $\sup_{\nu\in\mathcal{P}(E)} P\{ G_k^{(\nu)}(t)\} \le \varepsilon$, and for $\omega\notin G_k^{(\nu)}(t)$, $\rho(\pi_t^{k(\nu)},\pi_t^{(\nu)}) \le \eta$.

Finally for $k\ge N$ we have

$$\sup_{\nu\in\Gamma} E\{|F(\pi_t^{k(\nu)})-F(\pi_t^{(\nu)})|\} \le$$

$$\le \sup_{\nu\in\Gamma} \{ E \{ 1_{(G_k^{(\nu)}(t))^c}(\omega)\, 1_{\Gamma_t(\varepsilon)}(\pi_t^{(\nu)})\, 1_{\Gamma_t(\varepsilon)}(\pi_t^{k(\nu)})|F(\pi_t^{k(\nu)})-F(\pi_t^{(\nu)})|$$

$$+ 2\|F\| (P(G_k^{(\nu)}(t)) + P(\pi_t^{(\nu)}\in \Gamma_t(\varepsilon)) + P(\pi_t^{k(\nu)}\in \Gamma_t(\varepsilon))) \le \varepsilon(1+6\|f\|)$$

The proof of (21) is completed.

The next theorem provides an useful sufficient criterion under which a discrete time version of (21) holds.

Theorem 6. Assume

(a1) $g(w_k(z,y))|\Delta_k(z,y)|$ and $g(w(z,y))|\Delta(z,y)|$ are continuous and uniformly bounded as a functions of (z,y), (a2) for $f\in C(E)$, $\sup_{x\in E} |P^{(k)}(x,f) - P(x,f)| \longrightarrow 0$, as $k\longrightarrow \infty$, where $P^{(k)}(x,.)$ stands for a transition kernel of discrete time process $X^{(k)}$,

(a3) $g(w_k(z,y))|\Delta_k(z,y)| \longrightarrow g(w(z,y))|\Delta(z,y)|$, uniformly on compact sets from $E\times R^d$, as $k\longrightarrow \infty$,

(a4) for any compact set $\Gamma\subset\mathcal{P}(E)$

$$\int_{R^d} \left|\int_E g(w(z,u))|\Delta(z,u)|\nu P(dz) - \int_E g(w(z,u))|\Delta(z,u)|\nu P^{(k)}(dz)\right| du$$

$\longrightarrow 0$, as $k\longrightarrow \infty$, uniformly for $\nu\in\Gamma$,

(a5) $\sup_{z\in E} \int_{R^d} |g(w(z,u))|\Delta(z,u)| - g(w_k(z,u))|\Delta_k(z,u)||du \longrightarrow 0$, as $k\longrightarrow \infty$,

Then for any $n=1,2,\dots$, $F\in C(\mathcal{P}(E))$, $\Pi^{(k)n}(\nu,F) \longrightarrow \Pi^n(\nu,F)$, as $k\longrightarrow \infty$, uniformly on compact sets from $\mathcal{P}(E)$.

Proof. (Sketch) We point out only main steps which should be proved to obtain our result.

Step 1. For fixed $f\in C(E)$, the functions $(y,\nu)\longrightarrow S_k(f,y,\nu)$, $(y,\nu)\longrightarrow S(f,y,\nu)$ are continuous (S, S_k are given by (6) for processes X, $X^{(k)}$ respectively).

Step 2. For every $f\in C_0(E)$, and compact sets $K\subset R^d$, $\Gamma\subset\mathcal{P}(E)$

$$\sup_{y\in K}\sup_{\nu\in\Gamma} |S_k(f,y,\nu) - S(F,y,\nu)| \longrightarrow 0, \text{ as } k\longrightarrow \infty \tag{27}$$

Step 3. For given compact sets $K\subset R^d$, $\Gamma\subset\mathcal{P}(E)$, the set

$\bigcup_{k=1}^{\infty} S_k(.,K,\Gamma) \cup S(.,K,\Gamma)$ is compact in $\mathcal{P}(E)$

($S_k(.,K,\Gamma)=\{\nu\in\mathcal{P}(E)$: there exist $y\in K$, $\nu'\in\Gamma$, such that $\nu(.)=S_k(.,y,\nu')\}$)

Step 4. For every $\varepsilon>0$, and compact set $\Gamma\subset\mathcal{P}(E)$, there exists a compact set $K\subset R^d$, such that for each $\nu\in\Gamma$, $k=1,2,\dots$

$$\int_K \int_E g(w(z,u))|\Delta(z,u)|\nu P^{(k)}(dz)du \geq 1-\varepsilon \tag{28}$$

Step 5. For $F\in C(\mathcal{P}(E))$, Γ- compact set from $\mathcal{P}(E)$

$$\sup_{\nu\in\Gamma} |\Pi^{(k)}(\nu,F) - \Pi(\nu,F)| \longrightarrow 0, \text{ as } k\longrightarrow \infty \tag{29}$$

Step 6. For $n=1,2,\dots$, $F\in C(\mathcal{P}(E))$, $\Pi^{(k)n}(\nu,F) \longrightarrow \Pi^n(\nu,F)$, uniformly for ν from compact sets in $\mathcal{P}(E)$.

We finish the paper with an example for which all assumptions (a1)-(a5) are satisfied and by Theorem 4 we obtain an approximation of invariant measure of filtering process.

Example 5. Let

$$X_{n+1} = a(X_n) + \beta_n, \quad Y_n = b(X_n) + \eta_n \tag{30}$$

$$X^k_{n+1} = a_k(X^k_n) + \bar{\beta}^k_n, \quad Y_n = b_k(X^k_n) + \bar{\eta}^k_n \tag{31}$$

where a, a_k: $R^m\to R^m$, b, b_k: $R^m\to R^d$ are bounded continuous, (β_n), $(\bar{\beta}^k_n)$,

(η_n), $(\bar{\eta}_n^k)$ are mutually independent, sequences of i.i.d. Gaussian random variables, (β_n) and $(\bar{\beta}_n^k)$ as well as (η_n) and (η_n^k) have the same densities. Assume $\|a_k-a\| \to 0$, and $\|b_k-b\| \to \infty$, as $k\to\infty$. Then due to result pointed out in Example 4 and Theorem 6 of [5], the unique invariant measure μ_k of $X^{(k)}$ converges to invariant measure μ of uniformly ergodic Markov process X. Moreover (a1)-(a5) hold. The verification is left to the reader.

Appendix.

Let for $\phi_1,\dots,\phi_n \in C(E)$

$$\mathcal{T}_c(\phi_1,\dots,\phi_n) = \{\, F\in C(\mathcal{P}(E)):\ \exists_{p:R^n\to R,\ \text{convex continuous}}\ F(\nu) = p(\nu(\phi_1),\dots,\nu(\phi_n)) \text{ for } \nu\in\mathcal{P}(E)\}$$

$$\mathcal{T}(\phi_1,\dots,\phi_n) = \{\, F\in C(\mathcal{P}(E)):\ \exists_{p:R^n\to R,\ \text{continuous}}\ F(\nu) = p(\nu(\phi_1),\dots\nu(\phi_n)) \text{ for } \nu\in\mathcal{P}(E)\}$$

$$\mathcal{T}_c = \{\, F\in C(\mathcal{P}(E)):\ \exists_n\ \exists_{\phi_1,\dots,\phi_n\in C(E)},\ F\in\mathcal{T}_c(\phi_1,\dots,\phi_n)\ \}$$

$$\mathcal{T} = \{\, F\in C(\mathcal{P}(E)):\ \exists_n\ \exists_{\phi_1,\dots,\phi_n\in C(E)},\ F\in\mathcal{T}(\phi_1,\dots,\phi_n)\ \}$$

Proposition A1. Assume Φ, $\Psi\in\mathcal{P}(\mathcal{P}(E))$. If $\Phi(F)=\Psi(F)$, for every $F\in\mathcal{T}_c$, then $\Phi=\Psi$.

Proof. (Sketch). Step 1. Suppose $\Phi(F)=\Psi(F)$ for every $F\in\mathcal{T}_c(\phi_1,\dots,\phi_n)$. Then from the lattice version of Stone-Weierstrass theorem [9, Lemma 1.4.4C], we obtain $\Phi(F)=\Psi(F)$ also for $F\in\mathcal{T}(\phi_1,\dots,\phi_n)$.

Step 2. An analysis of the proof of Theorem 1.4.4.E [9] shows that for every $D\in C(\mathcal{P}(E))$, $K>0$, $\varepsilon>0$ and compact set $\Gamma\subset\mathcal{P}(E)$, there exists $F\in\mathcal{T}$, such that $\|F\|\le\|D\|+K$ and $|D(\nu)-F(\nu)|\le\varepsilon$, for $\nu\in\Gamma$.

Step 3. By step 1, $\Phi(F)=\Psi(F)$ for $F\in\mathcal{T}$. It remains to notice that the measures Φ and Ψ are tight and apply step 2.

References

[1] Azema J., Duflo M., Revuz D., Measure Invariante des Processus de Markov Recurrents, Sem. Prob. III, Lect. Notes Math. 88, Springer 1969, 24-33,

[2] Billingsley P., Convergence of Probability Measures, Wiley 1968,

[3] Duflo M., Revuz D., Proprietes asymptoptiques des probabilites de transition des processus de Markov recurrents, Ann. Inst. H. Poincare, Section B., Vol 5 (1969), 233-244,

[4] Ekeland I., Temam R., Convex Analysis and Variational Problems, North Holland 1976,

[5] Kartashov N.W., Criteria of uniform ergodicity and strong stability of Markov chains in general state space, Th. Prob. Math. Statistics 30 (1984), 65-81,

[6] Kunita H., Asymptotic Behavior of the Nonlinear Filtering Errors of Markov Processes, J. Mult. Anal. 1 (1971), 365-393,

[7] Kurtz T.G., Ocone D.L., Unique Characterization of Conditional Distributions in Nonlinear Filtering, Annals of Prob. 16 (1988), 80-107,

[8] Kushner H., Huang H., Approximate and Limit Results for Nonlinear

Filters with Wide Bandwidth Observation Noise, LCDS Report 84-36, Brown University 1984,

[9] Loomis L., An Introduction to Abstract Harmonic Analysis, Van Nostrand 1953,

[10] Mazziotto G., Stettner Ł., Szpirglas J., Zabczyk J., On Impulse Control with Partial Observation, to appear in SIAM J. Control Optimiz.,

[11] Stettner Ł., Zabczyk J., Optimal Stopping for Feller Processes, Preprint IMPAN No. 284, September 1983,

[12] Szpirglas J., Sur l'equivalence d'equations differentialles stochastiques a valeur measures intervenant dans le filtrage markovien non lineaire, Ann. Inst. H. Poincare, Section B, Vol 14 (1978), 33-59.

Polygonal Fields: A New Class of Markov Fields on the Plane

T.Arak and D.Surgailis

0. Introduction

The Markov property (Mp) of random fields can be briefed into the usual 'independence of the future and the past given the present'. However, it is not an obvious generalization of the Mp of random processes as the chronological notions ('the future','the past' and 'the present') do not refer to any time evolution. In fact the probabilists seemed unaware of the existence of Markov fields (Mf) with continuous parameter untill the middle of the fifties, when Paul Lévy [6] conjectured the Mp of the odd-parameter (Lévy's) Brownian motion. His conjecture was proved by McKean [7]. Later, Gaussian Mf were studied by a number of authors; in particular, the papers by Molchan [8] and Pitt [10] played important role (see Dynkin [3] and Röckner [11] for recent advances, and Rozanov [12] for general theory). We should like to mention also an interesting work by Kusuoka [5] about the Mp of solutions of linear SPDE (not necessarily Gaussian). Starting with the seventies, much interest in Mf was roused by the probabilistic approach in field theory [15]. However, physically interesting Mf must obey very strict symmetry conditions which make their treatment extremely difficult.

In all the cases mentioned above, the state space of Mf is continuum. A long standing problem in this context was the existence of Mf with finite state space and 'piecewise const ant' trajectories. In 1982, the first author [1] solved this problem by constructing a Mf on the plane which takes two values. A 'typical' trajectory of this field is shown in Figure 1 a). Recently, this construction was generalized by the authors [2] to random fields with any finite number of values (see Figure 1 b)). In view of the form of trajectories, we have called these fields polygonal fields (pf).

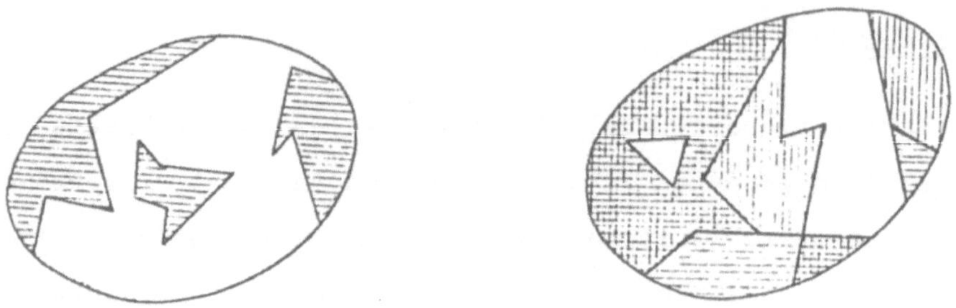

a) card $J = 2$ b) card $J = 4$

Figure 1

In this paper we want to make the reader familiar with the basic notions and ideas concerning pf. We present here also some new results about consistent pf with the sectional Mp which are not included in [2]. Open problems and possibilities of research are listed.

1. Definitions

Let be given a convex open bounded domain $T \subset \mathbb{R}^2$ and a finite set $J \subset \mathbb{Z}$. Denote by $\widetilde{\Omega}_T$ the set of all functions $\omega : T \to J$ such that

$$\omega(z) = \lim_{\varepsilon \to 0} \operatorname{ess\,sup}\{\omega(z') : |z' - z| < \varepsilon\}, \ \forall z \in T. \quad (1.1)$$

For any function $\omega \in \widetilde{\Omega}_T$, denote by $\partial\omega$ the set of its discontinuity points, i.e.

$$\partial\omega = \{z \in T: \limsup_{z' \to z} \omega(z') > \liminf_{z' \to z} \omega(z')\}.$$

Let $\mathcal{L}_T^n$ $(n \geq 1)$ denote the set of all collections $(\ell)_n \equiv (\ell_1, \ldots, \ell_n)$ of lines $\ell_i \subset \mathbb{R}^2$, $\ell_i \cap T \neq \emptyset$, $\ell_i \neq \ell_j$ $(i \neq j)$; $\mathcal{L}_T^1 \equiv \mathcal{L}_T$. By interval, or line segment, we mean a closed connected bounded subset I of a line $\ell \subset \mathbb{R}^2$ having strictly positive length $L(I) > 0$. (Sometimes we shall denote interval by $[\ell]$, indicating thus the line to which it belongs.) For any $(\ell)_n = (\ell_1, \ldots, \ell_n) \in \mathcal{L}_T^n$, consider the subset $\Omega_T(\ell)_n \subset \widetilde{\Omega}_T$, consisting of all functions ω such that for any $i = 1, \ldots, n$ there exists an interval $I_i \subset \ell_i$ such that

$$\partial\omega = \bigcup_{i=1}^n I_i \cap T. \quad (1.2)$$

Denote $\Omega_T^{(0)} = \{\omega \in \widetilde{\Omega}_T : \partial\omega = \emptyset\}$ and

$$\Omega_T^{(n)} = \bigcup_{(\ell)_n \in \mathcal{L}_T^n} \Omega_T(\ell)_n \quad (n \geq 1),$$

$$\Omega_T = \bigcup_{n=0}^{\infty} \Omega_T^{(n)}. \quad (1.3)$$

Introduce the σ-algebra $\mathcal{B}_T$ on Ω_T generated by the coordinate mappings $\omega \mapsto \omega(z) : \Omega_T \to J$, $z \in T$.

Remarks:

1.1. Note that card $\Omega_T(\ell)_n < \infty$ for any $n \geq 1$ and all $(\ell)_n \in \mathcal{L}_T^n$, as T is convex.

1.2. According to (1.2), for any $i=1,\dots,n$, $\partial\omega \cap \ell_i$ is connected up to a set of the (Lebesgue) length 0, i.e. $\partial\omega \cap \ell_i$ consists of a (connected) interval $I_i \cap T \subset \ell_i$ and eventually of some isolated points $\ell_i \cap \ell_j$, $j \neq i$.

1.3. One can introduce a metrizable topology in Ω_T such that the Borel σ-algebra coincides with $\mathcal{B}_T$ [2].

Let be given a finite non-atomic measure μ on $\mathcal{L}_T$. Denote

$$Z_T(A) = Z_{T,F,\mu}(A)$$

$$= \sum_{n=0}^{\infty} 1/n! \int_{\mathcal{L}_T^n} d^n\mu(\ell)_n \sum_{\omega \in \Omega_T(\ell)_n} \mathbb{1}_A(\omega)\, e^{-F(\omega)} \quad , A \in \mathcal{B}_T \tag{1.4}$$

where $d^n\mu(\ell)_n = d\mu(\ell_1) \dots d\mu(\ell_n)$ and $F(\omega)$ is a random variable (rv) taking values in $R \cup \{+\infty\}$. Denote

$$\Phi_{T,\mu} = \{F(\omega): \ Z_T \equiv Z_T(\Omega_T) < +\infty\}.$$

Definition 1.1. By polygonal field (pf) on T, corresponding to the measure μ and $F \in \Phi_{T,\mu}$, we mean the probability $P_T = P_{T,F,\mu}$ on Ω_T given by

$$P_T(A) = Z_T(A)/Z_T \quad , \quad A \in \mathcal{B}_T . \tag{1.5}$$

Remarks:

1.4. There is an analogy between pf and Gibbs field for continuous classical system (see e.g. Ruelle [13]). In fact, let $(z)_n = (z_1,\dots,z_n) \in T^n$, $z_i \neq z_j$ $(i \neq j)$, and let $\Omega'_T(z)_n$ consist of the single point measure $\omega' = \sum_{i=1}^n \delta_{z_i}$, where δ_z is the unite mass at z. Write

$$\Omega'_T = \bigcup_{n=0}^{\infty} \bigcup_{(z)_n} \Omega'_T(z)_n$$

for the set of all simple finite point measures ω' on T, with the topology of weak convergence and the corresponding Borel σ-algebra $\mathcal{B}'_T$. Let μ' be a finite non-atomic measure on T (e.g., the Lebesgue measure $\mu'(dz) = dz$), and let $F'(\omega')$ be a measurable function $\Omega'_T \to \mathbb{R} \cup \{+\infty\}$('potential') such that $Z'_T \equiv Z'_T(\Omega_T) < \infty$, where

$$Z_T(A) = \sum_{n=0}^{\infty} 1/n! \int_{T^n} d^n\mu'(z)_n \sum_{\omega' \in \Omega'_T(z)_n} \mathbf{1}_A(\omega')\, e^{-F'(\omega')}, \quad A \in \mathcal{B}'_T. \quad (1.6)$$

Gibbs field $P'_T = P'_{T,F',\mu'}$ on T corresponding to μ' and $F'(\omega')$ is defined by formula similar to (1.5):

$$P'_T(A) = Z'_T(A)/Z'_T, \quad A \in \mathcal{B}'_T. \quad (1.7)$$

Hovever, contrary to the case of Gibbs fields, the pf $P_{T,0,\mu}$ corresponding to the 'potential' $F(\omega) = 0$, is not trivial - the interaction enters via the basic condition (1.2) on admissible realizations. Clearly, $P'_{T,0,\mu'}$ is just the Poisson distribution on Ω'_T with the mean μ'.

1.5. Similarly as in the case of Gibbs fields, one can introduce pf with boundary conditions at ∂T (= the boundary of T), Such 'conditional pf' are important in the study of Mp of pf (see Sect. 2).

1.6. The existence of pf for general T, F and μ (i.e. the convergence of the series $Z_{T,F,\mu}$ (1.4)) is an open problem. However, sufficient conditions for the existence can be obtained, see Sect. 4.

1.7. The requirements for T to be convex and for μ to be atomless can be relaxed [2].

1.8. Definition 1.1 of pf can be generalized to dimensions higher than 2, but the Markov property of such fields seems doubtful, see Sect. 3.

2. Markov Property of Polygonal Fields

With any open set $U \subset T$, we associate the σ-algebra $\mathcal{B}_T^U \subset \mathcal{B}_T$, generated by $\omega \mapsto \omega(z): \Omega_T \to J$, $z \in U$.

Definition 2.1. A probability P on Ω_T will be called Markov field (Mf) if for any open sets $S, U \subset T$ such that $S \cup U = T$,

$$P(A \mid \mathcal{B}_T^U) = P(A \mid \mathcal{B}_T^{S \cap U}) \quad , \quad A \in \mathcal{B}_T^S . \tag{2.1}$$

Definition 2.2. A rv $F(\omega)$ is called additive if for any open sets $S, U \subset T$ such that $S \cup U = T$, there exist rv $F_S(\omega)$, $F_U(\omega)$, measurable with respect to $\mathcal{B}_T^S$, $\mathcal{B}_T^U$, correspondingly, and such that

$$F(\omega) = F_S(\omega) + F_U(\omega) \qquad P\text{-a.e.}$$

For Definitions 2.1, 2.2 see e.g. Nelson [9], Rozanov [12].

Theorem 1 [2]. Let $P_T = P_{T,F,\mu}$ be pf on T corresponding to an additive ('potential') $F \in \Phi_{T,\mu}$. Then P_T is Mf.

Let us explain the idea of the proof of Theorem 1.

According to the definition of pf $P_{T,F,\mu}$ and the result by Nelson [9] on the conservation of Mp by multiplicative change of measure, it suffices to consider the case $P_T = P_{T,0,\mu}$.

Roughly, the trajectories of this pf can be constructed as follows. Choose n lines $\ell_1, \dots, \ell_n$ intersecting T, distributed independently and identically according to $\mu(d\ell)$. Next, construct all (admissible) trajectories $\omega \in \Omega_T(\ell)_n$, taking into account the basic condition (1.2) about $\partial\omega$, and assigning to each of them equal weight.

Let be given an open set $U \subset T$ with the boundary ∂U (for simplicity, let us take U convex). One can construct $\omega \in \Omega_T^{(n)}$ in another way, namely, first by constructing $\omega' \in \Omega_U(\ell')_m$, $(\ell')_m = (\ell'_1, \dots, \ell'_m) \in \mathcal{L}_U^m$, $m \leq n$, and second, by completing ω' to a trajectory ω on T. For the second step, one chooses $n-m$ lines $\ell''_1, \dots, \ell''_{n-m} \in \mathcal{L}_{T\setminus U}$ according to $\mu(d\ell)$, independently of each other and independently of the previously chosen $\ell'_1, \dots, \ell'_m$, and constructs ω'' on $T\setminus U$ so that $\omega \in \Omega_T((\ell')_m \cup (\ell'')_{n-m})$, where $\omega(z) = \omega'(z)$ if $z \in U$, $= \omega''(z)$ if $z \in T\setminus U$.

Let $\partial\omega' = \bigcup_{i=1}^m I'_i \cap U$, $I'_i \subset \ell'_i$ (see Sect. 1). If an interval I'_i $(i=1,\dots,m)$ belongs to $U\setminus\partial_\varepsilon U$ for some $\varepsilon > 0$, where $\partial_\varepsilon U$ is ε-neighborhood of ∂U, then this interval I'_i together with the line $\ell'_i \supset I'_i$ shall have no effect on the choice of a suitable trajectory ω'' on $T\setminus U$, due to the connectedness of the intervals which form $\partial\omega = \partial\omega' \cup \partial\omega''$. But this means in fact the Mp of the pf P_T.

3. Polygonal Fields in Higher Dimensions

Let T, J, $\tilde{\Omega}_T$, $\mathcal{L}_T^n$ $(n \geq 1)$ satisfy the same conditions as in Sect. 1, with the difference that T is now a domain in $\mathbb{R}^d$ $(d \geq 2)$ and $\mathcal{L}_T$ is the set of $(d-1)$-dimensional hyperplanes $\ell \subset \mathbb{R}^d$ such that $\ell \cap T \neq \emptyset$. Similarly as in Sect. 1, one defines the set $\Omega_T(\ell)_n \subset \tilde{\Omega}_T$ $((\ell)_n \in \mathcal{L}_T^n)$ of functions $\omega : T \to J$ satisfying the condition (1.2), where each (face) I_i is a connected subset of ℓ_i of strictly positive (d-1)-dimensional Lebesgue measure. Let the probability space $(\Omega_T, \mathcal{B}_T)$ be defined analogously as in Sect. 1, and μ be a finite non-atomic measure on $\mathcal{L}_T$. The definition of pf $P_T = P_{T,F,\mu}$ on $T \subset \mathbb{R}^d$ is verbatim repetition of Definition 1.1.

Let us note, however, that the connectedness of the faces I_i in (1.2) is not sufficient for the Mp of pf in the case $d > 2$. In fact, consider the situation when (a) $I_i \cap \partial_\varepsilon U$ is disconnected for some $\varepsilon > 0$ and either (b) $I_i \cap U$ is connected, or (c) $I_i \cap U$ is disconnected. As $I_i \subset \ell_i$ is necessarily connected, the prediction problem for ω in $T \setminus U$ depends not only on the information (a) but rather on (b) or (c), for which the whole 'past' $\omega(z)$, $z \in U$ is needed.

This 'counterargument to the Mp' fails if one assumes each face $I_i \subset \partial\omega$ to be convex (in the case $d = 2$, connectedness and convexity of I_i coincide). On the other hand, convexity of I_i imposes a severe restriction on the form of trajectories of pf in the case $d > 2$.

4. Consistent Polygonal Fields

In view of eventual applications of pf (simulation, etc.), one of the major difficulties consists in finding the 'partition function' $Z_{T,F,\mu}$. Moreover, one might be interested in pf with the given marginal distribution and/or other characteristic properties. These problems can be solved for a special class of pf [2]. Below we present some new results on this subject, restricting our discussion to the case $\operatorname{card} J = 2$.

Let $\mathcal{G}$ be a family of convex open bounded domains $T \subset \mathbb{R}^2$, e.g. $\mathcal{G} = \mathcal{G}_{pol}$, where $\mathcal{G}_{pol}$ is the family of convex polygons. A family $(P_T)_{T\in\mathcal{G}}$ of probability measures P_T on $(\Omega_T, \mathcal{B}_T)$ is said consistent if for any $T, S \in \mathcal{G}$ such that $T \supset S$,

$$P_{T|\Omega_S} = P_S ,$$

where $P_{T|\Omega_S}$ denotes the restriction of P_T to Ω_S, i.e. $P_{T|\Omega_S}(A) = P_T(\pi_S^{-1}A)$, $A \in \mathcal{B}_S$ and $\pi_S : \Omega_T \to \Omega_S$ is given by $(\pi_S\omega)(z) = \omega(z)$, $z \in S$. By Kolmogorov's theorem, any consistent family $(P_T)_{T\in\mathcal{G}}$ determines random field $P = P_{\mathbb{R}^2}$ on $\mathbb{R}^2$ with values in J such that $P_{|\Omega_T} = P_T \quad \forall T \in \mathcal{G}$.

Below, let $J = \{+1,-1\}$ and $\mathcal{G} = \mathcal{G}_{pol}$. For any $T \in \mathcal{G}$, $j \in J$, $\omega \in \Omega_T(\ell_1,\dots,\ell_n)$ $(n \geq 0)$, let $\partial\omega_j = \partial T_j(\omega)$ denote the boundary of the sojourn set $T_j(\omega) = \{z \in T : \omega(z) = j\}$. Write $\partial\omega_{+1} = \partial\omega_+$, $\partial\omega_{-1} = \partial\omega_-$. By definition, $\partial\omega_+ \cap \partial\omega_- = \partial\omega$ and $\partial\omega_+ \cup \partial\omega_- = \partial\omega \cup \partial T$. The set $\partial\omega_j$ consists of a finite number of intervals $I_i^{(j)} \subset \bigcup_{k=1}^n \ell_k \cup \partial T$ intersecting only by their end points. Denote

$$L(\partial\omega_j) = \sum_{i=1}^{n(j)} L(I_i^{(j)})$$

the length (i.e. 1-dimensional Lebesgue measure) of $\partial\omega_j$, $j \in J$.

A point $z \in \partial\omega_j$ will be called node if there are two intervals $I_{i_1}^{(j)}, I_{i_2}^{(j)} \subset \partial\omega_j$, belonging to different lines and such that z is their (only) common point. We shall call such z simple if the pair $I_{i_1}^{(j)}, I_{i_2}^{(j)}$ is unique, and complex, if such a pair of intervals is not unique. Denote by $N(\partial\omega_j)$ the set of all simple nodes $z \in \partial\omega_j$ $(\omega \in \Omega_T)$. For any node $z \in N(\partial\omega_j)$, denote by $\alpha_j(z) = \alpha_j(z;\omega) \in (0,2\pi)$ the angle between the line segments $I_{i_1}^{(j)}$, $I_{i_2}^{(j)}$ measured on $T_j(\omega)$. Figure 2 illustrates these notions.

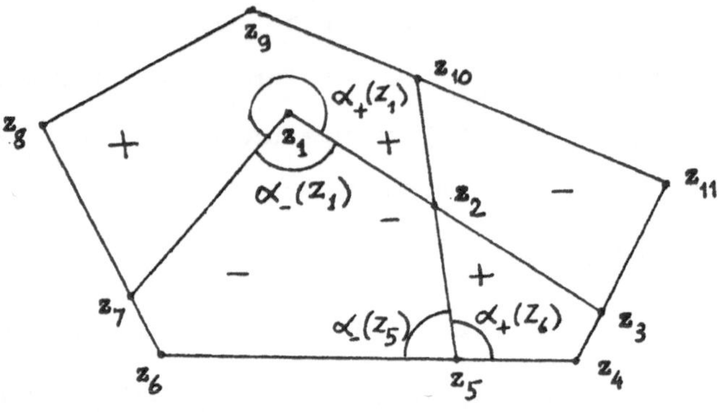

Figure 2. $N(\partial\omega_+) = \{z_1, z_3, z_4, z_5, z_7, z_8, z_9, z_{10}\}$,
$N(\partial\omega_-) = \{z_1, z_7, z_6, z_5, z_3, z_{10}, z_{11}\}$,
z_2 is complex node

Denote

$$n_j(\omega) = \operatorname{card}\{z \in N(\partial\omega_j) \cap T : \alpha_j(z) > \pi\}.$$

Consider the additive functional:

$$F(\omega) = \sum_{j\in J} \Big\{ \theta_j \, L(\partial\omega_{-j}) - n_j(\omega) \log \theta_j - (2\pi)^{-1} \sum_{z\in N(\partial\omega_j)} (\pi - \alpha_j(z;\omega)) \log\theta_j \Big\} \qquad (4.1)$$

if $\partial\omega_j$, $j \in J$ do not have complex nodes, $F(\omega) = +\infty$ if otherwise. Here, $\theta_{+1} = \theta_+$, $\theta_{-1} = \theta_-$ are strictly positive numbers and $L(\partial\omega_j)$, $n_j(\omega)$, $N(\partial\omega_j)$, $\alpha_j(z)$ are defined above.

For any line $\ell \subset R^2 = \{(t,y): t,y \in R\}$ introduce 'dynamic coordinates' (x,v), corresponding to the position at time $t = 0$ and the velocity of a particle moving on R with constant speed so that ℓ is its trajectory in time-space. (We shall write $\ell = \ell(x,v)$ below.) The measure $\mu(d\ell)$ invariant under the Euclidean transformations of the plane is given up to a constant factor by

$$\mu(d\ell) = dx\,\gamma(dv), \quad \ell = \ell(x,v), \qquad (4.2)$$

where

$$\gamma(dv) = dv/(1+v^2)^{3/2} . \tag{4.3}$$

Let $P_T = P_{T,F,\mu}$ be pf on $T \in \mathcal{G} = \mathcal{G}_{pol}$, corresponding to $F(\omega)$ (4.1) and μ (4.2). Denote $S(D) = \int_D dz$ $(D \subset \mathbb{R}^2)$.

Consider stationary Markov process $\eta = (\eta_t)_{t\in\mathbb{R}}$ with values in $J = \{+1,-1\}$ such that

$$\text{Prob}(\eta_t = j) = \theta_j/(\theta_+ + \theta_-) \tag{4.4}$$

and

$$\text{Prob}(\eta_{t+h} = j \mid \eta_t = i) = \theta_j h + o(h) , \tag{4.5}$$

$i,j \in J$, $i \neq j$. Let $\overset{d}{=}$ denote equality of (finite-dimensional) distributions.

Theorem 2.

(i) For any $T \in \mathcal{G} = \mathcal{G}_{pol}$, the pf P_T is well-defined and

$$Z_T = (\theta_+ + \theta_-) \exp\{\pi\theta_+\theta_- S(T)\} . \tag{4.6}$$

(ii) The family $(P_T)_{T\in\mathcal{G}}$ is consistent. The corresponding random field $P = P_{\mathbb{R}^2}$ on $\mathbb{R}^2$ is Euclidean invariant.

(iii) For any line $\ell \subset \mathbb{R}^2$,

$$(\omega(z) ; P)_{z\in\ell} \overset{d}{=} \eta .$$

Remarks:

4.1. From Theorem 2 (ii)-(iii) it follows that pf P_T enjoys the so- called sectional Markov property: for any line $\ell \in \mathcal{L}_T$, the sectional process $(\omega(z); P_T)_{z\in\ell\cap T}$ is Markov.

4.2. Theorem 2 can be generalized to measures μ which are only translation invariant, as well as to slightly more general additive functionals $F(\omega)$ allowing complex nodes.

4.3. Our paper [2] contains a similar result about consistent pf taking values in any finite set but under the additional condition that η is symmetric (for the case card$J = 2$, this means $\theta_+ = \theta_-$). In a forthcoming paper we extend these results to card$J \geqslant 2$ and reversible η .

4.4. In the case $\theta_+ = \theta_- \equiv \theta$, $F(\omega)$ (4.1) becomes

$$F(\omega) = 2\theta L(\partial\omega) + \theta L(\partial T) - (n(\omega) + 1)\log\theta ,$$

where $n(\omega) = n$, $\omega \in \Omega_T^{(n)}$. Imagine $\omega(z) = +1$ as white and $\omega(z) = -1$ as black colour. A typical trajectory $\omega \in \Omega_T$ in the case $\theta_+ = \theta_-$ looks symmetric in white and black, while in the case $\theta_+ \gg \theta_-$ it resembles a map of small convex black islands in the sea of white.

In the last Sect. 5 we explain the basic steps of the proof of Theorem 2. (The full proof will appear in the paper dealing with the case card$J \geqslant 2$.) The main idea of the proof, originatig from Arak [1], consists in a dynamic description of pf in terms of the evolution of a system of 1-dimensional particles, with death and birth. For simplicity, we consider below rectangular domains T . Also, the case $\theta_+ = \theta_-$ is relativly simpler and can be recommended for the first reading.

5. Dynamics of Particle System

Let $T = (t_0, t_1) \times (a,b) \subset \mathbb{R}^2 = \{(t,y): t,y \in \mathbb{R}\}$, $\bar{T} = T \cup \partial T$. The dynamics is described in $d_1) - d_3)$ below.

$d_1)$ A particle $z = (y,v)$ is characterized by its position $y \in \mathbb{R}$, velocity $v \in \mathbb{R}$ and environment $(i,j) \in J \times J$, $i \neq j$. Here, i is the environment above y and j is the environment below y , so that any particle is either $(^{+}_{-})$particle or $(^{-}_{+})$particle. The evolution $(z_t)_{t \in [t_0, t_1]}$ of a particle is determined by its initial coordinates $z_{t_0} = (y_0, v_0)$ and the evolution $(v_t)_{t \in [t_0, t_1]}$ of velocity (the environment remains constant during the evolution). The evolution of the velocity of $(^{i}_{j})$particle is described by a jump- type Markov process such that

$$\mathrm{Prob}(v_{t+h} \in du \mid v_t = v) = q_{ij}(v,du)h + o(h) , \tag{5.1}$$

where

$$q_{ij}(v,du) = \begin{cases} \theta_i |u-v| \gamma(du) , & \text{if } u<v , \\ \theta_j |u-v| \gamma(du) , & \text{if } u>v . \end{cases} \tag{5.2}$$

Thus, the length of the time interval between two successive jumps of v_t is distributed exponentially with the parameter

$$q_{ij}(v) = \int_{-\infty}^{\infty} q_{ij}(v,du) . \tag{5.3}$$

Trajectory $(y_t)_{t\in[t_0,t_1]}$ of $\binom{i}{j}$particle in the (t,y)-plane consists of line segments $[\ell_0], [\ell_1], \dots, [\ell_n]$ $(n \geq 0)$, where $\ell_0 = \ell(x_0, v_0)$, $x_0 = y_0 - v_0 t_0$ is determined by the initial coordinates (y_0, v_0) (see Figure 3). It follows from (5.1)-(5.3) that

$$\mathrm{Prob}(\ell_r \in d\ell_r , r=1,\dots,n \mid z_{t_0} = (y_0,v_0)) =$$
$$\exp\left\{ -\sum_{r=0}^{n} L([\ell_r])\, \rho_{ij}(\ell_r) + n_+ \log\theta_j + n_- \log\theta_i \right\} \prod_{r=1}^{n} \mu(d\ell_r) , \tag{5.4}$$

where n_+ (n_-) is the number of positive (resp., negative) jumps of v_t in the interval $[t_0, t_1]$, $n = n_+ + n_-$, and

$$\rho_{ij}(\ell) = q_{ij}(v)/(1+v^2)^{1/2} , \quad \ell = \ell(x,v) . \tag{5.5}$$

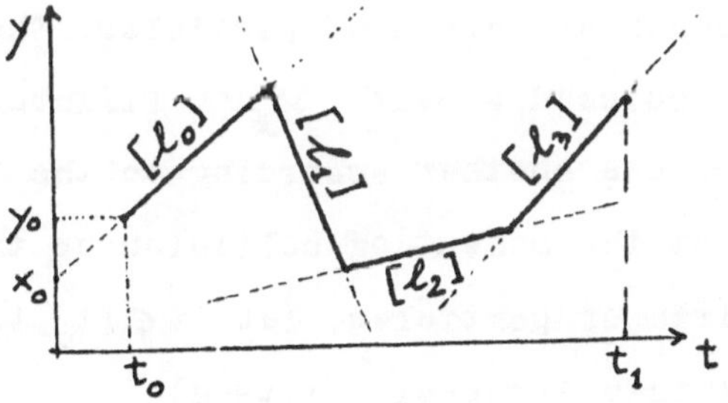

Figure 3

d_2) Consider a system $(z_1,\dots,z_n) \equiv \mathcal{X}$, $z_r = (y_r, v_r)$, with environments (i_r, j_r) , $r=1,\dots,n$, such that $a < y_1 < \dots < y_n < b$ and $i_r = -j_{r+1}$, $r=1,\dots, n-1$ (we shall call such systems of particles consistent). With any such system $\mathfrak{X}$, one can associate a function $\omega(\cdot;\mathfrak{X}): [a,b] \to J$ called the environment of $\mathfrak{X}$ (see Figure 4).

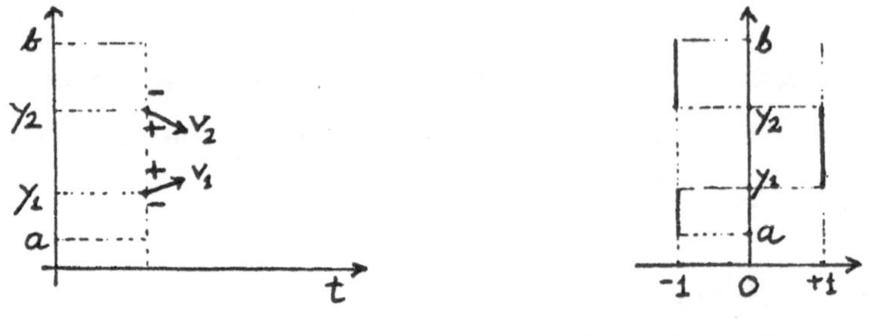

1) $\mathfrak{X} = (z_1, z_2)$ 2) $\omega(\cdot;\mathfrak{X})$

Figure 4

One can introduce a probability $\nu_{\mathfrak{X}}$ on the set $\mathcal{X}$ of all consistent systems of particles such that the distribution of $(\omega(y;\mathfrak{X}))_{y\in[a,b]}$ coincides with the distribution of the Markov process $(\eta_y)_{y\in[a,b]}$ (4.4)-(4.5), while the velocities v_1, v_2, ... of particles are mutually independent, independent of $\omega(\cdot;\mathfrak{X})$ and distributed according to $\gamma(dv)/\gamma(\mathbb{R})$.

d_3) Consider Markov evolution of a consistent system of particles $(\mathfrak{X}_t)_{t\in[t_0,t_1]}$, with death and birth of particles. The initial distribution of $\mathfrak{X}_{t_0}$ coincides with $\nu_{\mathfrak{X}}$. Different $\binom{i}{j}$particles evolve independently of one another according to the law described in d_1). Death occures at the moment of collision or the exit from T. To describe the birth of particles, let $t\in(t_0,t_1)$ and $\mathfrak{X}_{t-} \equiv \mathfrak{X} \in \mathcal{X}$. In a small time interval $(t, t+h)$,

b_1) with probability

$$\theta_{-\omega(b;\mathfrak{X})} |u| \gamma(du)\, \mathbb{1}_{(u<0)} h + o(h) \tag{5.6}$$

a new $\binom{i}{j}$particle $z=(y,v)$ is born at $y=b$ with $v\in du$, $i = -j = -\omega(b;\mathfrak{X})$;

b_2) with probability

$$\theta_{-\omega(a;\varkappa)}\, u\,\gamma(du)\,\mathbb{1}_{(u>0)}\, h + o(h) \tag{5.7}$$

a new $\binom{i}{j}$particle $z=(y,v)$ is born at $y=a$, with $v\in du$, $i=-j=\omega(a;\varkappa)$;

b_3) with probability

$$\theta_+\theta_-\,|u'-u''|\,\gamma(du')\,\gamma(du'')\,\mathbb{1}_{(u'>u'')}\,dy\,h + o(h) \tag{5.8}$$

two new particles $z'=(y',v')$, $z''=(y'',v'')$ with the environments (i',j'), (i'',j''), respectively, are born simultaneously at $y'=y''\in dy\subset(a,b)$, with $v'\in du'$, $v''\in du''$, $i'=-j'=i''=-j''=\omega(y;\varkappa)$.

Remarks:

5.1. Consider the Poisson line process on $\mathcal{L}_T$ with the mean (4.2). Then the right hand side of (5.1) (where $\theta_{\pm}=1$) is equal to the probability that a random line $\ell'=\ell(x',v')$ from this Poisson process intersects a fixed line $\ell=\ell(x,v)$ in the (small) time interval $(t,t+h)$ and $v'\in du$. Note also that (5.6) and (5.7) are particular cases of this probability, corresponding to $v=0$, $u<0$ and $v=0$, $u>0$, respectively. Similarly, (5.8) is related to the probability of intersection of two random lines, see (5.14) below.

5.2. The evolution (z_t) described in d_1), in the case $\theta_+=\theta_-=1$ resembles somewhat the evolution of a 'tagged' particle under ellastic collisions with random 'free' particles (Harris [4], Spitzer [14]). However, the two evolutions are actually different, due to the possibility of multiple collisions between the same particles.

It is clear that the evolution $(\varkappa_t)_{t\in[t_0,t_1]}$ described in d_1) - d_3) determines a random partition of $T=(t_0,t_1)\times(a,b)$ by sets with polygonal boundaries. The corresponding evolution $(\omega(\cdot;\varkappa_t))_{t\in[t_0,t_1]}$ of the environment of $\varkappa_t$ induces a probability measure (say, Q_T) on $(\Omega_T, \mathcal{B}_T)$. Our program now is as follows.

Step 1. To write Q_T as pf $P_{T,G,\mu}$, corresponding to some functional $G(\omega)$ and the same measure μ (4.2).

Step 2. To identify $G(\omega)$ with $F(\omega)$ (4.1) and consequently Q_T with P_T .

Step 3. To show how Steps 1-2 together with the Markov property of the evolution $(\mathcal{X}_t)$ and some symmetry properties of $F(\omega)$ yield the basic statements of Theorem 2.

From this program, Step 2 is the most involved, although in the case $\theta_+ = \theta_-$ it follows almost immediately.

Step 1. Let $\omega \in \Omega_T^{(n)}$ arise in the result of the evolution of $n_o + 2n_1 \leq n$ particles, n_o of which are born at ∂T and n_1 pairs of particles are born in T . Let

$$(t_j^o, y_j^o, v_j^o) \ (1 \leq j \leq n_o), \quad (t_j', y_j', v_j'), (t_j'', y_j'', v_j'') \ (1 \leq j \leq n_1) \tag{5.9}$$

denote the 'birth coordinates' of these particles (namely, the moment of birth, position and velocity at this moment, respectively). Of course, $t_j' = t_j''$ and $y_j' = y_j''$ $(1 \leq j \leq n_1)$. Note that in the case $\theta_+ = \theta_-$, the point process (5.9) is Poisson as the right hand sides of (5.6)-(5.8) do not depend on $\mathcal{X}$ and the process (η_t) is symmetric. Denote by

$$[\ell_j^o] \ (1 \leq j \leq n_o), \quad [\ell_j'], [\ell_j''] \ (1 \leq j \leq n_1) \tag{5.10}$$

the line segments, or the intervals of $\partial\omega$, corresponding to the initial parts of trajectories of the particles up to the first change of velocity or the death, if the velocity of a particle does not change. Consider the 'dynamic' coordinates (x,v) of lines ℓ_j^o , ℓ_j' , ℓ_j'' :

$$\ell_j^o = \ell(x_j^o, v_j^o) \ (1 \leq j \leq n_o), \quad \ell_j' = \ell(x_j', v_j'), \ \ell_j'' = \ell(x_j'', v_j'') \ (1 \leq j \leq n_1), \tag{5.11}$$

which are related to (5.9) by $x_j^o = y_j^o - v_j^o t_j^o$, $x_j' = y_j' - t_j' v_j'$, $x_j'' = y_j' - t_j' v_j''$. Hence

$$dy_j^o\,\gamma(dv_j^o) = dx_j^o\,\gamma(dv_j^o) = \mu(d\ell_j^o) \tag{5.12}$$

for $1 \leqslant j \leqslant n_o$, $t_j^o = t_o$ (i.e. for the particles born at $t = t_o$);

$$|v_j^o|\gamma(dv_j^o)dt_j^o = dx_j^o\,\gamma(dv_j^o) = \mu(d\ell_j^o) \tag{5.13}$$

for $1 \leqslant j \leqslant n_o$, $t_o < t_j^o < t_1$ (i.e. for the particles born at $y = a$ or $y = b$ after $t = t_o$, see (5.6)-(5.7));

$$|v_j' - v_j''|\gamma(dv_j')\,\gamma(dv_j'')dt_j'dy_j' = dx_j'dx_j''\,\gamma(dv_j')\,\gamma(dv_j'') = \mu(d\ell_j')\,\mu(d\ell_j'') \tag{5.14}$$

$1 \leqslant j \leqslant n_1$, see (5.8). From d_1)-d_3) and (5.12)-(5.14) we have that

$$Q_T(d\omega) = (\theta_+ + \theta_-)^{-1} \exp\{-G(\omega) - G_S\} \prod_{i=1}^{n} \mu(d\ell_i) \,, \tag{5.15}$$

Here, $G(\omega) = G_L(\omega) + G_N(\omega)$ and

$$\begin{aligned} G_L(\omega) = {} & \sum_{j\in J} L(\partial\omega_j \cap \{t = t_o\})\,\theta_{-j}\,\gamma(\mathbb{R}) \\ & + \sum_{j\in J} L(\partial\omega_j \cap \{y = a\})\,\theta_{-j} \int_0^{\infty} v\,\gamma(dv) \\ & + \sum_{j\in J} L(\partial\omega_j \cap \{y = b\})\,\theta_{-j} \int_{-\infty}^{0} |v|\,\gamma(dv) \\ & + \sum_{[\ell]\subset\partial\omega_\pm^\pm} L([\ell])\,\rho_{+,-}(\ell) \\ & + \sum_{[\ell]\subset\partial\omega_\mp^\mp} L([\ell])\,\rho_{-,+}(\ell) \,, \end{aligned} \tag{5.16}$$

where $\partial\omega_\pm^\pm$, $\partial\omega_\mp^\mp$ are the parts of $\partial\omega$, corresponding to the trajectories of ($\pm$)particles and ($\mp$)particles, respectively. Next,

$$\begin{aligned} -G_N(\omega) = {} & \log \theta_{\omega(t_o,a)} \\ & + n_o(\pm)\log\theta_+ + n_o(\mp)\log\theta_- \\ & + n_a(\pm)\log\theta_- + n_a(\mp)\log\theta_+ \\ & + n_b(\pm)\log\theta_+ + n_b(\mp)\log\theta_- \\ & + (n(\overset{+}{\underset{-}{\wedge}}) + n(\overset{-}{\underset{+}{\vee}}))\log\theta_+ + (n(\overset{-}{\underset{+}{\wedge}}) + n(\overset{+}{\underset{-}{\vee}}))\log\theta_- \\ & + (n(+\langle -) + n(-\langle +))\log(\theta_+\theta_-) \,, \end{aligned} \tag{5.17}$$

where $n_o\binom{i}{j}$, $n_a\binom{i}{j}$, $n_b\binom{i}{j}$ are the quantities of $\binom{i}{j}$particles born at $t=t_o$, $y=a$ and $y=b$, respectively; $n(i\langle j)$ is the number of pairs $\binom{i}{j},\binom{j}{i}$ of particles born simultaneously in T in which the velocity of the $\binom{i}{j}$particle is greater than the velocity of the $\binom{j}{i}$particle; finally, $n(\curlywedge^{i}_{j})$ (resp., $n(\curlyvee^{i}_{j})$) is the number of negative jumps (resp., the number of positive jumps) of velocity of $\binom{i}{j}$particles, $i,j\in J$, $i\neq j$. At last,

$$G_S = \theta_+ \theta_- S(T) \iint_{v'<v''} (v''-v')\,\gamma(dv')\gamma(dv'') = \pi\theta_+\theta_- S(T) . \quad (5.18)$$

Step 2. Write $F(\omega)$ (4.1) as $F_L(\omega)+F_N(\omega)$, where

$$F_L(\omega) = \sum_{j\in J} \theta_j\, L(\partial\omega_{-j}) , \quad (5.19)$$

$$-F_N(\omega) = \sum_{j\in J} \Big\{ n_j(\omega)\log\theta_j + (2\pi)^{-1} \sum_{z\in N(\partial\omega_j)} (\pi - \alpha_j(z))\log\theta_j \Big\}. \quad (5.20)$$

To prove Step 2, one has to verify

$$F_L(\omega) = G_L(\omega) \quad (5.21)$$

and

$$F_N(\omega) = G_N(\omega) \quad (5.22).$$

Note that (5.21)- (5.22) together with (5.15)-(5.18) imply Theorem 2 (i) for rectangular domains T . The verification of (5.21) is rather easy and we shall omit it, noting just that $\gamma(\mathbb{R})=1$ and $\int_{(0,\infty)} v\gamma(dv) = \int_{(-\infty,0)} |v|\gamma(dv) = 1$. Let us consider (5.22).

Let us introduce some terminology. Connected components of $\partial\omega_j$ will be called (j)contours ($j\in J$), and denoted by Γ_1, $\Gamma_2,\ldots$. A (j)contour Γ will be called internal if $\omega(z)=j$ for z near Γ inside it , and external, if $\omega(z)=j$ for z near Γ outside it (Figure 5).

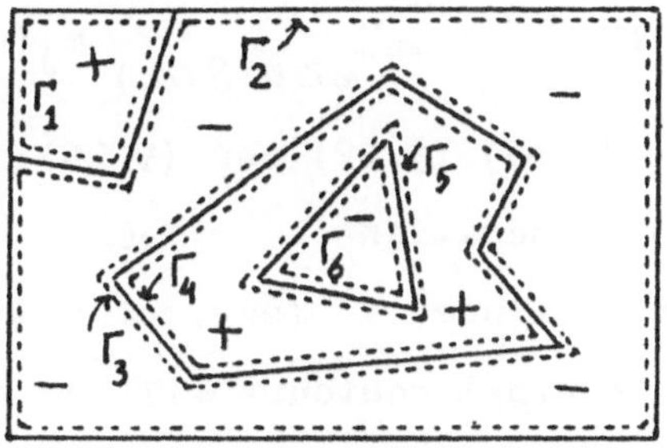

Figure 5. Γ_1 and Γ_4 are internal (+)contours, Γ_2 and Γ_6 are internal (−)contours, Γ_3 is external (−)contour and Γ_5 is external (+)contour

Consider the second sum on the right hand side of (5.20). Each node $z \in N(\partial\omega_j)$ can be attached to one and only one (j)contour and $\pi - \alpha_j(z)$ is just the angle at which this contour makes a turn around z . It is easy to observe that

$$(2\pi)^{-1} \sum_{z \in N(\partial\omega_j)} (\pi - \alpha_j(z)) = \operatorname{card}\{\Gamma\}_{j,\mathrm{int}} - \operatorname{card}\{\Gamma\}_{j,\mathrm{ext}} , \tag{5.23}$$

$j \in J$, where $\{\Gamma\}_{j,\mathrm{int}}$ and $\{\Gamma\}_{j,\mathrm{ext}}$ denote the sets of the internal (j)contours and external (j)contours, respectively.

Consider a scaling transformation:

$$\mathbb{R}^2 \ni z = (t,y) \longmapsto (z)_\varepsilon = (t, \varepsilon y) \in \mathbb{R}^2 \quad (\varepsilon > 0) . \tag{5.24}$$

This transformation maps $T = (t_0, t_1) \times (a,b)$ into $(T)_\varepsilon = (t_0, t_1) \times (\varepsilon a, \varepsilon b)$ and any (j)contour $\Gamma \subset \overline{T}$ into (j)contour $(\Gamma)_\varepsilon \subset (\overline{T})_\varepsilon$. Denote by $(\alpha_j(z))_\varepsilon$ the image of the angle $\alpha_j(z)$ ($z \in N(\partial\omega_j)$) under the transformation (5.24). Set

$$a_j(z) = a_j(z;\omega) = \lim_{\varepsilon \to 0} (\pi - (\alpha_j(z))_\varepsilon)/2\pi . \tag{5.25}$$

It is clear that $a_j(z)$ takes one of the values $+1/2$, $+1/4$, 0 , $-1/4$, $-1/2$. As the right hand side of (5.23) is invariant under the scaling,

$$- F_N(\omega) = \sum_{j\in J}\left\{ n_j(\omega)\log\theta_j + \sum_{z\in N(\partial\omega_j)} a_j(z)\log\theta_j \right\}. \quad (5.26)$$

For the identification of $G_N(\omega)$ (5.17) and (5.26), we use a procedure called 'transport operation'.

Fix a direction of travel along ∂T (say, clockwise). Then the direction of travel along internal contours will be also clockwise, and the direction of travel along external contours will be anti-clockwise.

Consider a line segment $[\ell]\subset\partial\omega_j$ between two adjacent nodes z_1, $z_2 \in N(\partial\omega_j)$. The direction of travel along $[\ell]$ will be called the transport direction if its projection on the t-axis is non-negative.

Imagine now that along each segment $[\ell]\subset\partial\omega_j$ in the transport direction there moves the 'cargo'

$$\frac{1}{2}\log\theta_j + \frac{1}{4}\log\theta_j\, \mathbb{1}_{(|v|=\infty)} \qquad (\ \ell = \ell(x,v)),\ (5.27)$$

which leaves one of the two adjacent nodes and comes to the other node.

Write $F_N(\omega)$ (5.26) and $G_N(\omega)$ (5.17) as sum of 'cargos' at nodes $z\in N(\partial\omega_j)$, $j\in J$:

$$- F_N(\omega) = \sum_{j\in J} \log\theta_j \sum_{z\in N(\partial\omega_j)} f_j(z)\ , \qquad (5.28)$$

$$- G_N(\omega) = \sum_{j\in J} \log\theta_j \sum_{z\in N(\partial\omega_j)} g_j(z)\ , \qquad (5.29)$$

In the result of the 'transport operation', $G_N(\omega)$ is transformed into

$$- G'_N(\omega) = \sum_{j\in J} \log\theta_j \sum_{z\in N(\partial\omega_j)} g'_j(z)\ . \qquad (5.30)$$

Of course, $G_N(\omega) = G'_N(\omega)$ as the 'transport operation' does not change the total sum of 'cargos' but only redistributes them between the nodes. We claim that $G'_N(\omega) = F_N(\omega)$, moreover,

$$g_j'(z) = f_j(z) \tag{5.31}$$

for any $z \in N(\partial \omega_j)$, $j \in J$.

Eq. (5.31) can be verified easily for each type of nodes (i.e. for nodes arising in the result of birth, death and evolution of particles). For example, consider the type shown in Figure 6.

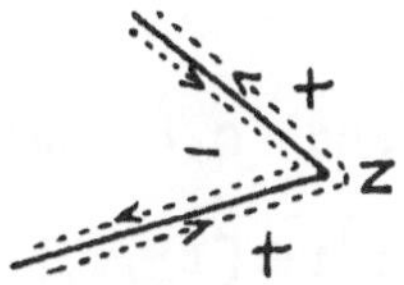

Figure 6

Here, $z \in N(\partial\omega_+) \cap N(\partial\omega_-)$ arises in the result of collision of $(\pm)$particle with $(\mp)$particle. By (5.17) and (5.29),

$$g_+(z) = g_-(z) = 0 . \tag{5.32}$$

After the 'transport operation', to z come the 'cargo' $\frac{1}{2}\log\theta_+$ along (+)contour, and the 'cargo' $\frac{1}{2}\log\theta_-$ along (-)contour, while no 'cargo' leaves z (along both contours, the direction of travel from z is the opposite to the transport direction). Hence

$$g_+'(z) = g_-'(z) = 1/2 . \tag{5.33}$$

As $\alpha_+(z) > \pi > \alpha_-(z)$, so $a_+(z) = -1/2$ and $a_-(z) = 1/2$ (see (5.25)). Consequently, according to (5.26) and (5.28),

$$f_+(z) = 1 + a_+(z) = 1/2 \ , \quad f_-(z) = a_-(z) = 1/2 . \tag{5.34}$$

From (5.32)-(5.34), it follows (5.31). This proves $G(\omega) = F(\omega)$ and consequently

$$Q_T = P_T . \tag{5.35}$$

Step 3. Let us show how the discussion above leads to the consistency of the pf $P_T = P_{T,F,\mu}$ with respect to the family of rectangular domains. Let $T=(t_o,t_1)\times(a,b)$, $T'=(t_o',t_1')\times(a',b')$ and $T\supset T'$. We want to show that

$$P_T\,|\,\Omega_{T'} = P_{T'} \,. \qquad (5.36)$$

Assume first that $a=a'$, $b=b'$, $t_o=t_o'$ and $t_1'<t_1$. By (5.35), $P_T=Q_T$ and $P_{T'}=Q_{T'}$. The relation

$$Q_T\,|\,\Omega_{T'} = Q_{T'}$$

follows from the fact that both Q_T and $Q_{T'}$ can be obtained from the same (Markov) evolution $(\mathcal{H}_t)_{t\in[t_o,t_1]}$: for $Q_{T'}$, this evolution should be observed up to time $t_1'<t_1$ only.

Assume now that $t_o=t_o'$, $t_1=t_1'$, $a=a'$ and $b'<b$. As $F(\omega)$ (4.1) and μ (4.2) are Euclidean invariant, we can interchange the coordinates t and y and use the same argument to prove (5.36). In a similar way, one proves (5.36) for $t_o<t_o'$, $t_1=t_1'$, $a=a'$, $b=b'$ and $t_o=t_o'$, $t_1=t_1'$, $b=b'$, $a<a'$. The general case $T\supset T'$ is a superposition of these four particular cases.

The discussion above yields also the statement (iii) of Theorem 2 for lines ℓ which are either horizontal or vertical. For example, let $T=(t_o,t_1)\times(a,b)$, $\ell=\{(t,y):\ t=t_o'\}$, $t_o<t_o'<t_1$. By (5.36),

$$(\omega(t_o',y);\ P_T)_{y\in(a,b)} \overset{d}{=} (\omega(t_o'+,y);\ P_{T'})_{y\in(a,b)} \,,$$

where $T'=(t_o',t_1)\times(a,b)\subset T$. As $P_{T'}=Q_{T'}$, so

$$(\omega(t_o'+,y);\ P_{T'})_{y\in(a,b)} \overset{d}{=} (\omega(y;\mathcal{H});\ \nu_{\mathcal{H}})_{y\in(a,b)}$$

$$\overset{d}{=} (\eta_t)_{t\in(a,b)}$$

according to the definition of the initial distribution $\nu_{\mathcal{H}}$ of our particle system, see d_2).

References

1. Arak, T.: On Markovian random fields with finite number of values. 4th USSR-Japan Symp. Probab.Theory and Math.Stat. Abstracts of communications. Tbilissi 1982.
2. Arak, T., Surgailis, D.: Markov fields with polygonal realizations (submitted to Probab. Theory Rel.Fields).
3. Dynkin, E.B,: Markov processes and random fields. Bull. Am. Math. Soc. 3 (3) , 975-999 (1980).
4. Harris, T.E.: Diffusion with 'collisions' between particles. J. Appl. Prob. 2 , 323-338 (1965).
5. Kusuoka, S.: Markov fields and local operators. J.Fac.Sci.Univ. Tokyo Ser. IA 26 , 199-212 (1979).
6. Lévy, P.: A special problem of Brownian motion and a general theory of Gaussian random functions. In: Proc. 3rd Berkeley Symp.Math.Stat. and Probab., 133-175. University of California Press: Berkeley 1955.
7. McKean, H.P.: Brownian motion with a several dimensional time. Theory Probab.Appl. 8 , 335-354 (1963).
8. Molchan, G.M.: Characterization of Gaussian fields with Markovian property. Soviet Math.Dokl. 197 , 784-787 (1971).
9. Nelson, E.: Probability theory and Euclidean field theory. In: Velo, G., Wightman, A. (ed.). Constructive quantum field theory. Lecture Notes Phys. 25 , 94-124. Springer: Berlin-Heidelberg-New York 1973.
10. Pitt, L.D.: A Markov property for Gaussian processes with a multidimensional parameter. Arch.Rat.Mech.Anal. 43 , 367-391 (1971)
11. Röckner, M.: Generalized Markov fields and Dirichlet forms. Acta Appl. Math. 3 , 285-311 (1985).
12. Rozanov, Yu.A. Markov random fields. Springer: Berlin-Heidelberg-New York 1982.

13. Ruelle, D. Statistical mechanics. Benjamin: New York 1969.

14. Spitzer, F.: Uniform motion with elastic collision of an infinite particle system. J. Math. Mech. 18 , 973-989 (1969).

15. Velo, G., Wightman, A. (ed.). Constructive quantum field theory. Lecture Notes Phys. 25 . Springer: Berlin-Heidelberg-New York 1973.

Arak Taivo

Tartu University
202400 Tartu
Vanemuise 46
Estonia, USSR

Surgailis Donatas

Institute of mathematics and cybernetics
232600 Vilnius
Akademijos 4
Lithuania, USSR

STRICTLY STATIONARY PROCESSES WITH
THE LINEAR PREDICTION PROPERTY*

Ngnyen Van Thu**
Universität Konstanz
Fakultät für Wirtschaftswissenschaften und Statistik
Postfach 55 60
D - 7750 Konstanz 1

*Dedicated to Professor Kazimierz Urbanik
for his coming sixtieth birthday.*

1. Introduction, Notation and Preliminaries

The purpose of this paper is to prove the moving average representation of strictly stationary processes with the linear prediction property. Moreover, a martingale approach to prediction problems for processes with infinite variance is introduced.

Suppose $(\Omega, F, F_t, \mathbb{P})$, $t \in T$, is a filtered probability space. The time index set T is assumed to be either $\mathbb{R} = (-\infty, \infty)$ or $\mathbb{Z} = \{0, \pm1, \pm2, \ldots\}$. The filtration $\{F_t\}$ is usually assumed to be increasing, i.e. for $s < t$

$$F_s \subset F_t.$$

Through the paper all terms like adapted, martingale, martingale differences, ... are referred to this fixed filtration $\{F_t\}$. Let $\{X_t\}$ be an adapted process such that $\{X_t\} \subset L^p$, $p \geq 1$, the Banach space of all p-integrable random variables with the usual norm $||\cdot||_p$. Let E_t be conditional expectation given F_t. For $x \in L^p$ the regression

---

* This research was completed during the author's stay at University of Konstanz under the Alexander von Humboldt fellowship.

** Permanent address: Institute of Mathematics, Hanoi,
P.O. Box 631, Bo-ho
Hanoi, Vietnam

$E_t x$ will be regarded as a prediction of x based on the full past information $\{F_s\}$ up to the time s = t. Denote by $[X_s]$ (or $[X_t]$) and $[X_s, s \le t]$ the closed subspaces of L^p generated by all random variables X_s, $s \in T$, and by the random variables X_s, $s \le t$, respectively.

We say that the process $\{X_s\}$ has a *linear prediction property* (shortly, lpp) or is a *lpp process*, if for every $t \in T$ E_t transforms $[X_s]$ onto $[X_s, s \le t]$.

Note that the above concept is similar to that of processes with the linear regression property which was introduced and investigated by Hardin [6].

Examples of lpp processes are the following:

(i) Gaussian, sub-Gaussian processes with mean zero;

(ii) processes whith the linear regression property (cf. Hardin [6]);

(iii) martingales;

(iv) strictly stationary processes in L^1 admitting a prediction in Urbanik's sense [15];

(v) stable processes with the right Wold decompositions and right innovations (cf. Cambanis, Hardin, Weron [1], Theorem 2.3, pp. 10).

The main aim of this paper is to study lpp strictly stationary (shortly, lppss) processes. Let $\{X_t\}$, $t \in T$, be such a process and let $\{S_t\}$ be its shift group defined by $S_t X_s = X_{t+s}$, $t, s \in T$. By stationarity of the process every S_t can be extended to a linear isometric operator in $[X_s]$ preserving probability distributions of random variables. Conversely, each such a group of operators in conjunction with a random variable x defines a strictly stationary process $X_t = S_t x$, $t \in T$ (cf. Doob [2], pp. 454).

Let $E_{-\infty}$ be the conditional expectation given $F_{-\infty} := \bigcap_{t=-\infty}^{\infty} F_t$. By a well known martingale convergence theorem we have $E_{-\infty}x = \lim_{t \to -\infty} E_t x$. Consequently, $E_{-\infty}$ maps $[X_s]$ into $[X_s]$. Moreover, on the space $[X_s]$ the following equations hold:

$$E_t = S_t E_0 S_{-t} \tag{1.1}$$

and

$$E_{-\infty} E_t = E_t E_{-\infty} = E_{-\infty} \qquad (1.2).$$

A lppss process $\{X_t\}$ is said to be *deterministic*, if $E_{-\infty}x = x$ for every $x \in [X_t]$ and *completely nondeterministic*, if $E_{-\infty}$ is identically zero on $[X_t]$.

Putting $X_t' = E_{-\infty}X_t$ and $X_t'' = (I - E_{-\infty}) X_t$ and taking into account (1.1) and (1.2) we infer that $\{X_t\}$ is the sum of two lppss processes $\{X_t'\}$ and $\{X_t''\}$, one deterministic and the other completely nondeterministic. Moreover, the spaces $[X_t']$ and $[X_t'']$ are (Jame) orthogonal in the sense that for any $x \in [X_t']$, $y \in [X_t'']$ and $\lambda \in \mathbb{R}$

$$|| x + \lambda y ||_p \geq || x ||_p \qquad (1.3).$$

For more information on the Jame orthogonality, see [8]. In the paper [1] Cambanis, Hardin and Weron have used the Jame orthogonality to investigate the Wold decompositions and innovations of stable processes and have obtained very interesting results. It should be noted that, if x and y are jointly stable random variables then (1.3) holds if and only if x and y form a martingale difference sequence.

The above decomposition of lppss processes into deterministic and completely nondeterministic components shows, in some sense, that the study of lppss processes can be reduced to that of completely nondeterministic processes. Hence in the sequel, unless otherwise stated, all lppss processes will be assumed to be completely nondeterministic.

The main result of this paper is to demonstrate and prove some moving average representations of lppss processes, both discrete time (Theorem 2.1) and continuous time (Theorem 4.1). We replace the orthogonality and stochastic independence relations in the Wiener-Kolmogorov and Urbanik prediction theory (cf. [9], [14], [15], [17], [18]) by a martingale relation which gives a new approach to prediction problems for strictly stationary processes with infinite variance. As a consequence we obtain a law of large numbers for lppss sequences (Corollary 2.1) which is closely related to Elton's law of large numbers for i.d. martingale differences [4]. The techniques developed by Urbanik in the standard papers [14] and [15] are widely exploited.

It should be noted that in the discrete time case our results are obtained in L^1. However, their continuous time counterpart can be proved only in L^p, $p > 1$.

2. lppss Sequences

Suppose $\{v_t\}$ is an adapted sequence ($t \in \mathbb{Z}$). Then it is called a martingale difference sequence (mds) or a sequence of martingale differences, if for every $t \in \mathbb{Z}$

$$E_t v_{t+1} = 0.$$

It is evident that every mds is a lpp sequence and a basic sequence in L^1. Moreover, every strictly stationary sequence of martingale differences is completely nondeterministic.

Theorem 2.1. Let $\{X_t\}$, $t \in \mathbb{Z}$, be a nonzero lppss sequence. Then there exists a strictly stationary sequence $\{v_t\}$ of martingale differences such that

$$[v_s,\ s \le 0] = [X_s,\ s \le 0] \tag{2.1}$$

and X_t is a moving average

$$X_t = \sum_{s=-\infty}^{0} \lambda_s v_{t+s} \tag{2.2},$$

the series converging in L^1 and almost surely, uniformly for $t \in \mathbb{Z}$.

Conversely, if $\{v_t\}$ is a strictly stationary sequence of martingale differences, then the moving average (2.2) is a (completely nondeterministic) lppss sequence provided the equation (2.1) holds.

Proof. Suppose $\{X_t\}$ is a lppss sequence. Putting

$$v_t = X_t - E_{t-1}X_t \tag{2.3}$$

and taking into account the formula (1.1) we get a strictly stationary sequence $\{v_t\}$ of martingale differences. Further, by the same procedure as in Urbanik ([14], Theorem 2) one can show that there is a sequence $\{\lambda_t\}$ of scalars such that

$$X_0 = \sum_{t=-\infty}^{0} \lambda_t v_t \tag{2.4}$$

which yields (2.2). Note that the series (2.2) is convergent in L^1

and almost surely. By stationarity of sequences $\{X_t\}$ and $\{v_t\}$ it follows easily that (2.2) is convergent uniformly in L^1 norm for $t \in \mathbb{Z}$.

To prove the uniform convergence with P.1 we denote

$$y_{t,n} = \sum_{s=n}^{0} \lambda_s v_{s+t} \qquad (t \in \mathbb{Z},\ n \le 0)$$

and observe that (2.4) is convergent with P.1 if and only if for every $\varepsilon > 0$

$$P\Big(\bigcup_{k=-\infty}^{0} |Y^0_{0,n+k} - X_0| > \varepsilon \Big) \xrightarrow[n \to -\infty]{} 0 \qquad (2.5)$$

(cf. Loève [10], pp.116). On the other hand, by strict stationarity of sequences $\{X_t\}$ and $\{v_t\}$, for every fixed n all events $\{ \bigcup_{k=-\infty}^{0} |Y_{t,n+k} - X_t| > \varepsilon \}$, $t \in \mathbb{Z}$, have the same probability which together with (2.5) implies the uniform convergence with P.1 of (2.2) for $t \in \mathbb{Z}$.

It remains to prove the converse statement of the theorem. But it can be done in the same way as in Urbanik ([14], Theorem 2). Thus the proof is complete.

As an application of the above theorem we get:

<u>Corollary 2.1.</u> (Law of large numbers. See also Doob [2], pp.465) If $\{X_s\}$ is a lppss sequence, then

(i) $$E\Big| \frac{1}{2n} \sum_{s=-n}^{n} X_s \Big| \xrightarrow[n \to \infty]{} 0$$

and

(ii) if $E|X_0| \log^+ |X_0| < \infty$ i.e. $X_0 \in$ LLog, then with P.1

$$\frac{1}{2n} \sum_{s=-n}^{n} X_s \xrightarrow[n \to \infty]{} 0.$$

Proof. (i) follows from the uniform convergence of (2.2) in L^1 (Theorem 2.1) and the law of large numbers in L^1 for i.d. mds $\{v_s\}$ (cf. Elton [4], Theorem 1). To prove (ii) we note that $X_0 \in$ LLog implies $E_{-1}X_0 \in$ LLog and therefore, by (2.3), $v_0 \in$ LLog. Now, by Theorem 2.1, (2.2) is convergent with P.1 uniformly for $t \in \mathbb{Z}$ and by the strong law of large numbers for i.d. mds $\{v_s\}$ (cf. Elton [4], Theorem 2) we get (ii).

<u>Corollary 2.2.</u> Let $\{X_t\}$ be a lppss (not necessarily completely

nondeterministic). Then there exists a norm $\|\!|\cdot|\!\|$ on $[X_s]$ invariant under shifts induced by $\{X_t\}$ and equivalent to the L^1-norm. Moreover, for every $x \in [X_t]$ and for every $t \in \mathbb{Z}$

$$\|\!| x - E_t x |\!\| = \inf \{ \|\!| x - y |\!\| : y \in [X_s, s \le t] \} .$$

Proof is the same as that in Urbanik ([14], Theorem 3) and is omitted.

Remark 2.1. If $\{X_t\}$ is a SαS (symmetric stable of order α) process one can prove that the moving average representation (2.2) is equivalent to the right Wold decomposition of $\{X_t\}$ in the sense of Cambanis, Hardin and Weron [1]. Therefore, in the case of stable and stationary processes, Theorem 2.1 and Theorem 2.3 in [1] are the same.

3. M-Stochastic Measures

Through this section we suppose that $p > 1$. Let $\mathfrak{B}$ denote the field of all bounded Borel subsets of $\mathbb{R}$ and $\mathfrak{B}_0$ the class of all bounded semiclosed intervals $(a,b]$.

A set function $M: \mathfrak{B} \longrightarrow L^p$ is said to be an *M-stochastic measure,* if

(i) for any sequence $\{A_n\}$ of disjoint sets in $\mathfrak{B}$ with union in $\mathfrak{B}$

$$M\left(\bigcup_{n=1}^{\infty} A_n\right) = \sum_{n=1}^{\infty} M(A_n),$$

the series converging in L^p;

(ii) for every $a \in \mathbb{R}$ the process $M((a,t])$, $t \ge a$, is a martingale.

Similarly, a function $N: \mathfrak{B}_0 \longrightarrow L^p$ is said to be an *M-interval function,* if

(i') for any $a < b < c$

$$N((a,c]) = N((a,b]) + N((b,c]);$$

(ii') for every $a \in \mathbb{R}$ the process $M((a,t])$, $t \ge a$, is a martingale;

(iii') for any $a < b$

$$\lim_{c \to b+} N((a,c]) = N((a,b]).$$

The terms *M-stochastic measure* and *M-interval function* are related with properties (ii) and (ii'). It should be noted that the restriction of an M-stochastic measure to $\mathfrak{B}_0$ is an M-interval function. We shall prove that every M-interval function can be extended to an M-stochastic measure.

Lemma 3.1. Let N be an M-interval function and $I_1, I_2, \ldots$ be disjoint intervals in $\mathfrak{B}_0$. Then for any sequence $\varepsilon_1, \varepsilon_2, \ldots$ of numbers 0 and 1 the following inequality holds:

$$\| \sum_{i=1}^{n} \varepsilon_i N(I_i) \|_p \le c_p \| \sum_{i=1}^{n} N(I_i) \|_p \qquad (3.1),$$

the constant c_p depending only on P. Consequently, if the union $\bigcup_{i=1}^{\infty} I_i$ is contained in an interval $I \in \mathfrak{B}_0$, then the series $\sum_{i=1}^{\infty} N(I_i)$ is convergent in L^P.

Proof. Given n one may assume, making a rearrangement if necessary, that $N(I_1), N(I_2), \ldots, N(I_n)$ is a mds. By virtue of the Burkhölder inequality for martingale transforms (cf. for example Meyer [11], Theorem 48, p.51) we get (3.1). Further, by (3.1) and by the Burkhölder inequalities for quadratic variations (cf. Meyer [11], Theorem 60, p.62), it follows that the series $\sum_{i=1}^{\infty} N(I_i)$ is convergent in L^P whenever $\bigcup_{i=1}^{\infty} I_i \subset I$, $I \in \mathfrak{B}_0$.

Theorem 3.2. Every M-interval function $N: \mathfrak{B}_0 \longrightarrow L^P$ has a unique extension to an M-stochastic measure on $\mathfrak{B}$.

Hint of the proof. Let $\mathfrak{B}_1$ denote the ring of all finite unions of intervals from $\mathfrak{B}_0$. Let [N] be the Banach space generated by random variables $N(I)$, $I \in \mathfrak{B}_0$. For a continuous functional y in [N] the real valued set function N defined by $N_y(.) = yN(.)$ is of finite variation on every bounded interval $I \in \mathfrak{B}_0$ (by (3.1)) which together with (iii') implies the countable additivity of N_y on $\mathfrak{B}_1$ (cf. Dunford-Schwartz [3], Theorem 14, pp.138, Lemma 16, pp. 140). Consequently N_y can be extended to a countably additive set func- tion on $\mathfrak{B}$. Therefore, the M-interval function N can be extended to an M-stochastic measure on $\mathfrak{B}$. Moreover, the extension is unique.

Given an increasing family of σ-fields $\{F_t\}$ let us define a projector valued function on $\mathfrak{B}_0$ by

$$A(I) = E_b - E_a \quad (3.2),$$

where $I = (a,b]$. Then,

$$\| A(I) \| \le 2 \quad (3.3),$$

$\|.\|$ being the operator norm in L^p.

Let $\mathcal{A}$ denote the set of all random variables x in L^p such that the function N_x defined by

$$N_x = A(.)x \quad (3.4)$$

is an M-interval function.

Lemma 3.2. $\mathcal{A}$ is a closed subspace of L^p.

Proof. It is easy to check that $\mathcal{A}$ is a linear space. Let $\{x_n\} \subset \mathcal{A}$ and $x_n \longrightarrow x$ in L^p. Obviously the function N_x defined by (3.3) satisfies (i') and (ii').

Suppose $a < b < c$. Then we have

$$N_x((a,c]) - N_x((a,b]) = N_x((b,c])$$

and by (3.3), for every n

$$\| N_x((b,c]) \|_p \le \| A((b,c])(x - x_n) \|_p + \| N_{x_n}((b,c]) \|_p$$

$$\le 2 \| x - x_n \|_p + \| N_{x_n}((b,c]) \|_p.$$

Therefore,

$$\lim_{c \longrightarrow b+} \| N_x((b,c]) \|_p \le 2 \| x - x_n \|_p.$$

Letting $n \longrightarrow \infty$ the last inequality yields

$$\lim_{c \longrightarrow b+} N_x((b,c]) = 0$$

or, equivalently,

$$\lim_{c \longrightarrow b+} N_x((a,c]) = N_x((a,b])$$

which shows that the condition (iii') is satisfied and hence $x \in \mathcal{A}$. Thus $\mathcal{A}$ is a closed subspace of L^p.

Lemma 3.3. Let M be an M-stochastic measure and let [M] denote the Banach space generated by all random variables M(E), $E \in \mathcal{B}$. Then the following inclusion holds:

$$[M] \subset \mathcal{A}$$

Proof. It follows from the closeness of $\mathcal{A}$ and from the fact that every M-stochastic measure is an M-interval function.

By Theorem 3.2 for every $x \in \mathcal{A}$ the M-interval function N_x defined by (3.4) can be extended to an M-stochastic measure. Therefore, the projector valued interval function A defined by (3.2) can be extended to a strongly countably additive operator valued measure. Let us denote the extension of A by the same symbol A.

Lemma 3.4. Let M be an M-stochastic measure. Then for any $G, H \in \mathcal{B}$

$$A(G)\ M(H) = M(G \cap H) \tag{3.5}.$$

Proof. Let G,H be intervals in $\mathcal{B}_0$. Then by (3.2) and by definition of M-stochastic measures the formula (3.5) holds. Hence it holds true for all $G, H \in \mathcal{B}_1$. By the parallel extension of measures A and M we infer that (3.5) holds for all $G, H \in \mathcal{B}$.

Lemma 3.5. Let I be a finite interval in $\mathcal{B}_0$ and let $\mathcal{B}(I)$ be the σ-field of all Borel subsets of I. Then

$$\varkappa(I) := \sup_{G \in \mathcal{B}(I)} \| A(G) \|_{\mathcal{A}} < \infty \tag{3.6},$$

the operator norm $\|.\|_{\mathcal{A}}$ being defined on $\mathcal{A}$.

Proof. It follows from the σ-additivity of A and from Corollary 2 in Dunford-Schwartz ([3], pp.319).

Remark 3.1. The constant $\varkappa(I)$ depends only on I, p and $\{F_t\}$.

Now combining Lemmas 3.4 and 3.5 we get the following:

Theorem 3.3. Let I be a finite interval and M an M-stochastic measure. Then for any G, $H \in \mathcal{B}(I)$ with $G \subset H$

$$\| M(G) \|_p \le \varkappa(I)\ \| M(H) \|_p \tag{3.7},$$

the constant $\varkappa(I)$ being given by (3.6).

Corollary 3.1. For every M-stochastic measure M there corresponds a nonnegative Borel measure μ on $\mathcal{B}$ such that for $G \in \mathcal{B}(I)$

$$\mu(G) \le \varkappa(I)\ \| M(G) \|_p \tag{3.8}$$

and

$$\| M(G) \|_p \longrightarrow 0 \quad \text{whenever} \quad \mu(G) \longrightarrow 0 \tag{3.9}$$

Proof. It follows from Theorem 3.3 and general results concerning vector valued measures (cf. Dunford-Schwartz [3], Lemma 4, pp.320, Lemma 5, pp.321).

Let M be an M-stochastic measure and μ its associated nonnegative measure (Corollary 3.1). Considering M as a vector valued measure one can define an integral of a scalar function f w.r.t. M in the same way as in Dunford-Schwartz ([3], p.323). Let us denote such an integral by $\int_{-\infty}^{\infty} f\, dM$. It should be noted that this integral has such a special property that the process $\int_{-\infty}^{t} f\, dM$, $t \in \mathbb{R}$, is a martingale.

By virtue of Corollary 3.1 and by the same method of Urbanik ([14], Theorem 2.2, p.241) one can prove the following:

Theorem 3.4. For every $x \in [M]$ there exists a unique M-integrable function f such that

$$x = \int_{-\infty}^{\infty} f\, dM$$

4. lppss process: Continuous time case.

Through this section we suppose that $p > 1$ and $\{X_t\}$ is a continuous lppss process, i.e.

$$\| X_t - X_s \|_p \xrightarrow[s \to t]{} 0 \tag{4.1}$$

Lemma 4.1. Let $\{X_t\}$, $t \in \mathbb{R}$, be a (completely nondeterministic) lppss process. Then,

$$[X_t] \subset \mathcal{A} \tag{4.2}$$

Proof. First, note that (4.1) implies the strong continuity of $\{S_t\}$. On the other hand, for any $a,b,t \in \mathbb{R}$ we have, by (1.1), the equation

$$\begin{aligned}(E_b - E_a)X_t &= S_b E_0 S_{-b} X_t - S_a E_0 S_{-a} X_t \\ &= S_a (S_{b-a} E_0 X_{t-b} - E_0 X_{t-a}) \\ &= S_a [(S_{b-a} E_0 X_{t-b} - S_{b-a} E_0 X_{t-a}) \\ &\quad + (S_{b-a} E_0 X_{t-a} - E_0 X_{t-a})].\end{aligned}$$

Hence it follows that

$$\| (E_b - E_a)X_t \|_p \le \| X_{t-b} - X_{t-a} \|_p + \| S_{b-a}E_0X_{t-a} - E_0X_{t-a} \|_p$$

$$\le \| X_{-b}-X_{-a} \|_p + \| S_{b-a}E_0X_{t-a}-E_0X_{t-a} \|_p .$$

From the last inequality and by the continuity of $\{X_t\}$ and $\{S_t\}$ we infer that

$$\lim_{b\to a} \| (E_b - E_a)X_t \| = 0,$$

which implies that for every $x \in [X_t]$

$$\lim_{b\to a} \| (E_b - E_a)x \| = 0 \qquad (4.3).$$

Finally, if we define an interval function N_x, $x \in [X_t]$, as in formulas (3.2) and (3.4) then, by (4.3) and by an easy argument, N_x is an M-interval function which implies that $x \in \mathcal{A}$ and consequently (4.2) holds.

Suppose $\{S_t\}$ is the shift group associated with a lppss process $\{X_t\}$. An M-stochastic measure M is said to be *$\{S_t\}$- homogenuous*, if for each set $G \in \mathcal{R}$ we have the equation

$$S_tM(G) = M(G + t) \quad , t \in \mathcal{R}.$$

Here $G + t$ denotes the set $\{u + t: u \in G\}$.

It should be noted that for every nonzero $\{S_t\}$-homogenuous M-stochastic measure M with the associated nonnegative measure μ (Corollary 3.1) the class of M-null sets coincides with the class of Lebesge measure zero sets.

The following theorem stands for a slight generalization of Theorem 4.2 by Urbanik ([15], pp. 251):

Theorem 4.1. Let $\{X_t\}$ be a nonzero lppss process. Then there exists an $[X_t]$-valued nontrivial $\{S_t\}$-homogenuous M-stochastic measure M and an M-integrable function f such that

$$[M(I): I \in \mathbb{R}_0 , I \subset (-\infty,0]] = [X_t , t \le 0] \qquad (4.4)$$

and

$$X_t = \int_{-\infty}^{t} f(u - t) M(du) , \qquad t \in \mathbb{R} \qquad (4.5).$$

Conversely, if M is a nontrivial $\{S_t\}$-homogenuous stochastic measure and f is M-integrable, then the process $\{X_t\}$ defined by

(4.5) is a lppss process provided (4.4) holds.

Proof. All that we need in the proof is Theorem 3.4, Lemma 4.1 and Urbanik's method in the proof of Theorem 4.2 in ([1], pp. 251).

Acknowledgements: I wish to thank Professor Dr. M. Kohlmann and PD Dr. N. Christopeit for inviting me to present some of these results at the Bad Honnef Conference. Warm thanks are also due to Mr. S. Dudey, Mrs. A. Krake and Mr. J. Monser for very carefully typing the manuscript.

References.

1. Cambanis, S., Hardin, C.D., Weron, A.: Innovations and Wold decompositions of stable sequences. Prob. Th. Rel. Fields 79, 1 -27 (1988).
2. Doob, J.L.: Stochastic process. New York - London (1967).
3. Dunford, N., Schwartz, J.T.: Linear operators. Part I: General theory. New York - London - Sydney (1957).
4. Elton, J.: Law of large numbers for identically distributed martingale differences. Ann. Prob. 9, No. 3, 405 - 412 (1981).
5. Hardin, C.D.: On the linearity of regression. Z. Wahrscheinlichkeitstheor. Verw. Geb. 61, 293 - 302 (1982).
6. Hardin, C.D.: On the spectral representation of symmetric stable processes. J. Multivariate Anal. 12, 385 - 401 (1982).
7. Hosoya, Y.: Harmonizable stable processes. Z. Wahrscheinlichkeitstheor. Verw. Geb. 60, 517 - 533 (1982).
8. James, R.C.: Orthogonality and linear functionals in a normed linear space. Trans. Am. Math. Soc. 61, 265 - 292 (1947).
9. Kolmogorov, A.N.: Stationary sequences in Hilbert space. Bull. Math. Uni. Moscow 2, 1 - 40 (1941).
10. Loève, M.: Probability theory I. 4th Edition. New York - Heidelberg - Berlin (1977).
11. Meyer, P.A.: Martingales and Stochastic integrals I. Lecture Notes in Math. 284 (1972).
12. Thu, N.V.: Prediction problems. Dissertationes Math. 163, 1 - 69 (1980).
13. Thu,N.V.: Prediction of Banach space valued strictly stationary sequences. Theor. Prob. Appl. 29, 334 - 344 (1984).
14. Urbanik, K.: Prediction of strictly stationary sequences. Colloq. Math. 12, 115 - 129 (1964).
15. Urbanik, K.: Some prediction problems for strictly stationary processes. In: Proc. Fifth Berkeley Symp. Math. Stat. Prob. vol. 2, Part I, pp. 235 - 258. University of California Press (1967).
16. Weron, A.: Harmonizable stable processes on groups: Spectral, ergodic and interpolation properties. Z. Wahrscheinlichkeitstheor. Verw. Geb. 68, 473 - 491 (1985).
17. Wiener, N., Masani, P.R.: The prediction theory of multivariate stochastic processes. Part I: Acta Math. 98, 111 -150; Part II: Acta Math. 99, 93-137 (1958).
18. Wold, H.: A study in the analysis of stationary time series. Stockholm: Almquist & Wicksell (1954).

Multiparameter Martingale and Markov Process

Eugene Wong
University of California, Berkeley

1. Introduction

One of the motivations in studying multiparameter martingales and Markov processes is to extend the results of filtering and detection problems in one dimension to the multidimensional (especially 2-dimensional) case. We begin with a brief review of these problems as follows:

Let $\{ \xi_t, \ t_0 \leq t \leq t_1 \}$ be a stochastic process representing the *observation* and assume that it has the form

$$\xi_t = S_t + N_t \tag{1}$$

where $\{ S_t, N_t \}$ are stochastic processes with known distributions that represent *signal* and *noise* respectively. The (causal) filtering problem deals with the evaluation of

$$\hat{S}_t = E(S_t \,|\, \xi_s, \ 0 \leq s \leq t) \tag{2}$$

while the detection problem concerns testing (1) as a hypothesis against the alternative that ξ_t consists of noise alone. In most cases, a likelihood ratio test would be used, and this involves the evaluation of the likelihood ratio

$$L_t = E_0\left(\frac{dP}{dP_0} \,\middle|\, \xi_s, \quad 0 \leq s \leq t \right) \tag{3}$$

where $\mathbf{P}$ is the probability measure associated with (1), and $\mathbf{P}_0$ is the probability measure under the hypothesis that ξ_t consists of noise only.

If we assume that the noise is a Gaussian white noise process, then the two problems: filtering and detection, are closely related, and one can show that the likelihood ratio can be expressed as

$$L_t = \exp\left\{ \int_0^t \hat{S}_\tau \, dX_\tau - \frac{1}{2} \int_0^t \hat{S}_\tau^2 \, d\tau \right\} \tag{4}$$

where $X_t = \int_0^t \xi_\tau \, d\tau$ and we have assumed that the white noise is normalized to have a unit spectral density. The first integral in (4) is an Ito integral. This formula, due to Duncan [DUN68,70] and Kailath [KAI69] is one of the highlights in applying martingale theory to problems in communication and control.

If a Markovian model is assumed for the signal S_t, then the problem of evaluating $\hat{S}_t$ can be formulated as a stochastic partial differential equation, viz., the Zakai equation [ZAK69]. If S_t is also Gaussian, then the problem can be reduced drastically, to a linear stochastic differential equation that is the Kalman filtering equation for continuous time.

Processing signals that depend on several parameters (space or space-time) is of considerable practical interest. Image processing, for example, involves signals and noise with a two dimensional parameter. In this particular case, the additive-noise model given by (1) is again appropriate if we take the observation to be the logarithm of the image intensity.

The research was supported by U.S. Army Research Office Grant DAAG29-85-K-0223

Both analytically and computationally, the problem of processing multiparameter signals is more difficult than its one-dimensional counterpart. Thus, there is strong motivation to seek the kind of simplification made possible by martingale theory and Markovian models in one dimension. Over the last two decades, there has been a significant effort in developing a theory of multiparameter martingales and Markov processes. Although this effort has met with incomplete success, the results that have been found are most interesting and suggest that much more is yet to come. The objective of this paper is to review some of these results, and to indicate some directions where future efforts might be usefully deployed.

2. Martingales

Let $\mathbb{R}_t^2$ denote the positive quadrant of the plane with a partial ordering defined by

$$t > s \iff t_i \geq s \quad i = 1, 2 \tag{5}$$

Let $(\Omega, \mathbf{F}, \mathbf{P})$ denote a probability space with a filtration $\{\mathbf{F}_t, \ t \in \mathbb{R}_t^2\}$ that satisfies

$$t > s \iff \mathbf{F}_t \supset \mathbf{F}_s \tag{6}$$

We assume the $\mathbf{F}_4$-condition of Cairoli and Walsh [CAI 75], namely,

$$\mathbf{F}_{t_1, \infty} \text{ and } \mathbf{F}_{\infty, t_2} \text{ are independent given } \mathbf{F}_{t_1, t_2}$$

We define a Wiener process on the plane as a zero-mean Gaussian process $\{ W_t, \ t \in \mathbb{R}_+^2 \}$ with

$$E W_t W_s = \min(t_1, s_1) \cdot \min(t_2, s_2) \tag{7}$$

The properties of W_t are most easily seen by viewing W as

$$W_t = \int_{A_t} \xi(s)\, ds \tag{8}$$

where $\xi(s)$ is a Gaussian white noise and A_t is the rectangle in $\mathbb{R}_t^2$ bound by the origin and the point $t = (t_1, t_2)$.

Wong and Zakai [WON74] introduced stochastic integrals of two different types with respect to the Wiener process and showed that processes defined by these integrals:

$$M_t^{(1)} = \int_{A_t} \psi_s \, W(ds) \tag{9}$$

$$M_t^{(2)} = \int_{A_t \times A_t} \psi_{s,s'} \, W(ds)\, W(ds') \tag{10}$$

were martingales. These integrals were found to form a basis for representing martingales generated by a Wiener process. Specifically, if $\mathbf{F}_t$ is the σ-field generated by $\{ W_s, \, s < t\}$ then every square-integrable $\mathbf{F}_t$-martingale has the representation

$$M_t = M_0 + M_t^{(1)} + M_t^{(2)} \tag{11}$$

where $M^{(1)}$ and $M_{(2)}$ are given by (9) and (10) respectively.

The papers [WON74] and [CAI75] provided a framework within which a considerable body of results on multi-parameter martingales has been developed. These include a formula of exponential type for the likelihood ratio [WON77] and a set of linear recursive filtering equations for signals that are Gaussian and Markovian with respect to rectangular boundaries [OGI81].

The definition of martingale and the ensuing results depend on the choice of partial ordering given by (15). In this sense multiparameter martingale theory is not "geometric." The partial ordering should be viewed as an artifact introduced to facilitate computation. In this respect, the situation is not significantly different from Ito's definition of stochastic integral in one dimension. It too is dependent on the choice of a "forward" direction and as such is not "geometric."

In one dimension the Ito differentiation formula provides much of the manipulative power of stochastic calculus, and its simplest version can be stated as follows. Let X_t be an *Ito process*, i.e., a process of the form

$$X_t = X_0 + \int_0^t \theta_s ds + \int_0^t \psi_s dW_s \tag{12}$$

where the last integral is a stochastic integral with respect to the one-parameter Wiener process W. Let $f: \mathbb{R} \to \mathbb{R}$ be a twice continuously-differentiable function. Then, $Y_t = f(X_t)$ is given by

$$Y_t = f(X_0) + \int_0^t f'(X_s)\, dX_s + \frac{1}{2}\int_0^t f''(X_s)\, d\langle X\rangle_s \tag{13}$$

where

$$dX_t = \theta_t dt + \psi_t dW_t \tag{14}$$

and

$$d\langle X\rangle_t = \psi_t^2 dt \tag{15}$$

It is worth noting that both dX_t and $d\langle X\rangle_t$ can be computed from X directly, and are not dependent on its representation (12).

The differentiation formula (13) has several interpretations. First, it is a statement of the *closure* of Ito processes under C^2 transformations. As such, the most appropriate form of it is:

$$f(X_t) = f(X_0) + \int_0^t \left[f'(X_s)\theta_s + \frac{1}{2} f''(X_s)\psi_s^2 \right] ds + \int_0^t f'(X_s)\psi_s dW_s \tag{16}$$

Additionally, it is also a statement of how differentiation of functions of an Ito process must be modified from ordinary calculus. The most appropriate expression of this statement is:

$$df(X_t) = f'(X_t)\, dX_t + \frac{1}{2} f''(X_t)\, d\langle X\rangle_t \tag{17}$$

where the fact that the differential is intrinsic, i.e., independent of the representation (12), is of considerable importance.

In higher dimensional parameter spaces, these statements are no longer the same. The counterpart to the "closure" statement is the following: [WON76] Let *weak semi-martingale* be a process of the form

$$X_t = X_0 + \int_{A_t} \phi_s W(ds) + \int_{A_t^2} \psi_{s,s'} W(ds)\, W(ds') + \int_{A_t} \theta_s ds + \int_{A_t^2} \alpha_{s,s'} W(ds)\, ds' + \int_{A_t^2} \beta_{s,s'} ds\, W(ds') \tag{18}$$

Let $f: \mathbb{R} \to \mathbb{R}$ be four times continuously differentiable. Then $f(X_t)$ is again of the form (18). The explicit expression for $f(X_t)$ that is the counterpart to (16) is complicated, and the complexity increases with dimensionality as one attempts to generalize to processes with higher dimensional parameters. A major reason for this is that the stochastic integrals (9) and (10) are integrals over "volumes" in $\mathbb{R}^n$, and are not the inverse to differentials in dimensions higher than one.

Unlike (16), the differential form (17) of Ito's formula should admit a simple generalization that in its appearance is both coordinate and dimension independent. To do this, we need to develop an exterior calculus, but one in which the martingale property is reflected. This is one of the motivations in the development of stochastic differential forms to be discussed in section 4.

3. Markov Processes

Lévy [LEV56] defined a multiparameter Markov process as follows: Let ∂D be a simply connected (n −1) surface dividing $\mathbb{R}^n$ into a bounded part D_- and an unbounded part D_+. A process $\{X_t,\ t \in \mathbb{R}^n\}$ is said to be Markov if X_t, $t \in D_+$ and $X_{t'}$, $t' \in D_-$ are conditionally independent given $\{X_s, s \in \partial D\}$. For $n > 1$, this is a rather restrictive condition, and it was shown in [WON68, 69] that no Gaussian, Markov, homogeneous and isotropic process could be Markov, unless the definition was relaxed to allow X to be a generalized process. In that case, a Gaussian generalized process with a covariance bilinear form given by

$$E\, X(\psi)\, \overline{X(\sigma)} = \int_{\mathbb{R}^n} \frac{\hat{\psi}(\nu)\, \bar{\hat{\sigma}}(\nu)}{1 + |\nu|^2}\, d\nu \tag{19}$$

is Markov. In (19), $\hat{\psi}$ and $\hat{\sigma}$ denote Fourier transforms. Indeed, except for scaling differences, this is the only example of isotropic-homogeneous Gauss-Markov process, and is widely known as the *free Euclidean field* [NEL73].

Generalized processes are usually defined as random functions parameterized by testing functions. As such, to define surface data: $\{X_s, S \in \partial D\}$ is difficult, though possible. A natural alternative is to introduce processes parameterized by k-dimensional sets, with $k < n$, in $\mathbb{R}^n$, and study Markovian properties for such processes. This is another motivation for introducing stochastic differential forms.

4. Stochastic Differential Forms [WON 87]

Intuitively, we want to define in a consistent way processes parameterized by k-dimensional sets in $\mathbb{R}^n$. We begin by considering *oriented k-rectangles* in $\mathbb{R}^n$ defined as follows: Let a_i denote an interval (left open, right closed) on the t_i axis of $\mathbb{R}^n$. Let $\sigma = a_{i_1} \wedge a_{i_2} \wedge \ldots \wedge a_{i_k}$ denote a rectangle with sides $a_{i_1}, a_{i_2}, \ldots, a_{i_k}$. The orientation of σ is positive if $\mathbf{i} = (i_1, i_2, \ldots, i_k)$ can be put into increasing order by an *even* permutation, and negative otherwise. We shall call $\mathbf{i}$ the direction of σ. A rectangular *k-chain* A is an algebraic sum

$$A = \sum_{\nu=1}^{m} \alpha_\nu \sigma_\nu \tag{20}$$

where $\alpha_\nu = \pm 1$ and σ_ν are oriented k-rectangles. We note that the boundary $\partial A = \sum \alpha_\nu\, \partial\sigma_\nu$ is a (k −1)-chain. A *random k-cochain* X is a random function defined on all *k*-chains such that

$$\begin{aligned} X(-A) &= -X(A) \\ X(A + B) &= X(A) + X(B) \end{aligned} \tag{21}$$

We note that a k-cochain is determined by its values on k-rectangles.

Chains can be used to approximate k-dimensional sets in $\mathbb{R}^n$ by introducing the *flat norm* $|A|^{\sim}$ as follows: Let $|\sigma|$ denote the k-dimensional volume of a k-rectangle σ. Let $|A|$ be defined by

$$\left|\sum \alpha_\nu \sigma_\nu\right| = \sum |\alpha_\nu|\, |\sigma_\nu| \tag{22}$$

and define

$$|A|^{\sim} = \inf \{\, |A - \partial B| + |B| \,\} \tag{23}$$

where the infimum is taken over all (k + 1)-chains B. It can be shown that $|\ |^{\sim}$ is a norm with

$$|\partial A|^{\sim} \leq |A|^{\sim} \leq |A| \tag{24}$$

Using the flat norm, we can approximate continuous paths by 1-chains and smooth surfaces by 2-chains [WHI57].

We can now define *stochastic differential k-forms* as random co-chains X that are continuous in probability with respect to the flat norm, and thus can be extended to all limits of k-chains. If ∂D is an (n −1) surface and X is a k-form, with $k \leq n-1$, the surface data of X on ∂D are easily defined. An interesting problem is the study of Markovian k-forms.

We note that a natural example of stochastic differential forms is the white noise process on $\mathbb{R}^n$, which can be defined as a zero-mean n-form η in $\mathbb{R}^n$ such that for n-rectangles σ and σ'

$$E\,\eta(\sigma)\,\eta(\sigma') = \pm \text{ volume }(\sigma \cap \sigma')$$

where the sign is + if σ and σ' are similarly oriented and −otherwise. If η is Gaussian and A_t is the rectangle bounded by the origin and the point t in $\mathbb{R}^n_+$, then

$$W_t = \eta(A_t)$$

is a Wiener process, the two-dimensional version of which was used in section 2.

Defining stochastic differential forms as above leads to a simple definition for *exterior derivative.* For any k-form $X(k \leq n-1)$, we define dX as a (k + 1)-form such that

$$(dX)(\sigma) = X(\partial\sigma) \tag{25}$$

for all oriented rectangles σ. In short, we use Stokes theorem to define the exterior derivative. Because of (24), continuity of dX with respect to the flat norm is assured and dX is guaranteed to be a stochastic differential form. Thus, differential forms are closed under exterior differentiation.

Square integrable stochastic differential forms are always random currents in the sense of Ito [ITO 56], but not conversely. Thus, a stochastic differential form admits any operation that is defined on random currents. However, such an operation may produce only a current, not differential form. The Hodge star operator on a form, for example, will in general produce only a current. Take the case of a white noise n-form η. If $*\eta$ were a 0-form, it must be an ordinary function such that

$$\eta(\sigma) = \int_\sigma (*\eta)\,dt \tag{26}$$

No such function exists.

The notion of differential form can be combined with that of martingale in a fruitful way. For ease of exposition, we restrict our discussion to the case of $\mathbb{R}^2$. The general case is treated in [WON87]. Consider a stochastic differential 1-form X. It can be evaluated on 1-rectangles in two directions, and these take the form of line segments:

$$\sigma_1 = (t_1, t_2) \rightarrow (t_1 + a, t_2) \quad \text{and} \quad \sigma_2 = (t_1, t_2) \rightarrow (t_1, t_2 + a)$$

respectively. We say X is a *1-martingale* if for all σ_1

$$E\left[X(\sigma_1)\,\middle|\,F_{t_1\infty}\right] = 0 \tag{27a}$$

a *2-martingale* if for all σ_2

$$E\left[X(\sigma_2)\,\middle|\,F_{\infty, t_2}\right] = 0 \tag{27b}$$

and a *martingale* if it is both a 1-martingale and a 2-martingale.

A 2-form M in $\mathbb{R}^2$ can be evaluated on a general rectangle: $\sigma = [t_1, t_1 + a] \times [t_2, t_2 + b]$. We say M is an i-martingale (i = 1, 2) if for all σ

$$E\left[M(\sigma)\,\middle|\,F^i_t\right] = 0 \tag{28}$$

where $F^i_t = F_{t_1,\infty}$ or F_{∞, t_2} for i = 1 or 2 respectively. Again, we say M is a martingale if it is both a 1-martingale and a 2-martingale.

A Gaussian white noise η is clearly a martingale 2-form. Let W be a Wiener process as defined in (7). Then its exterior derivative dW is a martingale 1-form. The definition for martingales given in section 2 is equivalent to one of defining a martingale 0-form as one whose exterior derivative is a martingale 1-form. This is consistent with the one-dimensional case where the martingale property is most naturally associated with the *increments* of a process rather than with the process itself.

One of the most interesting ways in which martingale differential forms can be used is in a bilinear operation that is at once a generalization of stochastic integral and a generalization of the *exterior product* for ordinary differential forms. Let X and Y be martingale k and r forms respectively with respect to a fixed filtration $\{\mathbf{F}_t,\ t \in \mathbb{R}^n\}$. Then we can define their exterior product $X \wedge Y$ as a martingale (k + r)-form. For example, suppose that m and M are 0-forms defined by two type-1 stochastic integrals:

$$m_t = \int_{A_t} \theta_s\, W(ds)$$

$$M_t = \int_{A_t} \psi_s\, W(ds)$$

Then, dm and dM are both martingale 1-forms, and $dm \wedge dM$ is a martingale 2-form that is related to a type-2 integral as follows:

$$\left(dm \wedge dM\right)(A_t) = \int_{A_t \times A_t} \left(\theta_s\, \psi_{s'} - \psi_s\, \theta_{s'}\right) W(ds)\, W(ds')$$

A natural and interesting question is whether the representation theorem for Wiener-martingales cited in section 2 can be re-expressed as representation theorems for martingale differential forms.

5. Markovian Random Currents

The notion of a random current was introduced in [ITO 56]. Let S^r denote the space of (ordinary) differential r-forms in $\mathbb{R}^n$ with coefficients in the Schwartz space of functions of rapid descent. Then, a *random k-current* X is defined as a continuous linear map of S^{n-k} into $L^2(\{\Omega, \mathbf{F}, \mathbf{P}\})$. Roughly speaking, a random k-current is a differential k-form with coefficients that are generalized n-parameter processes.

The space of random currents (of all order) is closed under both the exterior derivative

$$(dX)(\phi) = (-1)^k X(d\phi)$$

and the Hodge star operator

$$(*X)(\phi) = (-1)^{k(n-k)} X(*\phi)$$

If $\psi \in S^r$ and X is a random k-current, then the *exterior product* $X \wedge \psi$ can be defined as a (k + r)-current by

$$(X \wedge \psi)(\phi) = X(\psi \wedge \phi)$$

All stochastic k-forms, as defined in section 4, are random k-currents. Conversely, if $\psi \in S^{n-k}$ and X is a k-current, then $X \wedge \psi$ is a stochastic n-form. However, in general, for $r < n-k$, $X \wedge \psi$ is not a (k + r)-form.

We say a k-current is *localizable* if for all $\psi \in S^{n-k-1}$ and $\phi \in S^{k-1}$ both $X \wedge \psi$ and $*X \wedge \phi$ are (n −1)-forms. Given a (n −1)-surface ∂D, we take

$$\left\{(X \wedge \psi)(\partial D)\ ,\ (*X \wedge \phi)(\partial D)\ ;\ \psi \in S^{n-k-1},\ \phi \in S^{k-1}\right\}$$

to be the *surface data* of X on ∂D, and denote it by the abbreviated notation $X(\partial D)$. A localizable k-current X ($k \le n-1$) is said to be *Markov* if given an (n −1) surface ∂D that separates $\mathbb{R}^n$ into a bounded part D^- and an unbounded part D^+, $X(\sigma)$, $\sigma \subset D^+$ and $X(\sigma')$, $\sigma' \subset D^-$ are independent given $X(\partial D)$. This identifies an appropriate class of random functions for which the (simple) Markov property can be studied. The free Euclidean field, for example, is a Markov 0-current under our definition.

One of the simplest, yet interesting, classes of random currents is the class of isotropic and homogeneous currents. Let G denote the full group of isometries on $\mathbb{R}^n$ that preserve the Euclidean distance. A motion $g \in G$ induces a transformation τ_g on $\phi \in S^r$ which in turn induces a transformation T_g on k-forms X:

$$(T_g X)(\phi) = X(\tau_g^{-1} \psi)$$

A random k-current is said to be *isotropic and homogeneous* if for all $g \in G$

$$E\, X(\phi)\, \overline{X}(\psi) = E\left[(T_g X)(\phi)\,(T_g \overline{X})(\phi)\right]$$

It was shown in [ITO 56] that every isotropic and homogeneous k-current $(1 \leq k \leq n-1)$ is characterized by two Borel measures F_i and F_s on $[0, \infty)$. For $k = 0$ or n, only one such measure suffices. A natural question is: for a Gaussian isotropic and homogeneous k-current X, what must F_i and F_s be in order for X to be Markov?

Thus far, this question has only been partially answered: What we have been able to show is the following:

(1) For X to be Markov, F_i and F_s must be of the form:

$$F_i(d\lambda) = A\, \frac{\lambda^{n-1}\, d\lambda}{\alpha^2 + \lambda^2}$$

$$F_s(d\lambda) = B\, \frac{\lambda^{n-1}\, d\lambda}{\beta^2 + \lambda^2}$$

(2) If $\alpha^2 = \beta^2$, then A and B must be equal in which case X is indeed Markov.

A major open problem is to find examples of non-Gaussian isotropic and homogeneous Markov processes in $\mathbb{R}^n$. We think the introduction of stochastic differential forms may well aid this effort in several ways. First, it may allow "instantaneous" nonlinear operation to be defined on localizable currents. For example, "exterior product" would be such an operation. However, the martingale exterior product considered in section 4 is coordinate dependent and is unsuitable for the construction of Markovian currents.

Another approach is to use the exterior calculus available for currents to relate Markov processes to some basic elemental process such as "white noise." This may lead to a way of generating Markovian currents using stochastic differential equations of some kind.

6. Conclusions

In this paper we focus on the concepts of martingales and Markov processes as generalized to processes with a multidimensional parameter, and briefly review some of the known results on these two topics. We then introduce the recently developed notion of stochastic differential form and indicate how it can be related to martingales and Markov processes with a multidimensional parameter.

References

[CAI75] Cairoli, R. and J.B. Walsh, "Stochastic integrals in the plane," Acta. Math., *134*, (1975), pp. 111-183.

[DUN68] Duncan, T.E., "Evaluation of likelihood functions," Information and Control, *13*, (1968), pp. 62-74.

[DUN70] Duncan, T.E., "On the absolute continuity of measures," Ann. Math. Stat., *41*, (1970), pp. 30-38.

[ITO56] Ito, K., "Isotropic random current," Proc. 3rd Berkeley Symp. on Math. Stat. and Prob., pp. 125-132, Univ. Calif. Press, 1956.

[LEV56] Lévy, P., "A special problem of Brownian motion, and a general theory of Gaussian random function," Proc. 3rd Berkeley Symp. on Math. Stat. and Prob., pp. 133-175, Univ. Calif. Press, 1956.

[NEL73] Nelson, E., "Probability theory and Euclidean field theory," in *Constructive Quantum Field Theory,* Lecture notes in Physics, *25,* Springer-Verlag, New York, 1973.

[OGI81] Ogier, R. and E. Wong, "Recursive linear smoothing of two-dimensional random fields," IEEE Trans. Information Theory, *27,* (1981), pp. 77-83.

[WHI57] Whitney, H., *Geometric Integration Theory,* Princeton University Press, 1957.

[WON68] Wong, E. "Two dimensional random fields and representation of images," SIAM J. Appl. Math, *16,* (1968), pp. 756-770.

[WON69] ______ , "Homogeneous Gauss-Markov random fields," Ann. Math. Stat., *40,* (1969), pp. 1625-1634.

[WON74] Wong, E. and M. Zakai, "Martingales and stochastic integrals for processes with a multidimensional parameter," Z. Wahrscheinlichkeitstheorie, *29,* (1974), pp. 109-122.

[WON76] ______ , "Weak martingales and stochastic integrals in the plane," Ann. Prob., *4,* (1976), pp. 570-586.

[WON77] ______ , "Likelihood ratios and transformation of probability associated with two-parameter Wiener processes," Z. Wahrscheinlichkeitstheorie, *40,* (1977), pp. 283-308.

[WON87] ______ , "Multiparameter martingale differential forms," Prob. Th. and Rel. Fields, *74,* (1987) pp. 429-453.

[ZAK69] Zakai, M. "On the optimal filtering of diffusion processes," Z. Wahrscheinlichkeitstheorie, *11,* (1969), pp. 230-243.

Limit theorems for storage process with
the domain of attraction of a stable law

Keigo Yamada
Institute of Information Sciences and Electronics
University of Tsukuba
Tsukuba-shi, Ibaraki 305, Japan

1. Introduction

Let us consider a storage process X(t) which satisfies the following stochastic integral equation:

$$X(t)=X(0)-\int_0^t r(X(s))ds+\int_0^t\int_{R^+} f(X(s-),x)N(dsdx), \quad (1)$$

$X(0)\geq 0.$

Here $R^+=(0,\infty)$ and N(dsdx) is a Poisson random measuie of a point process p defined a complete probability space (Ω, F, P) with intensity measure $dsd\nu(x)$ and is independent of X(0). $r(\cdot)$ is a nonnegative continuous function on $[0,\infty)$ with r(0)=0. $f(\cdot,\cdot)$ defined on $[0,\infty)\times(0,\infty)$ is assumed to be Borel measurable and nonnegative. Furthermore if we assume that the measure $d\nu(x)$ defined on R^+ is finite (though we do not assume this in this paper), then it is known (Cinlar and Pinsky [1]) that equation (1) has a unique nonnegative solution.

The purpose of this paper is to show that under appropriate assumptions suitably normalized processes of X(t) converge weakly to a stable process or a reflecting stable process. In Yamada [2] the same problem was treated in the case when f(y,x)=x. It was shown in [2] that under the condition that input and output rates at the infinite level of the storage are equal and the second moment of the measure $d\nu(x)$ is finite, i.e., $\int_{R^+} x^2 d\nu(x)<\infty$, the limit of the suitably normalized processes of X(t) is a Bessel process. In this paper we consider the problem where the second moment of the Levy measure of the process X(t) at the infinite level of the storage is not necessarity finite. That is, defining the new measure $d\tilde{\nu}(x)=d\nu(g^{-1}(x))$ where $g(x)=f(\infty,x)$ $(=\lim_{y\to\infty} f(y,x))$ and $g^{-1}(x)$ is the inverse function of g(x), this second moment of $d\tilde{\nu}(x)$, $\int_{R^+} x^2 d\tilde{\nu}(x)$, is not necessarily finite. The basic condition we impose in considering our problem is that the measure $d\tilde{\nu}(x)$ is in the domain of attraction of a stable law. When the exponent α of the stable law is 2, it is shown, using the almost same technique as in [2], that the limit process under a suitable normalization is a Bessel process. Hence the main interest in this paper is in the case where the exponent α satisfies $0<\alpha<2$. It is shown that when $1<\alpha<2$, the limit process is a reflecting stable process and, when $0<\alpha<1$, the limit process is a stable process whose sample path is increasing a.s.. In this paper the case where $\alpha=1$ is excluded owing to some technical difficulties. In establishing these results, we use martingale theoretical results on point processes shch as presented in Jacod and Shiryaev [3]. Especially when we treat the case where $1<\alpha<2$, we use heavily the results and various techniques in Kasahara and Watanabe [4], which were developed to obtain some limit theorems for functionals of point processes.

We denote by $D([0,\infty),R^1)$ the space of right continuous functions $f: [0,\infty)\to R^1$ having left limits with the Skolohod topology ([3, VI]), and by $Y_n(t) \Rightarrow Y(t)$ the weak convergence of distributions of the processes Y_n to the distribution of the process Y. For any stochastic process $X(t)$, we denote by $\sigma_t(X)$ the σ-field generated by $X(s)$, $s\le t$.

2. Assumptions and results

Throughout the paper we shall assume, in addition to the conditions described in Introduction, the following assumptions:

(A1) The function $f(\cdot,\cdot)$ defined on $[0,\infty)\times R^+$ is such that for each $x>0$, $\lim_{y\to\infty} f(y,x)=g(x)$ exists, $g(x)$ is an increasing function which approaches infinity as x increases, and $f(y,x)\ge g(x)$ for all x and y. If we let $g^{-1}(x)$ be the inverse function of $g(\cdot)$, i.e., $g^{-1}(x)=\inf\{y;\ g(y)\ge x\}$, then

$$\lim_{x\to\infty,\, y\to\infty} \frac{f(y,g^{-1}(x))}{x} = 1. \tag{2}$$

Furthermore there exists a constant K such that

$$\frac{f(y,g^{-1}(x))}{x} \le K \tag{3}$$

for all $y\ge 0$ and $x>0$.

(A2) $\lim_{x\to\infty} r(x)=\bar{r}<\infty$ and $\bar{r}\ge r(x)$ for all $x\ge 0$.

As in Introduction, we define the measure $d\tilde{\nu}(x)$ by

$$d\tilde{\nu}(x)=d\nu(g^{-1}(x)).$$

Regarding this measure $d\tilde{\nu}(x)$, in view of the concept of the domain of attraction of a stable law, we consider the following two cases:

Case I There exists a regular varying function $\phi(\cdot)$ with index 1/2 such that as $n\to\infty$

$$\nu_n((x,\infty))\to 0 \quad \text{for any } x>0,$$

and

$$\int_{0<x<\varepsilon} x^2 d\nu_n(x)\to\sigma^2>0 \quad \text{for any } \varepsilon>0,$$

where $d\nu_n(x)=n\,d\tilde{\nu}(\phi(n)x)$, $n\ge 1$.

Note that we are assuming

$$\nu((x,\infty))<\infty \quad \text{and} \int_{0<x<\varepsilon} x^2 d\nu(x)<\infty$$

for any $x>0$ and $\varepsilon>0$.

<u>CaseII</u> There exists a regular varying function $\phi(\cdot)$ with index $1/\alpha$ $(0<\alpha<2)$ such that as $n\to\infty$,

$$\nu_n((x,\infty))\to C_+x^{-\alpha},\ x>0,\ C_+>0.$$

In <u>CaseI</u> and <u>CaseII</u> with $1<\alpha<2$, we assume the following:

(A3) $\int_{R^+} x d\tilde{\nu}(x)<\infty$

and

$$\bar{r} = \int_{R^+} x d\tilde{\nu}(x).$$

Furthermore we assume that

$$\lim_{x\to\infty} x(h(x)-r(x))=c<\infty$$

where $h(x)$ is defined as

$$h(x)=\int_{R^+} f(x,y)d\nu(y).$$

Now let us define a sequence of normalized peocesses $\{X_n(t),\ n\geq 1\}$ by

$$X_n(t)=\begin{cases} X(nt)/(\sigma\phi(n)) & \text{in CaseI} \\ X(nt)/\phi(n) & \text{in CaseII.}\end{cases}$$

Then we have the following result:

<u>Theorem 1</u> (a) Consider <u>CaseI</u> and assume (A1-2). We furhter assume the following condition which is stronger than (2):

$$\lim_{y\to\infty} \frac{f(y,g^{-1}(x))}{x} = 1 \tag{4}$$

uniformly with respect to $x>0$. We also asume that

$$\int_0^\infty x^2 d\tilde{\nu}(x)=\infty.$$

Then $X_n(t) \Rightarrow Y(t)$ in $D([0,\infty),R^1)$ where $Y(t)$ is a Bessel process with index 1 (i.e., a reflecting Brownion motion).

(b) Assume (A1-2) and suppose that

$$\int_{R^+} x^2 d\tilde{\nu}(x)<\infty.$$

Then with $\phi(x)=\sqrt{x}$, $X_n(t) \Rightarrow Y(t)$ in $D([0,\infty),R^1)$ where $Y(t)$ is a Bessel Process with index $\alpha=1+2kc$ where $k=\lim_{n\to\infty} n/(\phi(n))^2=1$. (Note that we are not assuming (4).)

Remark 1 It is easy to see that condition (4) is satisfied if we take, as $f(x,y)$, funcitons of the form $f_1(x)+f_2(y)$, $f_1(x)\ f_2(y)$ and composite functions of these forms such as $\sum_{i=1}^{m} f_i(x)k_i(y)+g(x)+h(y)$. Since the class of these functions commonly appears in applications, condition (4) seems not to be restrictive. We also note that these functions satisfy condition (2) and, if they are continuous, condition (3) is also satisfied.

To state our result corresponding to CaseII, we must define a reflecting stable process which will be the limit process of $X_n(t)$. Let

$$d\nu_0(x)=\alpha C_+ x^{-\alpha-1}dx, \quad x>0, \tag{5}$$

and let $N_0(dt,dx)$ be the random measure of a Poisson point process p_0 with its compensator $\hat{N}(dtdx)=dtd\nu_0(x)$ with respect to the filtration $\sigma_t(p_0)$. Now we consider CaseII and the following Skorohod equation:

$$Y(t)=\xi(t)+M_0(t) \tag{6}$$

where $M_0(t)$ is given by

$$M_0(t)=\int_0^t\int_{0<x<1} x\tilde{N}_0(dsdx)+\int_0^t\int_{x>1} xN_0(dsdx)-\int_0^t\int_{x>1} x\hat{N}_0(dsdx)$$

and $Y(t)$ and $\xi(t)$ are unknown nonnegative processes satisfying the condition that 1) $\xi(t)$ is an increasing continuous process with $\xi(0)=0$ and 2) $\xi(t)$ does not increase when $Y(t)>0$ and $Y(t-)>0$. We can show that the above equation (6) has a unique solution and $Y(t)$ is solved as

$$Y(t)=M_0(t)-\inf_{s\le t}(M_0(s)),$$

and the process $Y(t)$ is reflecting in the sense that $\int_0^t I(Y(s)=0)ds=0$ for any t (Tanaka [5]). We call the process $Y(t)$ as a reflecting stable process. Then we have the following

Theorem 2 Consider CaseII with $1<\alpha<2$ and assume (A1-3). Then $X_n(t) \Rightarrow Y(t)$ in $D([0,\infty),R^1)$ where $Y(t)$ is the reflecting stable process defined above.

The result corresponding to CaseII with $0<\alpha<1$ is stated in

Theorem 3 Consider CaseII with $0<\alpha<1$ and assume (A1-2). Then

$$X_n(t) \Rightarrow \int_0^t\int_{R^+} xN_0(dsdx)$$

in $D([0,\infty),R^1)$.

3. Outline of the proof of theorems

Since the proof of theorems is long, only the outline of the proof of Theorem 2 will be given here (see Yamada [6] for detailes). The normalized processes $\{X_n(t), n\geq 1\}$ can be written as

$$X_n(t)=X_n(0)+\frac{n}{\phi(n)}\int_0^t\{h(\phi(n)X_n(s))-r(\phi(n)X_n(s))\}ds+\int_0^t\int_{R^+}f_n(X_n(s-),x)x\tilde{N}_n(dsdx)$$

where

$$f_n(X_n(s-),x)=\frac{f(\phi(n)X_n(s-),g^{-1}(\phi(n)x))}{\phi(n)x}$$

and $N_n(dsdx)$ is the martingale measure of the random measure defined by

$$N_n(dsdx)=N(ndt,dg^{-1}(\phi(n)x)),$$

i.e., $\tilde{N}_n(dsdx)=N_n(dsdx)-\hat{N}_n(dsdx)$ with $\hat{N}_n(dsdx)$ being the compensator measure of $N_n(dsdx)$. Rewriting $X_n(t)$, we have

$$X_n(t)=X_n(0)+\xi_n(t)+\int_0^t\int_{0<x<1}f_n(X_n(s-),x)x\tilde{N}_n(dsdx)$$

$$+\int_0^t\int_{x\geq 1}f_n(X_n(s-),x)xN_n(dsdx)-\int_0^t\int_{x\geq 1}f_n(X_n(s-),x)x\hat{N}_n(dsdx)$$

$$\equiv X_n(0)+\xi_n(t)+Z_n^1(t)+Z_n^2(t)-Z_n^3(t)$$

where $\xi_n(t)$, $Z_n^1(t)$, $Z_n^2(t)$ and $Z_n^3(t)$ are obviously defined. We can show the tightness of the family of processes $\{X_n(t),\ \xi_n(t),\ Z_n^1(t),\ Z_n^2(t),\ Z_n^3(t),\ n\geq 1\}$, and let $(Y(t), \xi(t), Z_1(t), Z_2(t), Z_3(t))$ be any weak limits of these processes. Then we can identify these weak limits as follows. First the condition of Case II implies that $d\nu_n(x)$ converges vaguely to $d\nu_0(x)$ and, then this implies that $\hat{N}_n(dsdx)\Rightarrow\hat{N}_0(dsdx)$ and $N_n(dsdx)\Rightarrow N_0(dsdx)$ ([4]). While we can show that $X(t)\to\infty$ in probability as $t\to\infty$, and this implies that $f_n(X_n(s-),x)\to 1$ in probability as $n\to\infty$. Conbining this fact with the above result, i.e., $\hat{N}_n(dsdx)\Rightarrow\hat{N}_0(dsdx)$ and $N_n(dsdx)\Rightarrow N_0(dsdx)$, we can show that

$$Z_n^1(t)\Rightarrow\int_0^t\int_{0<x<1}x\tilde{N}_0(dsdx),$$

$$Z_n^2(t)\Rightarrow\int_0^t\int_{x\geq 1}xN_0(dsdx),$$

$$Z_n^3(t)\Rightarrow\int_0^t\int_{x\leq 1}x\hat{N}_0(dsdx)$$

in $D([0,\infty),R^1)$ by using the technique in [4]. Thus we have

$$Y(t)=Y(0)+\xi(t)+\int_0^t\int_{0<x<1}x\tilde{N}_0(dsdx)+\int_0^t\int_{x\geq 1}xN_0(dsdx)-\int_0^t\int_{x\geq 1}x\hat{N}_0(dsdx). \tag{7}$$

Now if we can show that 1) $\xi(t)$ is an increasing continuous process with $\xi(0)=0$ and 2) $\xi(t)$ does not increase when $Y(t)>0$ and $Y(t-)>0$, then equation (7) is nothing but the Skorohod equation given in Theorem 2 and we conclude that the limit process $Y(t)$ is the reflecting stable process. The argument for 1) and 2) is as follows. Since $\xi_n(t)$ is continuous and increasing, 1) is trivial. As for 2), suppose that $Y(t)>0$ and $Y(t-)>0$. Then there exists an ε $(=\varepsilon(\omega))$ such that $\inf_{t-\varepsilon\le s\le t+\varepsilon} Y(s)>0$. This means that since we may assume that $X_n(t)$ converges to $Y(t)$ in the Skorohod topology a.s., there exists a constant $\delta(\omega)>0$ such that $\inf_{t-\varepsilon\le s\le t+\varepsilon} X_n(s)>\delta(\omega)>0$ for all sufficiently large n. Then for sufficiently large n we have

$$0 \le \xi_n(t+\varepsilon)-\xi_n(t-\varepsilon)$$

$$\le \frac{n}{(\phi(n))^2}\frac{1}{\delta(\omega)}\int_{t-\varepsilon}^{t+\varepsilon}\phi(n)X_n(s)\{h(\phi(n)X_n(s)-r(\phi(n)X_n(s))\}ds.$$

By letting n tend to infinity in the above, condition (A3) and the fact that $n/(\phi(n))^2\to 0$ lead us to the conclusion that $\xi(t+\varepsilon)-\xi(t-\varepsilon)=0$, i.e., $\xi(t)$ does not increase at t.

Acknowledgement

The author is indebted to M. Tsuchiya and Y. Kasakara for their helpful comments.

References

[1] Cinlar, E. and Pinsky, M. (1972). On dams with additive inputs and a general release. J. Appl. Probab. 9 422-429

[2] Yamada, K. (1984). Diffusion approximation for storage processes with general release rules. Math. Oper. Res. 9 459-470

[3] Jacod, J. and Shiryaev, A.N. (1987). Limit Theorems for Stochastic Processes. Springer-Verlag

[4] Kasahara, Y. and Watanabe, S. (1986). Limit theorems for point processes and their functionals. J. Math. Soc. Japan, 38 543-574

[5] Tanaka, H. (1971). Construction of one-dimensional reflecting Markov processes by stochastic differential equations. Lecture Notes of Research Institute of Mathematical Sciences, No. 112, 120-133 (in Japanese)

[6] Yamada, K. (1988). Limit theorems for storage process with domain of attraction of a stable law, preprint

Lecture Notes in Control and Information Sciences

Edited by M. Thoma and A. Wyner

Vol. 81: Stochastic Optimization
Proceedings of the International Conference, Kiew, 1984
Edited by I. Arkin, A. Shiraev, R. Wets
X, 754 pages, 1986.

Vol. 82: Analysis and Algorithms of Optimization Problems
Edited by K. Malanowski, K. Mizukami
VIII, 240 pages, 1986.

Vol. 83: Analysis and Optimization of Systems
Proceedings of the Seventh International Conference of Analysis and Optimization of Systems
Antiba, June 26-27, 1986
Edited by A. Bensoussan, J. L. Lions
XVI, 901 pages, 1986.

Vol. 84: System Modelling and Optimization
Proceedings of the 12th IFIP Conference
Budapest, Hungary, September 2–6, 1985
Edited by A. Prékopa, J. Szelezsán, B. Strazicky
XII, 1046 pages, 1986.

Vol. 85: Stochastic Processes in Underwater Acoustics
Edited by Charles R. Baker
V, 205 pages, 1986.

Vol. 86: Time Series and Linear Systems
Edited by Sergio Bittanti
XVII, 243 pages, 1986.

Vol. 87: Recent Advances in System Modelling and Optimization
Proceedings of the IFIP-WG 7/1 Working Conference
Santiago, Chile, August 27-31, 1984
Edited by L. Contesse, R. Correa, A. Weintraub
IV, 199 pages, 1987.

Vol. 88: Bruce A. Francis
A Course in H_∞ Control Theory
XI, 156 pages, 1987.

Vol. 88: Bruce A. Francis
A Course in H_∞ Control Theory
X, 150 pages, 1987.
Corrected - 1st printing 1987

Vol. 89: G. K. H. Pang/A. G. J. McFarlane
An Expert System Approach to Computer-Aided Design of Multivariable Systems
XII, 223 pages, 1987.

Vol. 90: Singular Perturbations and Asymptotic Analysis in Control Systems
Edited by P. Kokotovic, A. Bensoussan, G. Blankenship
VI, 419 pages, 1987.

Vol. 91 Stochastic Modelling and Filtering
Proceedings of the IFIP-WG 7/1 Working Conference
Rome, Italy, Decembre 10-14, 1984
Edited by A. Germani
IV, 209 pages, 1987.

Vol. 92: L. T. Grujić, A. A. Martynyuk, M. Ribbens-Pavella
Large-Scale Systems Stability Under Structural and Singular Perturbations
XV, 366 pages, 1987.

Vol. 93: K. Malanowski
Stability of Solutions to Convex Problems of Optimization
IX, 137 pages, 1987.

Vol. 94: H. Krishna
Computational Complexity of Bilinear Forms
Algebraic Coding Theory and Applications to Digital Communication Systems
XVIII, 166 pages, 1987.

Vol. 95: Optimal Control
Proceedings of the Conference on Optimal Control and Variational Calculus
Oberwolfach, West-Germany, June 15-21, 1986
Edited by R. Bulirsch, A. Miele, J. Stoer and K. H. Well
XII, 321 pages, 1987.

Vol. 96: H. J. Engelbert/W. Schmidt
Stochastic Differential Systems
Proceedings of the IFIP-WG 7/1 Working Conference
Eisenach, GDR, April 6-13, 1986
XII, 381 pages, 1987.

Lecture Notes in Control and Information Sciences

Edited by M. Thoma and A. Wyner

Vol. 117: K.J. Hunt
Stochastic Optimal Control Theory
with Application in Self-Tuning Control
X, 308 pages, 1989.

Vol. 118: L. Dai
Singular Control Systems
IX, 332 pages, 1989

Vol. 119: T. Başar, P. Bernhard
Differential Games and Applications
VII, 201 pages, 1989

Vol. 120: L. Trave, A. Titli, A. M. Tarras
Large Scale Systems:
Decentralization, Structure Constraints
and Fixed Modes
XIV, 384 pages, 1989

Vol. 121: A. Blaquière (Editor)
Modeling and Control of Systems
in Engineering, Quantum Mechanics,
Economics and Biosciences
Proceedings of the Bellman Continuum
Workshop 1988, June 13–14, Sophia Antipolis, France
XXVI, 519 pages, 1989

Vol. 122: J. Descusse, M. Fliess, A. Isidori,
D. Leborgne (Eds.)
New Trends in Nonlinear Control Theory
Proceedings of an International
Conference on Nonlinear Systems,
Nantes, France, June 13–17, 1988
VIII, 528 pages, 1989

Vol. 123: C. W. de Silva, A. G. J. MacFarlane
Knowledge-Based Control with
Application to Robots
X, 196 pages, 1989

Vol. 124: A. A. Bahnasawi, M. S. Mahmoud
Control of Partially-Known
Dynamical Systems
XI, 228 pages, 1989

Vol. 125: J. Simon (Ed.)
Control of Boundaries and Stabilization
Proceedings of the IFIP WG 7.2 Conference
Clermont Ferrand, France, June 20–23, 1988
IX, 266 pages, 1989

Vol. 126: N. Christopeit, K. Helmes
M. Kohlmann (Eds.)
Stochastic Differential Systems
Proceedings of the 4th Bad Honnef Conference
June 20–24, 1988
IX, 342 pages, 1989